Groups of Circle Diffeomorphisms

CHICAGO LECTURES IN MATHEMATICS SERIES

Editors: Spencer J. Bloch, Peter Constantin, Benson Farb, Norman R. Lebovitz, Carlos Kenig, and J. P. May

Other *Chicago Lectures in Mathematics* titles available from the University of Chicago Press

SIMPLICIAL OBJECTS IN ALGEBRAIC TOPOLOGY, *by J. Peter May* (1967, 1993)

FIELDS AND RINGS, SECOND EDITION, *by Irving Kaplansky* (1969, 1972)

LIE ALGEBRAS AND LOCALLY COMPACT GROUPS, *by Irving Kaplansky* (1971)

SEVERAL COMPLEX VARIABLES, *by Raghavan Narasimhan* (1971)

TORSION-FREE MODULES, *by Eben Matlis* (1973)

STABLE HOMOTOPY AND GENERALISED HOMOLOGY, *by J. F. Adams* (1974)

RINGS WITH INVOLUTION, *by I. N. Herstein* (1976)

THEORY OF UNITARY GROUP REPRESENTATION, *by George V. Mackey* (1976)

COMMUTATIVE SEMIGROUP RINGS, *by Robert Gilmer* (1984)

INFINITE-DIMENSIONAL OPTIMIZATION AND CONVEXITY, *by Ivar Ekeland and Thomas Turnbull* (1983)

NAVIER-STOKES EQUATIONS, *by Peter Constantin and Ciprian Foias* (1988)

ESSENTIAL RESULTS OF FUNCTIONAL ANALYSIS, *by Robert J. Zimmer* (1990)

FUCHSIAN GROUPS, *by Svetlana Katok* (1992)

UNSTABLE MODULES OVER THE STEENROD ALGEBRA AND SULLIVAN'S FIXED POINT SET CONJECTURE, *by Lionel Schwartz* (1994)

TOPOLOGICAL CLASSIFICATION OF STRATIFIED SPACES, *by Shmuel Weinberger* (1994)

LECTURES ON EXCEPTIONAL LIE GROUPS, *by J. F. Adams* (1996)

GEOMETRY OF NONPOSITIVELY CURVED MANIFOLDS, *by Patrick B. Eberlein* (1996)

DIMENSION THEORY IN DYNAMICAL SYSTEMS: CONTEMPORARY VIEWS AND APPLICATIONS, *by Yakov B. Pesin* (1997)

HARMONIC ANALYSIS AND PARTIAL DIFFERENTIAL EQUATIONS: ESSAYS IN HONOR OF ALBERTO CALDERÓN, *edited by Michael Christ, Carlos Kenig, and Cora Sadosky* (1999)

A CONCISE COURSE IN ALGEBRAIC TOPOLOGY, *by J. P. May* (1999)

TOPICS IN GEOMETRIC GROUP THEORY, *by Pierre de la Harpe* (2000)

EXTERIOR DIFFERENTIAL SYSTEMS AND EULER-LAGRANGE PARTIAL DIFFERENTIAL EQUATIONS, *by Robert Bryant, Phillip Griffiths, and Daniel Grossman* (2003)

RATNER'S THEOREMS ON UNIPOTENT FLOWS, *by Dave Witte Morris* (2005)

Groups of Circle Diffeomorphisms

ANDRÉS NAVAS

THE UNIVERSITY OF CHICAGO PRESS   CHICAGO AND LONDON

ANDRÉS NAVAS is an associate researcher at the University of Santiago in Chile.

The University of Chicago Press, Chicago 60637
The University of Chicago Press, Ltd., London
©2011 by The University of Chicago
All rights reserved. Published 2011
Printed in the United States of America
20 19 18 17 16 15 14 13 12 11   1 2 3 4 5

ISBN-13: 978-0-226-56951-2 (cloth)
ISBN-10: 0-226-56951-9 (cloth)

Navas, Andrés.
    Groups of circle diffeomorphisms / Andrés Navas.
        p.   cm. — (Chicago lectures in mathematics series)
    ISBN-13: 978-0-226-56951-2 (cloth : alk. paper)
    ISBN-10: 0-226-56951-9 (cloth : alk. paper)   1. Diffeomorphisms.   2. Group actions
(Mathematics)   3. Manifolds (Mathematics)   I. Title.   II. Series: Chicago lectures in
mathematics.
    QA613.N393   2011
    514'.72—dc22   2010026628

TO ROCÍO AND NACHITO,
AND YAMILETH . . .

Contents

Introduction xi
Acknowledgments xv
Notation and General Definitions xvii

1   Examples of Group Actions on the Circle 1
    1.1 The Group of Rotations 1
    1.2 The Group of Translations and the Affine Group 2
    1.3 The Group PSL$(2, \mathbb{R})$ 4
        1.3.1 PSL$(2, \mathbb{R})$ as the Möbius group 4
        1.3.2 PSL$(2, \mathbb{R})$ and the Liouville geodesic current 7
        1.3.3 PSL$(2, \mathbb{R})$ and the convergence property 9
    1.4 Actions of Lie Groups 11
    1.5 Thompson's Groups 12
        1.5.1 Thurston's piecewise projective realization 15
        1.5.2 Ghys-Sergiescu's smooth realization 18

2   Dynamics of Groups of Homeomorphisms 23
    2.1 Minimal Invariant Sets 23
        2.1.1 The case of the circle 23
        2.1.2 The case of the real line 29
    2.2 Some Combinatorial Results 30
        2.2.1 Poincaré's theory 30
        2.2.2 Rotation numbers and invariant measures 37
        2.2.3 Faithful actions on the line 39
        2.2.4 Free actions and Hölder's theorem 43
        2.2.5 Translation numbers and quasi-invariant measures 49
        2.2.6 An application to amenable, orderable groups 58

2.3 Invariant Measures and Free Groups   63

    2.3.1 A weak version of the Tits alternative   63

    2.3.2 A probabilistic viewpoint   69

3   Dynamics of Groups of Diffeomorphisms   80

  3.1 Denjoy's Theorem   80

  3.2 Sacksteder's Theorem   90

    3.2.1 The classical version in class $C^{1+\text{Lip}}$ 91

    3.2.2 The C^1 version for pseudogroups   96

    3.2.3 A sharp C^1 version via Lyapunov exponents   103

  3.3 Duminy's First Theorem: On the Existence of Exceptional
     Minimal Sets   110

    3.3.1 The statement of the result   110

    3.3.2 An expanding first-return map   112

    3.3.3 Proof of the theorem   116

  3.4 Duminy's Second Theorem: On the Space of
     Semiexceptional Orbits   119

    3.4.1 The statement of the result   119

    3.4.2 A criterion for distinguishing two different ends   123

    3.4.3 End of the proof   127

  3.5 Two Open Problems   130

    3.5.1 Minimal actions   130

    3.5.2 Actions with an exceptional minimal set   141

  3.6 On the Smoothness of the Conjugacy between Groups
     of Diffeomorphisms   146

    3.6.1 Sternberg's linearization theorem and C^1 conjugacies   147

    3.6.2 The case of bi-Lipschitz conjugacies   152

4   Structure and Rigidity via Dynamical Methods   158

  4.1 Abelian Groups of Diffeomorphisms   158

    4.1.1 Kopell's lemma   158

    4.1.2 Classifying Abelian group actions in class C^2 164

    4.1.3 Szekeres's theorem   165

    4.1.4 Denjoy counterexamples   171

    4.1.5 On intermediate regularities   182

  4.2 Nilpotent Groups of Diffeomorphisms   192

    4.2.1 The Plante-Thurston Theorems   192

    4.2.2 On growth of groups of diffeomorphisms   195

    4.2.3 Nilpotence, growth, and intermediate regularity   202

4.3 Polycyclic Groups of Diffeomorphisms   209

4.4 Solvable Groups of Diffeomorphisms   211

   4.4.1 Some examples and statements of results   211

   4.4.2 The metabelian case   216

   4.4.3 The case of the real line   220

4.5 On the Smooth Actions of Amenable Groups   223

5 Rigidity via Cohomological Methods   226

5.1 Thurston's Stability Theorem   226

5.2 Rigidity for Groups with Kazhdan's Property (T)   233

   5.2.1 Kazhdan's property (T)   233

   5.2.2 The statement of the result   240

   5.2.3 Proof of the theorem   243

   5.2.4 Relative property (T) and Haagerup's property   251

5.3 Superrigidity for Higher-Rank Lattice Actions   253

   5.3.1 Statement of the result   253

   5.3.2 Cohomological superrigidity   256

   5.3.3 Superrigidity for actions on the circle   262

Appendix A   Some Basic Concepts in Group Theory   267

Appendix B   Invariant Measures and Amenable Groups   269

References   273

Index   287

Introduction

The theory of dynamical systems concerns the quantitative and qualitative study of the orbits of a map or the flow associated to a vector
field. If the map is invertible or the vector field is regular, these systems
may be thought of as actions of the group $\mathbb{Z}$ or $\mathbb{R}$, respectively. From this
point of view, classical dynamics may be considered a particular subject in
the general theory of group actions.

From an algebraic point of view, the theory of group actions may be considered a "nonlinear" version of that of group representations. In this view,
the internal structure of the groups is intended to be revealed by looking at
their actions on nice spaces, preferably differentiable manifolds. Several
algebraic notions become natural, and many topological and/or analytical
aspects of the underlying spaces turn out to be relevant. In recent years, this
approach has been very fruitful, leading (sometimes quite unexpectedly)
to dynamical proofs of some results of a purely algebraic nature.

Without any doubt, it would be too ambitious (and perhaps impossible)
to provide a complete treatment of the general theory of group actions on
manifolds. It is therefore natural to restrict the study to certain groups or
manifolds. In this book we follow this path by studying group actions on the
simplest closed manifold, namely, the circle. In spite of the apparent simplicity of the subject, it turns out to be highly nontrivial and quite extensive.
Its relevance lies mainly in its connections with many other branches, such
as low-dimensional geometry and topology (including foliation theory),
mathematical logic (through the theory of orderable groups), mathematical physics (through the cohomological aspects of the theory), etc. In this
spirit, this book has been conceived as a text where people who have encountered some of the topics contained here in their research may see
them integrated into an independent theory.

There already exists a very nice and complete survey by Ghys [88] of the theory of groups of circle *homeomorphisms*. Unlike [88], here we mainly focus on the theory of groups of circle *diffeomorphisms*, which is essentially different in many aspects. Nevertheless, even in the context of diffeomorphisms, this text is still incomplete. We would have liked to add at least one section on the theory of small denominators (including recent solutions of both Moser's problem on the simultaneous conjugacy to rotations of commuting diffeomorphisms by Fayad and Khanin [79] and Frank's problem on distortion properties for irrational Euclidean rotations by Avila [4]), to extend the treatment of Sacksteder's theorem by providing a discussion of the so-called level theory [46, 51, 114], to include two small sections on groups of real-analytic diffeomorphisms and piecewise affine homeomorphisms, respectively, to develop the notion of topological entropy for group actions on compact metric spaces by focusing on the case of one-dimensional manifolds [95, 121, 131, 248], to confront the dynamical and cohomological aspects involved in the study of the so-called Godbillon-Vey class (or Bott-Virasoro-Thurston cocycle) providing a proof of Duminy's "third" theorem [93, 119, 122, 125, 238], and to explore the representations of fundamental groups of surfaces [91, 94]. Moreover, since we mainly focus on actions of discrete groups, we decided not to include some relevant topics, for example, simplicity properties and diffusion processes for the groups of interval and circle diffeomorphisms (see [10] and [154, 155], respectively).

This book begins with a brief section in which we establish most of the notation and recall some general definitions. More elaborate concepts, for instance, group amenability, are discussed in appendixes.

Chapter 1 studies some simple but relevant examples of groups that act on the circle. After recalling some fundamental properties of the rotation group, the affine group, and the Möbius group, it treats the general case of Lie group actions on the circle and concludes by discussing the very important Thompson's groups.

In Chapter 2, we study some fundamental results about the dynamics of groups of interval and circle homeomorphisms. In the first part, we discuss some of their combinatorial aspects, for example, Poincaré's theory of rotation number and its relation to invariant probability measures. We next provide necessary and sufficient conditions for the existence of faithful actions on the interval and of free actions on the interval and the circle. As an application, we give a dynamical proof of a recent and beautiful result due to Witte-Morris asserting that left-orderable, amenable groups are locally indicable. At this point, it would have been natural to treat Ghys's characterization of faithful actions on the circle in terms of bounded cohomology. However, we decided not to treat this topic, mainly because it is well developed in [88], and furthermore because, although it is very important for describing continuous actions on the circle, its relevance for the smooth theory is smaller. In the second part of Chapter 2, we treat a result due to Margulis, which may be thought of as a weak form of the so-called Tits alternative for groups of circle homeomorphisms. Although we do not give Margulis' original proof of this result, we develop an alternative one due to Ghys that is more suitable for a probabilistic interpretation in the context of random walks on groups.

In Chapter 3, we collect several dynamical results that require some degree of smoothness, in general C^2. First, we extensively study the most important one, namely Denjoy's theorem. Next we present some closely related results, for example, Sacksteder's and Duminy's theorems. Then we discuss two of the major open problems of the theory, namely the zero Lebesgue measure conjecture for minimal invariant Cantor sets and the ergodicity conjecture for minimal actions. Finally, we treat the smoothness properties of conjugacies between group actions. At this point, it would have been natural to treat some particular properties in smoothness higher than C^2. Unfortunately, although there exist several interesting and promising results in this direction (see, for example, [38, 50, 244]), it still seems to be impossible to collect them in a systematic and coherent way.

Chapter 4 corresponds to a tentative description of the dynamics of groups of diffeomorphisms of one-dimensional manifolds based on some relevant algebraic information. It begins with the case of Abelian and nilpotent groups, where Denjoy's theorem and the well-known Kopell's lemma become relevant. After a digression concerning growth of groups of diffeomorphisms, it continues with the case of polycyclic and solvable groups and concludes by showing the difficulties encountered in the case of amenable groups.

Chapter 5 concerns obstructions to smooth actions for groups satisfying particular cohomological properties. It begins with the already-classical Thurston's stability theorem for groups of C^1 diffeomorphisms of the interval. It then passes to a rigidity theorem for groups having the so-called Kazhdan property (T) (roughly, Kazhdan group actions on the circle have a finite image). The chapter concludes with a closely related superrigidity result for actions of irreducible higher-rank lattices. These last two theorems (due to the author) may be viewed as natural (but still nondefinitive) generalizations of a series of results concerning obstructions to actions on one-dimensional spaces of higher-rank simple and semisimple Lie groups, in the spirit of the seminal works by Margulis and the quite inspiring Zimmer's program [81, 259].

We have made an effort to make this book mostly self-contained. Although most of the results treated here are very recent, the techniques involved are, in general, elementary. We have also included a large list of complementary exercises where we pursue some topics a little further or briefly explain some related results. However, we must alert the reader that these "exercises" may vary drastically in level of difficulty. Indeed, in many cases the small results that are presented in them do not appear in the literature. This is also the case of certain sections of the book. No doubt the most relevant case is that of § 3.4, where we give the original proof of a theorem proved by Duminy (more than 30 years ago) on the existence of infinitely many ends for semiexceptional leaves of transversely C^2 codimension-one foliations. The pressing need to publish Duminy's brilliant proof of this remarkable result (for which an alternative reference is [47]) was an extra motivation to the author to write this book.

Acknowledgments

The original version (in Spanish) of this text was prepared for a mini-course in Antofagasta, Chile. Subsequently, enlarged and revised versions were published in the series *Monografías del IMCA* (Peru) and *Ensaios Matemáticos* (Brazil). This translation arose from the necessity of making this text accessible to a larger audience. I would like to thank Juan Rivera-Letelier for his invitation to the II Workshop on Dynamical Systems (2001), for which the original version was prepared, and Roger Metzger for his invitation to IMCA (2006), where part of this material was presented. I would also like to thank both Étienne Ghys and Maria Eulália Vares for motivating me and allowing me to publish the revised Spanish version in Brazil, as well as all the people who strongly encouraged me to prepare this English version.

This work was partially supported by CONICYT and PBCT via the Research Network on Low Dimensional Dynamics. I would also to acknowledge the support of both the UMPA Department of the École Normale Supérieure de Lyon, where the idea of writing this text was born during my PhD thesis, and the Institut des Hautes Études Scientifiques, where the original notes started taking their definitive form while I benefited from a one-year postdoctoral position.

This work owes much to many of my colleges. Several remarks spread throughout the text and the content of some examples and exercises were born during fruitful and quite stimulating discussions. It is thus a pleasure to thank Sylvain Crovisier (Proposition 4.2.25), Albert Fathi (Exercise 5.3.15), Tsachik Gelander (Exercise 5.3.16), Adolfo Guillot (Exercise 3.3.8), Carlos Moreira (Exercise 3.1.4), Pierre Py (Remark 5.2.25), Takashi Tsuboi (Exercise 3.6.15), Dave Witte-Morris (Proposition 5.2.21), and Jean-Christophe Yoccoz (Exercises 1.3.11 and 5.2.26). It is also a pleasure to thank Levon Beklaryan, Rostislav Grigorchuk, Leslie Jiménez, Eduardo Jorquera, Jan Kiwi, Yoshifumi Matsuda, Daniel Pons, Juan Rivera-Letelier, Pierre Paul Romagnoli, Eugenio Trucco, and Dave Witte-Morris for their many corrections to this and earlier versions of the text, as well as Marina Flores for her help with the English translation.

Finally, it is a very great pleasure to thank Étienne Ghys for his encouragement and for numerous fascinating discussions on the subject of this book and many other aspects of mathematics.

Santiago de Chile, May 28, 2009

Notation and General Definitions

We will commonly denote the circle by S^1. As usual, we will consider the counterclockwise orientation on it. We will denote by $]a, b[$ the open interval from a to b according to this orientation. Notice that if b belongs to $]a, c[$, then $c \in]b, a[$ and $a \in]c, b[$. We will sometimes write $a < b < c < a$ for these relations. One defines similarly the intervals $[a, b]$, $[a, b[$, and $]a, b]$. The distance between a and b is the shorter of the lengths of the intervals $]a, b[$ and $]b, a[$. We will denote this distance by $dist(a, b)$ or $|a - b|$. We will also use the notation $|I|$ for the length of an interval I (on the circle or the real line). The Lebesgue measure of a measurable subset A of either S^1 or $\mathbb{R}$ will be denoted by $Leb(A)$.

Unless the contrary is said, in this text we will deal only with orientation-preserving maps. The group of (orientation-preserving) circle homeomorphisms will be denoted by $\text{Homeo}_+(S^1)$, and for $k \in \mathbb{N} \cup \{\infty\}$, we will denote the subgroup of C^k diffeomorphisms by $\text{Diff}_+^k(S^1)$. For $\tau \in]0, 1[$ we will also deal with the group $\text{Diff}_+^{1+\tau}(S^1)$ of circle diffeomorphisms whose derivatives are Hölder continuous of exponent τ (or just τ-Hölder continuous), that is, for which there exists a constant $C > 0$ such that $|f'(x) - f'(y)| \leq C|x - y|^\tau$ for all x and y.

The group of circle diffeomorphisms that have a Lipschitz derivative will be denoted by $\text{Diff}_+^{1+\text{Lip}}(S^1)$. Finally, the notation $\text{Diff}_+(S^1)$ will be employed when the involved regularity is clear from the context or is irrelevant.

We will sometimes see the real line as the universal covering of the circle through the maps $x \mapsto e^{ix}$ or $x \mapsto e^{2\pi i x}$, depending on whether we parameterize S^1 by $[0, 2\pi]$ or $[0, 1]$. In general, we will consider the first of these parameterizations. By a slight abuse of notation, for each orientation-preserving circle homeomorphism f, we will commonly denote also by f (sometimes by F) each of its lifts to the real line. Therefore, $f : \mathbb{R} \to \mathbb{R}$

will be an increasing continuous function such that for every $x \in \mathbb{R}$, either $f(x + 2\pi) = f(x) + 2\pi$ or $f(x + 1) = f(x) + 1$, depending on the chosen parameterization. We will denote by $\widetilde{\mathrm{Homeo}}_+(S^1)$ the group of homeomorphisms of the line obtained as lifts of circle homeomorphisms. The circle rotation of angle θ will be denoted by R_θ. Once again, notice that θ is an angle either in $[0, 2\pi]$ or $[0, 1]$, depending on the parameterization.

In many cases we will deal directly with subgroups of $\mathrm{Homeo}_+(S^1)$ or $\mathrm{Diff}_+(S^1)$. However, we will also consider representations of a group Γ in the group of circle homeomorphisms or diffeomorphisms. In other words, we will deal with homomorphisms Φ from Γ into $\mathrm{Homeo}_+(S^1)$ or $\mathrm{Diff}_+(S^1)$. These representations, which we may also think of as actions, will commonly be denoted by Φ. For simplicity, we will generally identify the element $g \in \Gamma$ with the map $\Phi(g)$. To avoid any confusion, we will denote the identity transformation by Id, and the neutral element in the underlying group by id.

Let us recall that in general, an action Φ of a group Γ on a space M is *faithful* if for every $g \neq id$, the map $\Phi(g)$ is not the identity of M. The action is *free* if for every $g \neq id$, one has $\Phi(g)(x) \neq x$ for all $x \in$ M. The **orbit** of a point $x \in$ M is the set $\{\Phi(g)(x) : g \in \Gamma\}$. More generally, the orbit of a subset $A \subset$ M is $\{\Phi(g)(x) : x \in A, \ g \in \Gamma\}$. If M is endowed with a probability measure μ, then Φ is **ergodic** with respect to μ if every measurable set A that is **invariant** (that is, $\Phi(g)(A) = A$ for every $g \in \Gamma$) has either null or total μ-measure. This is equivalent to saying that every measurable function $\phi : $ M $\rightarrow \mathbb{R}$ satisfying $\phi \circ \Phi(g) = \phi$ for every $g \in \Gamma$ is μ-almost surely constant. If the measure μ is invariant (that is, $\mu(\Phi(g)(A)) = \mu(A)$ for every measurable set A and every $g \in \Gamma$), this is also equivalent to saying that μ cannot be written as a nontrivial convex combination of two invariant probability measures.

Two elements f and g in $\mathrm{Homeo}_+(S^1)$ are **topologically conjugate** if there exists a circle homeomorphism h such that $h \circ f = g \circ h$ (in this case we say that hfh^{-1} is the **conjugate** of f by h). Similarly, two actions Φ_1 and Φ_2 of a group Γ by circle homeomorphisms are topologically conjugate if there exists $h \in \mathrm{Homeo}_+(S^1)$ such that $h \circ \Phi_1(g) = \Phi_2(g) \circ h$ for every $g \in \Gamma$. In most of the cases we will avoid the use of the symbol $\circ$ when we are composing maps. The non-Abelian free group on n generators will be denoted by $\mathbb{F}_n$. To avoid confusion, we will denote the one-dimensional torus by $\mathbb{T}^1$, thus emphasizing the group structure on it (which identifies with that of $(\mathbb{R} \mod 1, +)$, or $SO(2, \mathbb{R})$). Similarly, we will sometimes write $(\mathbb{R}, +)$ (resp. $(\mathbb{Z}, +)$) instead of $\mathbb{R}$ (resp. $\mathbb{Z}$) to emphasize the corresponding additive group structure.

Examples of Group Actions on the Circle

1.1 The Group of Rotations

The rotation group $SO(2, \mathbb{R})$ is the simplest group acting transitively by circle homeomorphisms. Up to topological conjugacy, it may be characterized as the group of the homeomorphisms of S^1 that preserve a probability measure having properties similar to those of Lebesgue measure. Recall that the **support** of a measure is the complement of the largest open set with null measure. Hence the measure has **total support** if the measure of every nonempty open set is positive.

Proposition 1.1.1. *If* Γ *is a subgroup of* $\mathrm{Homeo}_+(S^1)$ *that preserves a probability measure having total support and no atoms, then* Γ *is topologically conjugate to a subgroup of* $SO(2, \mathbb{R})$.

Proof. The measure μ on S^1 given by the hypothesis induces in a natural way a σ-finite measure $\tilde{\mu}$ on the real line satisfying $\tilde{\mu}([x, x+2\pi]) = 1$ for all $x \in \mathbb{R}$. Let us define $\varphi : \mathbb{R} \to \mathbb{R}$ by $\varphi(x) = 2\pi\tilde{\mu}([0, x])$ if $x \geq 0$, and $\varphi(x) = -2\pi\tilde{\mu}([x, 0])$ if $x < 0$. It is easy to check that φ is a homeomorphism. For an arbitrary element $g \in \Gamma$, let us fix a lift $\tilde{g}$ such that $\tilde{g}(0) > 0$. For $y > 0$, the point $\varphi\tilde{g}\varphi^{-1}(y)$ coincides with

$$2\pi\tilde{\mu}([0, \tilde{g}\varphi^{-1}(y)]) = 2\pi\tilde{\mu}([0, \tilde{g}(0)]) + 2\pi\tilde{\mu}([\tilde{g}(0), \tilde{g}\varphi^{-1}(y)])$$
$$= 2\pi\tilde{\mu}([0, \tilde{g}(0)]) + 2\pi\tilde{\mu}([0, \varphi^{-1}(y)]).$$

Therefore,

$$\varphi\tilde{g}\varphi^{-1}(y) = 2\pi\tilde{\mu}([0, \tilde{g}(0)]) + y,$$

and a similar argument shows that the same holds for $y \le 0$. Thus $\varphi \tilde{g} \varphi^{-1}$ is the translation by $2\pi \tilde{\mu}([0, \tilde{g}(0)])$. The map φ is 2π-periodic and hence induces a homeomorphism of S^1. The preceding computation then shows that after conjugation by this homeomorphism, each $g \in \Gamma$ becomes the rotation by angle $2\pi \tilde{\mu}([0, g(0)]) \mod 2\pi$. $\qquad\qquad\square$

Compact topological groups satisfy the hypothesis of the preceding proposition. Indeed, if Γ is a compact group, we may consider the (normalized) Haar measure dg on it. If Γ acts by circle homeomorphisms, we define a probability measure μ on the Borel sets of S^1 by letting

$$\mu(A) = \int_\Gamma Leb(gA) \, dg.$$

This measure μ is invariant under Γ and has total support and no atoms. We thus conclude the following:

Proposition 1.1.2. *Every compact subgroup of* $\mathrm{Homeo}_+(S^1)$ *is topologically conjugate to a subgroup of* $\mathrm{SO}(2, \mathbb{R})$.

1.2 The Group of Translations and the Affine Group

Group actions on the circle with a global fixed point may be thought of as actions on the real line. Because of this, to understand general actions on S^1, we first need to understand actions on $\mathbb{R}$.

An example of a nice group of homeomorphisms of the real line is the affine group $\mathrm{Aff}_+(\mathbb{R})$. To each element $g(x) = ax + b$ $(a > 0)$ in this group we associate the matrix

$$\begin{pmatrix} a & b \\ 0 & 1 \end{pmatrix} \in \mathrm{GL}_+(2, \mathbb{R}).$$

Via this correspondence, the group $\mathrm{Aff}_+(\mathbb{R})$ identifies with a subgroup of $\mathrm{GL}_+(2, \mathbb{R})$.

Recall that a **_Radon measure_** is a (nontrivial) measure defined on the Borel sets of a topological space that is finite on compact sets. An example of a Radon measure is the Lebesgue measure. Since this measure is preserved up to a multiplicative constant by each element in $\mathrm{Aff}_+(\mathbb{R})$, the following definition becomes natural.

Definition 1.2.1. Let v be a Radon measure on the real line and Γ a subgroup of $\mathrm{Homeo}_+(\mathbb{R})$. We say that v is **quasi-invariant** under Γ if for every $g \in \Gamma$ there exists a positive real number $\kappa(g)$ such that $g^*(v) = \kappa(g) \cdot v$ (that is, for each Borel set $A \subset \mathbb{R}$, one has $v(g(A)) = \kappa(g) \cdot v(A)$).

As an analogue to Proposition 1.1.1, we have the following characterization of the affine group.

Proposition 1.2.2. *Let Γ be a subgroup of $\mathrm{Homeo}_+(\mathbb{R})$. If there exists a Radon measure having total support and no atoms that is quasi-invariant under Γ, then Γ is conjugate to a subgroup of the affine group.*

Proof. Let us define $\varphi : \mathbb{R} \to \mathbb{R}$ by $\varphi(x) = v([0, x])$ if $x \geq 0$, and $\varphi(x) = -v([x, 0])$ if $x < 0$. If $g \in \Gamma$ and $x \geq 0$ are such that $g(x) \geq 0$ and $g(0) \geq 0$, then

$$\varphi(g(x)) = v([0, g(x)]) = \kappa(g)v([g^{-1}(0), x])$$
$$= \kappa(g)v([0, x]) + \kappa(g)v([g^{-1}(0), 0])$$
$$= \kappa(g)\varphi(x) - \kappa(g)\varphi(g^{-1}(0)),$$

and therefore,

$$\varphi g \varphi^{-1}(x) = \kappa(g)x - \kappa(g)\varphi(g^{-1}(0)).$$

If we look at all other cases, it is not difficult to check that the last equality actually holds for every $x \in \mathbb{R}$ and every $g \in \Gamma$, which concludes the proof. $\qquad\square$

The affine group contains the group of translations of the real line, and a similar argument to that of the proof of Proposition 1.1.1 shows the following:

Proposition 1.2.3. *Let Γ be a subgroup of $\mathrm{Homeo}_+(\mathbb{R})$. If Γ preserves a Radon measure having total support and no atoms, then Γ is topologically conjugate to a subgroup of the group of translations.*

Exercise 1.2.4. Define the **logarithmic derivative** (of the derivative) of a C^2 diffeomorphism $f : I \subset \mathbb{R} \to J \subset \mathbb{R}$ as $LD(f)(x) = (\log(f'))'(x)$. Prove that $LD(f) \equiv 0$ if and only if f is the restriction of an element of $\mathrm{Aff}_+(\mathbb{R})$. From the equality

$$\log((f \circ g)')(x) = \log(g')(x) + \log(f')(g(x)), \tag{1.1}$$

deduce the cocycle relation

$$LD(f \circ g)(x) = LD(g)(x) + g'(x) \cdot LD(f)(g(x)).$$

1.3 The Group PSL(2, $\mathbb{R}$)

1.3.1 PSL(2, $\mathbb{R}$) as the Möbius group

We denote by $\mathbb{D}$ the **Poincaré disk**, that is, the unit disk endowed with the hyperbolic metric

$$\frac{4du}{(1 - |u|^2)^2}, \quad u \in \mathbb{D}.$$

The group of nonnecessarily orientation-preserving diffeomorphisms of $\mathbb{D}$ that preserve this metric coincides with the group of conformal diffeomorphisms of $\mathbb{D}$, and it contains the Möbius group as a subgroup of index 2. This means that the only (orientation-preserving) diffeomorphisms $g : \mathbb{D} \to \mathbb{D}$ for which the equality

$$\frac{2\|Dg(u)(\zeta)\|}{1 - |g(u)|^2} = \frac{2\|\zeta\|}{1 - |u|^2}$$

holds for every $u \in \mathbb{D}$ and every vector $\zeta \in T_u(\mathbb{D}) \sim \mathbb{R}^2$ are those that in complex notation may be written in the form

$$g(z) = e^{i\theta} \cdot \frac{z - a}{1 - \bar{a}z}, \quad \theta \in [0, 2\pi], \ a \in \mathbb{C}, \ |a| < 1, \ z \in \mathbb{D}.$$

The Möbius group is denoted by $\mathcal{M}$. Each of its elements induces a real-analytic diffeomorphism of the circle (identified with the boundary $\partial\mathbb{D}$).

Let us now consider the map $\varphi(z) = (z + i)/(1 + iz)$. Notice that $\varphi(S^1) = \mathbb{R} \cup \{\infty\}$; moreover, the image of $\mathbb{D}$ under φ is the upper half plane $\mathbb{R}^2_+$, which, endowed with the induced metric, corresponds to the **hyperbolic plane** $\mathbb{H}^2$. In complex notation, the action of each element of $\mathcal{M}$ on $\mathbb{H}^2$ is of the form $z \mapsto (a_1 z + a_2)/(a_3 z + a_4)$ for some a_1, a_2, a_3, and a_4 in $\mathbb{R}$ such that $a_1 a_4 - a_2 a_3 = 1$. For $a_3 = 0$, we obtain an affine transformation, thus showing that $\text{Aff}_+(\mathbb{R})$ is a subgroup of $\mathcal{M}$.

To each element $z \mapsto (a_1 z + a_2)/(a_3 z + a_4)$ in $\mathcal{M}$ we associate the matrix

$$\begin{pmatrix} a_1 & a_2 \\ a_3 & a_4 \end{pmatrix} \in \text{SL}(2, \mathbb{R}). \tag{1.2}$$

An easy computation shows that the matrix associated to the composition of two elements in $\mathcal{M}$ corresponds to the product of the matrices associated to these elements. Moreover, two matrices M_1 and M_2 in $SL(2,\mathbb{R})$ induce the same element of $\mathcal{M}$ if and only if they coincide or $M_1 = -M_2$. In this way, the Möbius group naturally identifies with the projective group $PSL(2,\mathbb{R})$.

The action of $PSL(2,\mathbb{R})$ satisfies a remarkable property of transitivity and rigidity: given two triples of cyclically ordered points on S^1, say, (a,b,c) and (a',b',c'), there exists a unique element $g \in PSL(2,\mathbb{R})$ that sends a, b, and c into a', b', and c', respectively. In particular, if g fixes three points, then $g = Id$.

The elements of $\mathcal{M} \sim PSL(2,\mathbb{R})$ may be classified according to their fixed points on $S^1 \subset \overline{\mathbb{D}}$. Remark that to find these points in the upper half-plane model, we need to solve the equation

$$\frac{a_1 z + a_2}{a_3 z + a_4} = z. \tag{1.3}$$

A simple analysis shows that there are three cases:

(i) When $|a_1 + a_4| < 2$, the solutions of (1.3) are different (conjugate) points of the complex plane. Therefore, in the Poincaré disk model, the map g has no fixed point on the unit circle. Actually, the map g is conjugate to a rotation.

(ii) When $|a_1 + a_4| = 2$. the solutions of (1.3) coincide and are situated on the real line. Hence in the Poincaré disk model, g fixes a unique point on the circle.

(iii) When $|a_1 + a_4| > 2$. there exist two distinct solutions of (1.3), which are also on the real line. Therefore, the map g fixes two points on the circle, one of them attracting and the other one repelling.

Notice that $|a_1 + a_4|$ corresponds to the absolute value of the trace of the corresponding matrix. (Although the function $M \mapsto a_1 + a_4$ is not well defined on $PSL(2,\mathbb{R})$, there is no ambiguity in the definition of its absolute value.) Figures 1, 2, and 3 illustrate the cases (i), (ii), and (iii), respectively. In case (i) we say that the element is ***elliptic***, in case (ii) it is ***parabolic*** if it does not coincide with the identity, and in case (iii) it is ***hyperbolic***.

Exercise 1.3.1.   Prove that any two hyperbolic elements of $PSL(2,\mathbb{R})$ are topologically conjugate. Show that the same holds for parabolic elements, but not for elliptic ones.

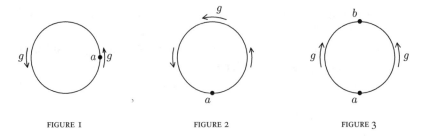

FIGURE I   FIGURE 2   FIGURE 3

There is another way to view the action of PSL(2, ℝ) on the circle. Recall that the ***projective space*** $\mathbb{PR}^1$ is the set of lines through the origin in the plane. This space naturally identifies with the circle parameterized by the interval $[0, \pi]$, since every such line is uniquely determined by the angle $\alpha \in [0, \pi[$ that it makes with the x-axis. Since a linear map sends lines into lines and fixes the origin, it induces a map from $\mathbb{RP}^1$ into itself; moreover, the maps induced by two matrices in $GL_+(2, \mathbb{R})$ coincide if and only if these matrices represent the same element in PSL(2, ℝ), that is, if any of them is a scalar multiple of the other one. Actually, if we consider $s = \cot(\alpha)$ as a parameter, then the action of (1.2) on $\mathbb{PR}^1$ is given by $s \mapsto (a_1 s + a_2)/(a_3 s + a_4)$. Although this view of the action of $\mathbb{RP}^1$ is perhaps simpler than the one arising from extensions of isometries of the Poincaré disk, the latter is more suitable as a source of motivation: in a certain sense, most groups of circle diffeomorphisms tend to have a "negative curvature" behavior.

Despite the fact that the dynamics of each element in PSL(2, ℝ) is very simple, the global structure of its subgroups is not uniquely determined by the individual dynamics. More precisely, for a subgroup Γ of Homeo$_+(S^1)$, the condition that each of its elements is topologically conjugate to an element of PSL(2, ℝ) does not imply that Γ itself is conjugate to a subgroup of PSL(2, ℝ), even when the orbits are dense [145] and the elements are real-analytic diffeomorphisms [182].

Exercise 1.3.2. The ***Schwarzian derivative*** $S(f)$ of a C^3 diffeomorphism $f : I \subset \mathbb{R} \to J \subset \mathbb{R}$ is defined by

$$S(f) = \frac{f'''}{f'} - \frac{3}{2}\left(\frac{f''}{f'}\right)^2.$$

Show that $S(f)$ is identically zero if and only if f is the restriction of a Möbius transformation, that is, a map of the form $x \mapsto (ax + b)/(cx + d)$. Prove also the cocycle relation

$$S(f \circ g)(x) = S(g)(x) + (g'(x))^2 \cdot S(f)(g(x)). \tag{1.4}$$

Exercise 1.3.3. Show the following formulae for the Schwarzian derivative of C^3 diffeomorphisms between intervals in the line:

$$
\begin{aligned}
S(g)(y) &= 6 \lim_{x \to y} \left[\frac{g'(x)g'(y)}{(g(x) - g(y))^2} - \frac{1}{(x - y)^2} \right] \\
&= 6 \lim_{x \to y} \frac{\partial^2}{\partial y \partial x} \log \left(\frac{g(x) - g(y)}{x - y} \right),
\end{aligned}
\tag{1.5}
$$

$$-\frac{1}{2\sqrt{dg/dx}} S(g) = \frac{d^2}{dx^2} \left(\frac{1}{\sqrt{dg/dx}} \right). \tag{1.6}$$

Exercise 1.3.4. A *projective structure* on the circle is given by a system of local coordinates $\varphi_i : I_i \to S^1$ such that the changes of coordinates $\varphi_j^{-1} \circ \varphi_i$ are restrictions of Möbius transformations.

(i) Check that the coordinates ϕ_1 and ϕ_2 with inverses $\alpha \mapsto \tan(\alpha)$ and $\alpha \mapsto \cot(\alpha)$, respectively, define a projective structure on S^1 (to be referred to as the **canonical** projective structure).

(ii) Show that given a projective structure on S^1, the Schwarzian derivative of a diffeomorphism $f : S^1 \to S^1$ is well defined as a **quadratic differential**. In other words, given systems of coordinates $\varphi_1, \bar{\varphi}_1, \varphi_2$, and $\bar{\varphi}_2$ that are compatible with the prescribed projective structure, for every x in the domain of φ_1, one has

$$S(\bar{\varphi}_2^{-1} \circ f \circ \bar{\varphi}_1)(x) = ((\bar{\varphi}_1^{-1} \circ \varphi_1)')^2 \cdot S(\varphi_2^{-1} \circ f \circ \varphi_1)(\bar{\varphi}_1^{-1} \circ \varphi_1(x)).$$

1.3.2 PSL(2, ℝ) and the Liouville geodesic current

Recall that each geodesic in the Poincaré disk is uniquely determined by its endpoints on the circle, which are necessarily different. Hence the space of geodesics may be naturally identified with the quotient space $S^1 \times S^1 \setminus \Delta$ under the equivalence relation that identifies the pairs (s, t) and (t, s), where $s \neq t$. A **geodesic current** is a Radon measure defined on this space of geodesics. It may be thought of as a measure L defined on the Borel subsets of $S^1 \times S^1$ that are disjoint from the diagonal Δ, which is finite on compact subsets of $S^1 \times S^1 \setminus \Delta$ and which satisfies the symmetry condition

$$L([a, b] \times [c, d]) = L([c, d] \times [a, b]), \qquad a < b < c < d < a. \qquad (1.7)$$

Proposition 1.3.5. *The diagonal action of* $\mathrm{PSL}(2, \mathbb{R})$ *on* $S^1 \times S^1 \setminus \Delta$ *preserves the geodesic current*

$$Lv = \frac{ds\,dt}{4\sin^2\left(\frac{s-t}{2}\right)}.$$

Proof. Let us first recall that the cross-ratio of four points e^{ia}, e^{ib}, e^{ic}, and e^{id} in S^1 is defined as

$$[e^{ia}, e^{ib}, e^{ic}, e^{id}] = \frac{(e^{ia} - e^{ic})(e^{ib} - e^{id})}{(e^{ia} - e^{id})(e^{ic} - e^{ib})}.$$

One easily checks that cross-ratios are invariant under Möbius transformations. Conversely, if a circle homeomorphism preserves cross-ratios, then it belongs to the Möbius group. Now notice that for $a < b < c < d < a$, the measure $Lv([a, b] \times [c, d])$ equals

$$\int_c^d \int_a^b \frac{ds\,dt}{4\sin^2(\frac{s-t}{2})} = \int_c^d \left[-\frac{\cos(\frac{s-t}{2})}{2\sin(\frac{s-t}{2})} \right]_{s=a}^{s=b} dt$$

$$= \int_c^d \frac{1}{2} \left[\cot\left(\frac{a-t}{2}\right) - \cot\left(\frac{b-t}{2}\right) \right] dt$$

$$= \log\left(\left| \frac{\sin(\frac{b-d}{2})\sin(\frac{a-c}{2})}{\sin(\frac{b-c}{2})\sin(\frac{a-d}{2})} \right| \right).$$

Since $\left|\sin(\frac{x-y}{2})\right| = \frac{|e^{ix} - e^{iy}|}{2}$, this yields

$$Lv([a, b] \times [c, d]) = \log(|[e^{ia}, e^{ib}, e^{ic}, e^{id}]|) = \log([e^{ia}, e^{ib}, e^{ic}, e^{id}]),$$

where the last inequality follows from the fact that the cross-ratio of cyclically ordered points on the circle is a *positive* real number. Since the action of $\mathrm{PSL}(2, \mathbb{R})$ on S^1 preserves cross-ratios, it also preserves the measure Lv. $\qquad \square$

The measure Lv, called the **Liouville measure**, satisfies the equality

$$e^{-Lv([a,b]\times[c,d])} + e^{-Lv([b,c]\times[d,a])} = 1 \qquad (1.8)$$

for all $a < b < c < d < a$. Indeed,

$$e^{-Lv([a,b]\times[c,d])} + e^{-Lv([b,c]\times[d,a])}$$

$$= \frac{1}{[e^{ia}, e^{ib}, e^{ic}, e^{id}]} + \frac{1}{[e^{ib}, e^{ic}, e^{id}, e^{ia}]}$$

$$= \frac{(e^{ia} - e^{id})(e^{ib} - e^{ic})}{(e^{ia} - e^{ic})(e^{ib} - e^{id})} + \frac{(e^{ib} - e^{ia})(e^{ic} - e^{id})}{(e^{ib} - e^{id})(e^{ic} - e^{ia})}$$

$$= \frac{(e^{ia} - e^{id})(e^{ib} - e^{ic}) - (e^{ib} - e^{ia})(e^{ic} - e^{id})}{(e^{ia} - e^{ic})(e^{ib} - e^{id})} = 1.$$

As we will next see, property (1.8) characterizes the Liouville measure.

Given a geodesic current L, we denote the group of circle homeomorphisms preserving L by Γ_L. For instance, it is easy to check that $\Gamma_{Lv} = $ PSL(2, $\mathbb{R}$). In general, the group Γ_L is very small (it is "generically" trivial). However, there exists a very simple condition that ensures that it is topologically conjugate to PSL(2, $\mathbb{R}$). The following result may be considered analogous for the Möbius group to Propositions 1.1.1, 1.2.2, or 1.2.3. We refer to [19] for its proof (see also Exercise 1.3.11).

Proposition 1.3.6. *If L is a geodesic current satisfying property (1.8), then Γ_L is conjugate to PSL(2, $\mathbb{R}$) by a homeomorphism sending L into Lv.*

Exercise 1.3.7. Using the invariance of the Liouville geodesic current under Möbius transformations, show that if $f : I \to \mathbb{R}$ is a C^1 local diffeomorphism satisfying

$$f'(x)f'(y) = \frac{(f(x) - f(y))^2}{(x - y)^2}$$

for all $x \neq y$ in I, then f is of the form $x \mapsto (ax + b)/(cx + d)$ (compare (1.5)).

1.3.3 PSL(2, $\mathbb{R}$) and the convergence property

A sequence (g_n) of circle homeomorphisms has the **convergence property** if it contains a subsequence (g_{n_k}) satisfying one of the following properties:

(i) There exist a and b in S^1 (not necessarily different) such that g_{n_k} converges pointwise to b on $S^1 - \{a\}$, and $g_{n_k}^{-1}$ converges pointwise to a on $S^1 \setminus \{b\}$;

(ii) There exists $g \in \text{Homeo}_+(S^1)$ such that g_{n_k} converges to g and $g_{n_k}^{-1}$ converges to g^{-1} on the circle.

A subgroup Γ of $\mathrm{Homeo}_+(S^1)$ has the convergence property if every sequence of elements in Γ satisfies this property above. Notice that the convergence property is invariant under topological conjugacy.

One easily checks that every subgroup of $\mathrm{PSL}(2, \mathbb{R})$ has the convergence property (see Exercises 1.3.3 and 1.3.11). Conversely, a difficult theorem due to Casson and Jungreis, Gabai, Gëhring and Martin, Hinkkanen, and Tukia [53, 84, 86, 116, 246] asserts that this property characterizes (up to topological conjugacy) the subgroups of $\mathrm{PSL}(2, \mathbb{R})$.

Theorem 1.3.8. *A group of circle homeomorphisms is topologically conjugate to a subgroup of* $\mathrm{PSL}(2, \mathbb{R})$ *if and only if it satisfies the convergence property.*

It is not difficult to show that for discrete subgroups of $\mathrm{Homeo}_+(S^1)$, the convergence property is equivalent to the property that the action on the space of ordered triples of points in S^1 is free and properly discontinuous [246].

Exercise 1.3.9. Prove directly from the definition that if Γ satisfies the convergence property and $g \in \Gamma$ fixes three points on S^1, then $g = Id$.

Exercise 1.3.10. A circle homeomorphism g is C-***quasi-symmetric*** if for all $a < b < c < d < a$ for which $[a, b, c, d] = 2$, one has $1/C \leq [g(a), g(b), g(c), g(d)] \leq C$. Prove that if Γ is a ***uniformly quasi-symmetric*** subgroup of $\mathrm{Homeo}_+(S^1)$ (that is, all of its elements are C-quasi-symmetric with respect to the same constant C), then Γ satisfies the convergence property. Conclude that Γ is topologically conjugate to a subgroup of $\mathrm{PSL}(2, \mathbb{R})$.

Remark. According to a difficult result due to Markovic [160], under the preceding hypothesis the group Γ is conjugate to a subgroup of $\mathrm{PSL}(2, \mathbb{R})$ by a *quasi-symmetric* homeomorphism.

Exercise 1.3.11. Let L be a geodesic current satisfying $L([a, a] \times [b, c]) = 0$ for all $a < b \leq c < a$ and $L([a, b[\times]b, c]) = \infty$ for all $a < b < c < a$ (notice that the Liouville measure satisfies these properties). Prove that Γ_L has the convergence property (see also [181]). Conclude that Γ_L is topologically conjugate to a subgroup of $\mathrm{PSL}(2, \mathbb{R})$. Remark, however, that this conjugacy does not necessarily send L to the Liouville geodesic current, since property (1.8) is invariant under conjugacy and is not necessarily satisfied by the elements of Γ_L.

1.4 Actions of Lie Groups

The object of this section is to show that among locally compact groups that act by circle homeomorphisms, those that may provide new phenomena are the discrete ones (which we think of as zero-dimensional Lie groups). This is the reason that we will mostly consider actions on the circle of *discrete* groups.

Recall that a deep (and already-classical) result by Montgomery and Zippin [171] asserts that a locally compact topological group is a Lie group if and only if it has no "small compact subgroups"; that is, there exists a neighborhood of the identity that contains no nontrivial compact subgroups.

Proposition 1.4.1. *Every locally compact subgroup of* $\mathrm{Homeo}_+(S^1)$ *is a Lie group.*

Proof. Let Γ be a locally compact subgroup of $\mathrm{Homeo}_+(S^1)$. The set

$$V = \{g \in \Gamma : dist(x, g(x)) < 2\pi/3\}$$

is a neighborhood of the identity in Γ. We will show that V contains no nontrivial compact subgroups, which, because of the Montgomery-Zippin theorem, implies the proposition.

Let Γ_0 be a compact subgroup of $\mathrm{Homeo}_+(S^1)$ contained in V. For each $f \in \Gamma$, let $\tilde{f} \in \widetilde{\mathrm{Homeo}}_+(S^1)$ be the (unique) lift of f such that $dist(\tilde{f}(x), x) < 2\pi/3$ for all $x \in \mathbb{R}$. One readily checks that $\widetilde{gh} = \tilde{g}\tilde{h}$ for all g and h in Γ_0. Therefore, Γ_0 embeds into the group $\widetilde{\mathrm{Homeo}}_+(S^1)$. Now $\widetilde{\mathrm{Homeo}}_+(S^1)$ is torsion-free, although Proposition 1.1.2 implies that every nontrivial compact subgroup of $\mathrm{Homeo}_+(S^1)$ has torsion elements. This implies that Γ_0 must be trivial. $\square$

Exercise 1.4.2. In order to avoid referring to Proposition 1.1.2 in the preceding proof, prove the following lemma due to Newman [192] (see also [143]): If f is a nontrivial finite-order homeomorphism of the sphere S^n (normalized so that its diameter is 1), then there exists $i \in \mathbb{N}$ such that $dist(f^i, Id) > 1/2$.

Hint. If the opposite inequality holds for all i, then each orbit is contained in a hemisphere. We may then define a continuous map $bar : S^n \to S^n$ by associating to x the "barycenter" $bar(x)$ of its orbit inside the corresponding hemisphere. This map satisfies $bar(f(x)) = bar(x)$ for all x, from which one easily deduces that the order of f divides its topological degree. Nevertheless, since f is homotopic to the identity, its topological degree is equal to 1.

Remark. From the preceding argument one easily deduces that $dist(f, Id) > 1/2k$, where k is the order of f.

It is not very difficult to obtain the classification of transitive actions of *connected* Lie groups on one-dimensional manifolds; see [88]. Up to topological conjugacy, the complete list consists of

(i) the action of $(\mathbb{R}, +)$ by translations of the line;
(ii) the action of the rotation group $SO(2, \mathbb{R})$ on the circle;
(iii) the action of the affine group $\text{Aff}_+(\mathbb{R})$ on the line;
(iv) the action of the group $PSL_k(2, \mathbb{R})$ whose elements are the lifts of the elements of $PSL(2, \mathbb{R})$ to the k-fold covering of S^1 (which is topologically a circle); and
(v) the action of the group $\widetilde{PSL}(2, \mathbb{R})$ whose elements are the lifts to the real line of the elements of $PSL(2, \mathbb{R})$.

In a certain sense, this classification says that there exist only three distinct types of geometries on one-dimensional manifolds: Euclidean, affine, and projective (compare [159]). The classification of (faithful) nontransitive actions of connected Lie groups follows from the preceding one. Indeed, the orbits of these actions correspond to points or whole intervals. Therefore, denoting by $\text{Fix}(\Gamma)$ the set of global fixed points of the action, on each connected component of the complement of $\text{Fix}(\Gamma)$ we obtain an action given by a surjective homomorphism from Γ into $(\mathbb{R}, +)$, $SO(2, \mathbb{R})$, $\text{Aff}_+(\mathbb{R})$, $PSL_k(2, \mathbb{R})$, or $\widetilde{PSL}(2, \mathbb{R})$.

1.5 Thompson's Groups

For simplicity, in this section we will use the parameterization of the circle by the interval $[0, 1]$ via the map $x \mapsto e^{2\pi i x}$. Let us consider the group of homeomorphisms $\tilde{f} : \mathbb{R} \to \mathbb{R}$ satisfying the following conditions:

(i) $\tilde{f}(0) = 0$;

(ii) there exists a sequence $\ldots x_{-1} < x_0 < x_1 < \ldots$ (diverging in both directions) of dyadic rational numbers such that each of the restrictions $\tilde{f}|_{[x_i, x_{i+1}]}$ is affine with derivative an integer power of 2; and

(iii) $\tilde{f}(x + 1) = \tilde{f}(x) + 1$ for all $x \in \mathbb{R}$.

Each such $\tilde{f}$ induces a homeomorphism f of $[0, 1]$ by letting $f(0) = 0$, $f(1) = 1$, and $f(s) = \tilde{f}(s) \pmod{1}$ for $s \in \,]0, 1[$. We obtain in this way a group of homeomorphisms of $[0, 1]$. This group was first introduced by Thompson and is commonly denoted by F.

Thompson's group F has several remarkable properties, which are not always easy to prove. First of all, F admits the finite presentation

$$F = \langle f, g : [fg^{-1}, f^{-1}gf] = [fg^{-1}, f^{-2}gf^2] = id \rangle,$$

where $[\cdot, \cdot]$ denotes the commutator between two elements, and f and g are the homeomorphisms whose graphs are depicted in Figures 4 and 5, respectively.

Since every nontrivial homeomorphism of the interval has infinite order, F is torsion-free. From the proof of Theorem 1.5.1 later, it will be evident that F has free Abelian subgroups of infinite index. Moreover, the abelianized quotient F/[F, F] is isomorphic to $\mathbb{Z} \times \mathbb{Z}$. To show this, it suffices to take, for each $[h] \in$ F/[F, F], the value of the derivative of h at the endpoints (notice that these values do not depend on the representative of h, since $[f, g]'(0) = [f, g]'(1) = 1$ for all f and g in F). Then, taking the logarithm in base 2 of these values, we obtain a homomorphism from F/[F, F] into $\mathbb{Z} \times \mathbb{Z}$. Actually, it can be easily checked that this homomorphism is an isomorphism. (One may also use the nontrivial fact that the derived group [F, F] is simple [44].)

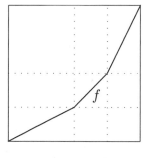

FIGURE 4

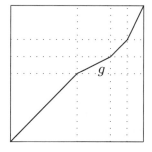

FIGURE 5

Further information concerning Thompson's group F may be found, for instance, in [44]. In particular, in the former reference one may find a discussion of the relevant problem of the *amenability* of F (see Appendix B for the notion of amenability of groups). According to Exercise B.6, one of the main difficulties in this is the fact that F does not contain free subgroups on two generators (see, however, Exercise 2.3.3). This is a corollary of a much more general and nice result due to Brin and Squier [34] that we reproduce below. We remark that if F is nonamenable, then this would lead to the first example of a finitely presented, *torsion-free*, nonamenable group that does not contain $\mathbb{F}_2$. (We point out that a finitely presented, nonamenable group not containing $\mathbb{F}_2$ but having torsion has been constructed by Ol'shanski and Sapir in [199].)

Theorem 1.5.1.   *The group* $\mathrm{PAff}_+([0,1])$ *of piecewise affine homeomorphisms of* $[0,1]$ *does not contain free subgroups on two generators.*

Proof.   Suppose for a contradiction that f and g in $\mathrm{PAff}_+([0,1])$ generate a free group. For each $h \in \mathrm{PAff}_+([0,1])$, let us denote by $supp_0(h)$ the *open support* of h, that is, the set of points in $[0,1]$ that are not fixed by h. The set $I = supp_0(f) \cup supp_0(g)$ may be written as the union of finitely many open intervals $I_1, \ldots, I_n$. Notice that the closure of the set $supp_0([f,g])$ is contained in I, since in a neighborhood of each endpoint x_0 of each I_i the maps f and g are of the form $x \mapsto \lambda(x - x_0) + x_0$, and hence they commute.

Among the nontrivial elements $h \in \langle f, g \rangle$ such that $\overline{supp_0(h)}$ is contained in I, let us choose one, say, h_0, such that the number of connected components of I that intersect $\overline{supp_0(h_0)}$ is minimal. Let $]a, b[$ be one of the connected components of the intersection, and let $[c, d]$ be an interval that is contained in the interior of $]a, b[$ and contains $supp_0(h_0) \cap]a, b[$. If x belongs to $]a, b[$, then the orbit of x under the group generated by f and g is contained in $]a, b[$; moreover, the supremum of this orbit is a point that is fixed by f and g and hence coincides with b. One then deduces the existence of an element $\bar{h} \in \langle f, g \rangle$ sending the interval $[c, d]$ to the right of d. In particular, the restrictions of h_0 and $\bar{h}h_0\bar{h}^{-1}$ to $[a, b]$ commute, and they generate a subgroup isomorphic to $\mathbb{Z} \times \mathbb{Z}$. On the other hand, h_0 and $\bar{h}h_0\bar{h}^{-1}$ do not commute in $\langle f, g \rangle \sim \mathbb{F}_2$, since otherwise they would generate a subgroup isomorphic to $\mathbb{Z}$. The commutator between h_0 and $\bar{h}h_0\bar{h}^{-1}$ is then a nontrivial element whose open support does not intersect $]a, b[$, and so it intersects fewer components of I than the open support of h_0. However, this contradicts the choice of h_0. $\square$

If we consider the homeomorphisms $\tilde{f}$ of the real line that satisfy only the properties (ii) and (iii) corresponding to the lifts of the elements in F, and such that $\tilde{f}(0)$ is a dyadic rational number, then, after passing to the quotient, we obtain a group G of circle homeomorphisms. This group is infinite, has a finite presentation, and is simple. In fact, G was the first example of a group satisfying these three properties simultaneously. (This was one of the original motivations that led Thompson to introduce these groups.)

1.5.1 Thurston's piecewise projective realization

In order to understand Thompson's groups better, we will give two alternative definitions in this section. One is based on Thompson's original work, and the other follows an idea due to Thurston. We begin with some definitions.

A (nondegenerate) *dyadic tree* T is a finite union of closed *edges* (that is, homeomorphic copies of the unit interval, including its endpoints or *vertices*) such that

(i) there exists a marked vertex, called the *root* of the tree and denoted by σ;

(ii) each vertex different from the root is the final point of either one or three edges, while the root is the final point of either two edges or no edge (the latter case appears only when the tree is degenerate: it contains no edge and is reduced to a single point, namely, the root); and

(iii) T is connected.

If a vertex is the final point of three edges, then they may be labeled Υ_d, Υ_l, and Υ_r, according to whether they point down, to the left, or to the right, respectively. The edges starting from σ are labeled Υ_l and Υ_r. A *leaf* of the tree is a vertex v that is the final point of a single edge. The root σ will also be considered a leaf of the degenerate tree. The set of all the leaves of a dyadic tree T will be denoted by $lv(T)$. Notice that there exists a natural cyclic order for the leaves of a dyadic tree, and the notion of *first leaf* may also be defined naturally.

Given a dyadic tree T and a leaf $p \in lv(T)$, we will say that a dyadic tree T' "germinates" from T at the leaf p if T' is the union of T and two edges starting from p. Notice that the number of leaves of the new tree equals that of the original tree plus 1.

Let us now consider two trees T_1 and T_2 having the same number of leaves. We will say that a map from $lv(T_1)$ to $lv(T_2)$ is G-admissible if it preserves the cyclic order between the leaves, and we will say that it is F-admissible if it in addition sends the first leaf of T_1 into the first leaf of T_2. We will define an equivalence relation for G-admissible maps, leaving the task of adapting the definition to the case of F-admissible maps to the reader.

Let us consider a G-admissible map $g : lv(T_1) \rightarrow lv(T_2)$ and a leaf p of T_1. Let T_1' and T_2' be dyadic trees germinating from T_1 and T_2 at p and $g(p)$, respectively. Let us define the map $g' : lv(T_1') \rightarrow lv(T_2')$ by $g'(q) = g(q)$ if q is a leaf of T_1 different from p, and by $g'(p_1) = p_1'$ and $g'(p_2) = p_2'$, where $p_1 \neq p$ and $p_2 \neq p$ are, respectively, the vertices of the edges Υ_l and Υ_r starting from p (and analogously for p_1' and p_2' in relation to $g(p)$). The map g' will be called a ***germination*** of g.

In general, given two G-admissible maps $g : lv(R_1) \rightarrow lv(S_1)$ and $h : lv(R_2) \rightarrow lv(S_2)$, we will say that g is G-equivalent to h if there exists a finite sequence $g_0 = g, g_1, \ldots, g_n = h$ of G-admissible maps such that for each $k \in \{1, \ldots, n\}$, either g_k is a germination of g_{k-1}, or g_{k-1} is a germination of g_k. Let us denote by G the set of G-admissible maps modulo this equivalence relation.

We will now proceed to define a group structure on F and G. Again, we will give the definition only for G, since that for F is analogous. Fix two elements f and g in G. It is not difficult to verify that there exist dyadic trees R, S, and T such that in the class of f and in that of g there exist maps—which we will still denote by f and g, respectively—such that $g : lv(R) \rightarrow lv(S)$ and $f : lv(S) \rightarrow lv(T)$. We then define the element $fg \in G$ as the class of the map

$$fg : lv(R) \rightarrow lv(T).$$

The reader can easily check that this definition does not depend on the chosen representatives and that, endowed with this product, G becomes a group. Figure 6 illustrates the composition of two elements in G. Notice that the neutral element is the class of the map sending the root (viewed as the unique leaf of the degenerate tree) into itself.

We now explain the relationship between the groups G and F previously defined and those acting on the circle and the interval, respectively. To do this, to each vertex of a dyadic tree we associate a ***dyadic interval*** in $[0, 1]$ (i.e., an interval of type $[i/2^n, (i+1)/2^n]$, with $i \in \{1, \ldots, 2^n\}$) as follows:

(i) To the root we associate the interval $[0, 1]$.

(ii) If to the vertex p that is not a leaf we have associated the interval $[a, b]$, then with p_1 and p_2 we associate the intervals $[a, (a+b)/2]$ and $[(a+b)/2, b]$,

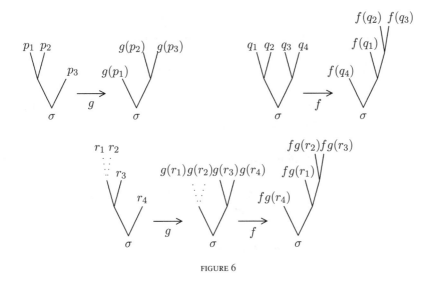

FIGURE 6

respectively, where $p_1 \neq p$ and $p_2 \neq p$ are the final vertices of the edges Υ_l and Υ_r starting from p.

For each $g \in G$, we choose a representative $g : lv(\mathcal{T}_1) \to lv(\mathcal{T}_2)$, and we associate to it the circle homeomorphism sending affinely the interval corresponding to each leaf p of $\mathcal{T}_1$ into the interval corresponding to the leaf $g(p)$. It is not difficult to check that this definition does not depend on the chosen representative of g. Thus we obtain a group homomorphism from the recently defined group G into the group G acting on the circle, and one is easily convinced that this homomorphism is, in fact, an isomorphism.

To conclude this section, we show that G embeds into $\mathrm{Diff}_+^{1+\mathrm{Lip}}(S^1)$. The proof that we give is based on an idea due to Thurston. Following a construction due to Ghys and Sergiescu, we will show in the next section that a stronger result holds: G is topologically conjugate to a subgroup of the group of C^∞ circle diffeomorphisms.

Thurston's idea uses finite partitions of S^1 given by the Farey sequence instead of partitions into dyadic intervals. In other words, to each vertex of a dyadic tree we associate a subinterval of $[0, 1]$ in the following way:

(i) To the root we associate the interval $[0, 1]$.
(ii) If to the vertex p we have associated the interval $[a/b, c/d]$ and p is not a leaf, then to p_1 and p_2 we associate the intervals $[a/b, (a+b)/(c+d)]$ and

$[(a+b)/(c+d), c/d]$, respectively. Here $p_1 \neq p$ and $p_2 \neq p$ are the vertices different from p of the edges Υ_l and Υ_r starting from p.

Then for each $g \in G$, we choose a representative $g : lv(\mathcal{T}_1) \to lv(\mathcal{T}_2)$, to which we associate the circle homeomorphism sending the interval associated to each leaf p of $\mathcal{T}_1$ into the interval associated to $g(p)$ by a (uniquely determined) map in $PSL(2, \mathbb{Z})$. As in the previous case, everything is well defined up to the equivalence relation defining the group structure on G. Actually, one can exhibit the elements in $PSL(2, \mathbb{Z})$ used in this defini-tion. Indeed, it is not difficult to verify by induction that if the interval associated to some vertex is $[a/b, c/d]$, then $bc - ad = 1$. Therefore, the unique map in $PSL(2, \mathbb{Z})$ sending $I = [a/b, c/d]$ into $J = [a'/b', c'/d']$ is $\gamma_{I,J} = \gamma_J \circ \gamma_I^{-1}$, where

$$\gamma_I(x) = \frac{(c-a)x + a}{(d-b)x + b}, \qquad \gamma_J(x) = \frac{(c'-a')x + a'}{(d'-b')x + b'}.$$

Notice that $\gamma_I'(x) = 1/((d-b)x + b)^2$, and therefore

$$\gamma_{I,J}'\left(\frac{a}{b}\right) = \left(\frac{b}{b'}\right)^2, \qquad \gamma_{I,J}'\left(\frac{a'}{b'}\right) = \left(\frac{d}{d'}\right)^2.$$

These equalities show that for each $g \in G$, the associated piecewise $PSL(2, \mathbb{Z})$ circle homeomorphism is actually of class $C^{1+\text{Lip}}$; indeed, the values of the derivatives to the left and to the right of the "break points" coincide.

We have thus constructed a new action of G on the circle, this time by $C^{1+\text{Lip}}$ diffeomorphisms. It is not very difficult to show that the resulting group of piecewise $PSL(2, \mathbb{Z})$ maps and that of piecewise dyadically affine maps are topologically conjugate.

1.5.2 Ghys-Sergiescu's smooth realization

A remarkable (and at first glance surprising) property of Thompson's group G is that it can be realized as a group of C^∞ circle diffeomorphisms. We mention, however, that the general problem of knowing what are the subgroups of $PAff_+([0, 1])$ and $PAff_+(S^1)$ sharing this property is wide open. This problem is particularly interesting (both from the dynamical and algebraic viewpoints) for the groups studied by Stein in [232] and their natural analogs acting on the circle.

Following (part of) [96], we will associate a representation of G in $\text{Homeo}_+(S^1)$ to each homeomorphism $H : \mathbb{R} \to \mathbb{R}$ satisfying the following properties:

(i) For each $x \in \mathbb{R}$ one has $H(x+1) = H(x) + 2$.
(ii) $H(0) = 0$.

Notice that the function $H(x) = 2x$ satisfies these two properties; the associated representation will correspond to the canonical action of G by piecewise affine circle homeomorphisms.

For the construction, let us first introduce some notation. Let us denote by $\mathbb{Q}_2(\mathbb{R})$ the group of dyadic rational numbers (which may be thought of as a subgroup of the translation group). By $\text{Aff}_+(\mathbb{Q}_2, \mathbb{R})$ we will denote the group of affine transformations of the real line that preserve the set of the dyadic rationals, and by $\text{PAff}_+(\mathbb{Q}_2, \mathbb{R})$ we will denote the group of homeomorphisms that are piecewise dyadically affine. Similarly, $\mathbb{Q}_2(S^1)$ will denote the group of dyadic rational rotations of the circle, and $\text{PAff}_+(\mathbb{Q}_2, S^1)$ will denote the group of piecewise dyadically affine circle homeomorphisms. Recall, finally, that for each $a \in \mathbb{R}$, the translation by a is denoted T_a.

Lemma 1.5.2. *The correspondence* $\Phi_H : \mathbb{Q}_2(\mathbb{R}) \to \text{Homeo}_+(\mathbb{R})$ *sending* $p/2^q$ *into* $H^{-q} T_p H^q$ *is well defined and is a group homomorphism.*

Proof. To show that the definition does not lead to a contradiction, we need to check that for all integers $q \geq 0$ and p, one has $\Phi_H(p/2^q) = \Phi_H(2p/2^{q+1})$, that is,

$$H^{-q} T_p H^q = H^{-(q+1)} T_{2p} H^{q+1}.$$

To do this, notice that this equality is equivalent to $H(x+p) = H(x) + 2p$, which follows directly from property (i). To show that Φ_H is a group homomorphism, remark that

$$\Phi_H\left(\frac{p}{2^q} + \frac{p'}{2^q}\right) = H^{-q} T_{p+p'} H^q$$
$$= H^{-q} T_p H^q H^{-q} T_{p'} H^q = \Phi_H\left(\frac{p}{2^q}\right) \Phi_H\left(\frac{p'}{2^q}\right). \quad \square$$

Lemma 1.5.3. *The homomorphism Φ_H of the preceding lemma extends to a homomorphism from $\mathrm{Aff}_+(\mathbb{Q}_2, \mathbb{R})$ into $\mathrm{Homeo}_+(\mathbb{R})$ by*

$$\begin{pmatrix} 2^n & p/2^q \\ 0 & 1 \end{pmatrix} \longmapsto \Phi_H\left(\frac{p}{2^q}\right) \circ H^n.$$

Proof. The claim follows directly from the equality

$$H \circ \Phi_H\left(\frac{p}{2^{q+1}}\right) = \Phi_H\left(\frac{p}{2^q}\right) \circ H. \qquad \square$$

The extension of the homomorphism Φ_H to $\mathrm{Aff}_+(\mathbb{Q}_2, \mathbb{R})$ will also be denoted by Φ_H. The definition for each $g \in \mathrm{PAff}_+(\mathbb{Q}_2, \mathbb{R})$ is a little more subtle. Let us fix a strictly increasing sequence $(a_n)_{n \in \mathbb{Z}}$ of dyadic rational numbers without accumulation points, as well as a sequence of elements $h_n \in \mathrm{Aff}_+(\mathbb{Q}_2, \mathbb{R})$, in such a way that for all $n \in \mathbb{Z}$, one has

$$g|_{[a_n, a_{n+1}]} = h_n|_{[a_n, a_{n+1}]}.$$

If we define $b_n = \Phi_H(a_n)(0)$, then it is easy to see that the sequence $(b_n)_{n \in \mathbb{Z}}$ is also strictly increasing and does not have accumulation points.

Proposition 1.5.4. *If to each $g \in \mathrm{PAff}_+(\mathbb{Q}_2, \mathbb{R})$ we associate the map that on each interval $[b_n, b_{n+1}[$ coincides with $\Phi_H(h_n)$, then we obtain a homomorphism from $\mathrm{PAff}_+(\mathbb{Q}_2, \mathbb{R})$ into $\mathrm{Homeo}_+(\mathbb{R})$ that extends Φ_H.*

Proof. The fact that the map associated to each g is well defined (i.e., it does not depend on the choice of the a_n's) follows readily from the definition, as well as the fact that the map associated to each $g \in \mathrm{Aff}_+(\mathbb{Q}_2, \mathbb{R})$ coincides with $\Phi_H(g)$. To prove that the map associated to each $g \in \mathrm{PAff}_+(\mathbb{Q}_2, \mathbb{R})$ is a homeomorphism, we need to check the continuity at each point b_n, which reduces to showing that

$$\Phi_H(h_n)(b_n) = \Phi_H(h_{n-1})(b_n).$$

Notice that this equality is equivalent to

$$\Phi_H(h_n T_{a_n})(0) = \Phi_H(h_{n-1} T_{a_n})(0),$$

that is,

$$\Phi_H(T_{-a_n} h_{n-1}^{-1} h_n T_{a_n})(0) = 0. \tag{1.9}$$

Now since g is continuous, we have $h_n(a_n) = h_{n-1}(a_n)$. Therefore, $T_{-a_n} h_{n-1}^{-1} h_n T_{a_n}$ is an element of $\text{Aff}_+(\mathbb{Q}_2, \mathbb{R})$ that fixes the origin, that is, a map f of the form $x \mapsto 2^k x$. The desired equality (1.9) then follows from $\Phi_H(f) = H^k$ and from the fact that by property (ii), H fixes the origin. □

Notice that from property (i) it follows that $\Phi_H(p) = T_p$ for every integer p. Therefore, Φ_H induces an injective homomorphism (which we will still denote by Φ_H) from G into $\text{Homeo}_+(S^1)$.

Proposition 1.5.5. *Assume that for some positive integer r or for $r = \infty$, the map H is a C^r diffeomorphism satisfying the following condition:*

(iii)$_r$ $H'(0) = 1$ and $H^{(i)}(0) = 0$ for all $i \in \{2, \ldots, r\}$.

Then the image $\Phi_H(G)$ is contained in the group of C^r circle diffeomorphisms.

Proof. Using the notation of the preceding proposition, we need to show that for each $i \in \{1, \ldots, r\}$, one has

$$\Phi_H(h_n)^{(i)}(b_n) = \Phi_H(h_{n-1})^{(i)}(b_n).$$

But this follows from the fact that the Taylor series expansion of order r of the map $\Phi_H(T_{-a_n} h_{n-1}^{-1} h_n T_{a_n}) = H^k$ coincides with that of the identity. □

Notice that when $r = \infty$, property (iii)$_r$ cannot be satisfied by any real-analytic diffeomorphism. In fact, the groups F and G cannot act faithfully on S^1 by real-analytic diffeomorphisms. (Since G is a simple group, this implies that every action of G by real-analytic diffeomorphisms of the circle is trivial.) This may be shown in many distinct ways, but it will appear evident after § 4.4: Thompson's group F contains solvable subgroups of arbitrary length of solvability, while every solvable group of real-analytic diffeomorphisms of the (closed) interval is metabelian.

In § 2.1.1 we will deal again with some dynamical aspects of the preceding realization. To conclude this chapter, let us point out that the dyadic feature of the preceding arguments is not essential. Indeed, for each integer $m \geq 2$, an analogous construction starting with a map H satisfying $H(x + 1) = H(x) + m$ for each $x \in \mathbb{R}$ leads to an *m-adic Thompson's group*.

From an algebraic point of view, the case $m = 2$ is special in relation to the automorphism group; we refer the reader to [31, 32] for further information on this very interesting topic. For a complete survey of the recent progress (especially on cohomological aspects) on Thompson's groups, see [225].

CHAPTER TWO

Dynamics of Groups
of Homeomorphisms

2.1 Minimal Invariant Sets

2.1.1 The case of the circle

Recall that a subset Λ of S^1 is *invariant* under a subgroup Γ of $\mathrm{Homeo}_+(S^1)$ if $g(x) \in \Lambda$ for every $x \in \Lambda$ and every $g \in \Gamma$. A compact invariant set Λ is *minimal* if its only closed invariant subsets are the empty set and Λ itself.

Theorem 2.1.1. *If Γ is a subgroup of $\mathrm{Homeo}_+(S^1)$, then one (and only one) of the following possibilities occurs:*

(i) *There exists a finite orbit.*

(ii) *All the orbits are dense.*

(iii) *There exists a unique minimal invariant compact set that is homeomorphic to the Cantor set (and that is contained in the set of accumulation points of every orbit).*

Proof. The family of nonempty closed invariant subsets of S^1 is ordered by inclusion. Since the intersection of nested compact sets is (compact and) nonempty, Zorn's lemma allows us to deduce the existence of a minimal nonempty closed invariant set Λ. The boundary $\partial\Lambda$ and the set Λ' of the accumulation points of Λ are closed invariant sets contained in Λ. By the minimality of Λ, one of the following possibilities occurs:

(i) Λ' is empty; in this case Λ is a finite orbit.

(ii) $\partial\Lambda$ is empty; in this case $\Lambda = S^1$, and therefore all the orbits are dense.

(iii) $\Lambda = \Lambda' = \partial\Lambda$; in this case Λ is a closed set with empty interior and having no isolated point; in other words, Λ is homeomorphic to the Cantor set.

We will show that in the last case, Λ is contained in the set of accumulation points of every orbit, which clearly implies its uniqueness. Let x and y be arbitrary points in S^1 and Λ, respectively. We need to show that there exists a sequence (g_n) of elements in Γ such that $g_n(x)$ converges to y. For $x \in \Lambda$, this follows from the minimality of Λ. If $x \in S^1 \setminus \Lambda$, let us consider the interval $I =]a, b[$ contained in $S^1 \setminus \Lambda$ such that both a and b belong to Λ and $x \in I$. Since the orbit of a is dense in Λ, and since Λ does not have isolated points, there must exist a sequence (g_n) in Γ for which $g_n(a)$ tends to y in such a way that the intervals $g_n(I)$ are two-by-two disjoint. The length of $g_n(I)$ must go to zero, and thus $g_n(x)$ converges to y. This concludes the proof. $\qquad\square$

Exercise 2.1.2. Prove that if a group of circle homeomorphisms has finite orbits, then all of them have the same cardinality.

In the case where all the orbits are dense, the action is said to be *minimal*. If there exists a minimal invariant Cantor set, this set is called an *exceptional minimal set*. To understand this case better, it is useful to introduce the following terminology.

Definition 2.1.3. A circle homeomorphism f is *semiconjugate* to g if there exists a continuous degree-1 map $\varphi : S^1 \to S^1$ whose lifts to $\mathbb{R}$ are nondecreasing functions and such that $\varphi f = g\varphi$. Similarly, a group action Φ_1 by circle homeomorphisms is semiconjugate to an action Φ_2 if there exists φ satisfying the preceding properties and such that $\varphi\Phi_1(g) = \Phi_2(g)\varphi$ for every element g in the acting group.

The map φ may be noninjective; if it is injective, the semiconjugacy is in fact a conjugacy. Notice that if a group Γ acts on S^1 with an exceptional minimal set Λ, then, replacing the closure of each connected component of $S^1 \setminus \Lambda$ by a point, we obtain a topological circle S^1_Λ on which Γ acts in a natural way by homeomorphisms. The original action is semiconjugate to the induced minimal action on S^1_Λ.

Remark 2.1.4. The relation of semiconjugacy is not an equivalence relation. The equivalence relation that it generates is sometimes called

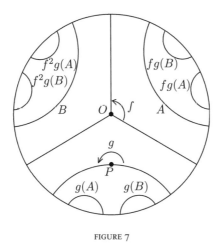

FIGURE 7

monotonic equivalence. More precisely, two actions Φ_1 and Φ_2 are mono-tonically equivalent if there exists an action Φ that is semiconjugate to both Φ_1 and Φ_2. The interested reader may find more on this notion in [43].

It is not difficult to construct homeomorphisms of S^1 that admit exceptional minimal sets. Indeed, every subset of S^1 homeomorphic to the Cantor set may be exhibited as the exceptional minimal set of a circle homeomorphism (see § 2.2.1). We now show an example of a group of real-analytic circle diffeomorphisms having an exceptional minimal set. This group is generated by two Möbius transformations f and g. The diffeomorphism f is the rotation $R_{2\pi/3}$ (centered at the origin $O = (0, 0)$). The diffeomorphism g is also elliptic and corresponds to a rotation of angle π centered at a point $P \in \mathbb{D}$ situated at a Euclidean distance larger than $2 - \sqrt{3}$ from O (see Figure 7). Equivalently, g is the "hyperbolic reflection" with respect to the geodesic passing through P and perpendicular to the geodesic joining this point to O.

The acting group in this example coincides with the ***modular group***, which admits the finite presentation $\Gamma = \langle f, g : g^2 = f^3 = id \rangle$. The standard injection of Γ into $PSL(2, \mathbb{R})$ is obtained by identifying f with $R_{2\pi/3}$ and g with the rotation by π centered at the point $(0, \sqrt{3} - 2) \in \mathbb{D}$. Via this injection, the modular group identifies with $PSL(2, \mathbb{Z})$, and the corresponding action is minimal.

A small perturbation of a "piecewise affine version" of the preceding example allows us to produce an action by C^∞ diffeomorphisms having

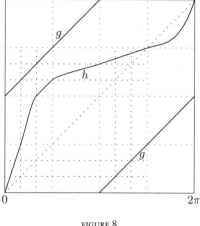

the triadic Cantor set as an exceptional minimal set. To do this, it suffices to consider a circle diffeomorphism h satisfying the following conditions:

(i) Its restriction to $[0, \pi/6]$ (resp. $[\pi, 3\pi/2]$) is affine, with derivative 3 (resp. $1/3$).
(ii) For all $x \in [\pi/3, \pi/2]$, $h(x) = x + 2\pi/3$.
(iii) It is defined in a coherent way on the remaining intervals so that $(hg)^3 = Id$, where g denotes the rotation by an angle π (see Figure 8).

Letting $f = hg$, we obtain the desired action.

It is very interesting to notice that although the example illustrated by Figure 7 already appears in works by Klein and Poincaré, for many years this example was "forgotten", and actually the "first example" of a group of C^∞ circle diffeomorphisms having an exceptional minimal set is commonly attributed to Sacksteder [221]. His example corresponds to (a slight modification of) the one illustrated below...

In the case of **Fuchsian groups** (that is, discrete subgroups of $PSL(2, \mathbb{R})$), a particular terminology for Theorem 2.1.1 is sometimes used. If there exists a finite orbit, then the group is called **elementary**. If the orbits are dense, the group is said to be of **first kind**. Finally, in the case of an exceptional minimal set, the group is of **second kind** [135]. An interesting family of examples of groups of second kind is the **Schottky groups**. These correspond to groups generated by hyperbolic elements g_0 and g_1 in $PSL(2, \mathbb{R})$ for which there exist disjoint intervals I_0, I_1, J_0, and J_1 in S^1 such that $g_i(I_i \cup J_i \cup J_{1-i}) \subset I_i$ and $g_i^{-1}(I_i \cup I_{1-i} \cup J_{1-i}) \subset J_i$ for $i \in \{0, 1\}$.

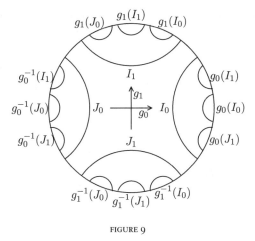

FIGURE 9

Figure 9 illustrates the corresponding combinatorics of the dynamics. It is easy to check that $\langle g_0, g_1 \rangle$ acts in a way that admits an exceptional minimal set, namely,

$$\Lambda = \bigcap_{g \in \langle g_0, g_1 \rangle} g(\overline{I_0} \cup \overline{I_1} \cup \overline{J_0} \cup \overline{J_1}).$$

Moreover, the group generated by g_0 and g_1 is free (see § 2.3.1).

Exercise 2.1.5. Give examples of finitely generated subgroups of $\mathrm{Diff}_+^\infty(S^1)$ admitting an exceptional minimal set Λ so that the orbits of all points in $S^1 \setminus \Lambda$ are dense.

Remark. Using a result of Hector, it is possible to show that if Γ is a (nonnecessarily finitely generated) group of real-analytic circle diffeomorphisms having an exceptional minimal set, then none of its orbits is dense (see, for example, [182]).

Exercise 2.1.6. Give an example of a (non–finitely generated) group of real-analytic circle diffeomorphisms whose action is minimal and such that all of its finitely generated subgroups admit finite orbits (see Example 3.1.7 in case of problems with this).

We close this section with another important example of a group of C^∞ circle diffeomorphisms having an exceptional minimal set, namely, Thompson's group G. Indeed, in § 1.5.2, to each homeomorphism H :

$\mathbb{R} \to \mathbb{R}$ satisfying properties (i) and (ii), we associated a homomorphism $\Phi_H : G \to \mathrm{Homeo}_+(S^1)$, which takes values in $\mathrm{Diff}^r_+(S^1)$ when H is a C^r diffeomorphism satisfying condition (iii)$_r$. It is fairly clear that the groups $\Phi_H(G)$ are topologically conjugate to (the canonical inclusion of) G (in $\mathrm{Homeo}_+(S^1)$). However, according to the following proposition, some of them are not conjugate to G.

Proposition 2.1.7. *If the homeomorphism $H : \mathbb{R} \to \mathbb{R}$ satisfies properties* (i) *and* (ii) *of § 1.5.2 and has at least two fixed points, then the group $\Phi_H(G)$ admits an exceptional minimal set.*

Proof. If a and b are fixed points of H, then property (i) implies that they belong to the same "fundamental domain". In other words, the open interval in the real line whose endpoints are a and b injectively projects into an open interval I of the circle satisfying $\bar{H}^n(I) = I$ for every $n \geq 0$ (where $\bar{H}$ denotes the degree-2 map of S^1 induced by H; see Figure 10). On the other hand, for each $n \geq 0$, the set $\bar{H}^{-n}(I)$ is the union of a family consisting of 2^n disjoint open intervals, and $\bar{H}^{-n}(I) \subset \bar{H}^{-m}(I)$ for all $m \geq n$. Therefore, the union $\cup_{n \geq 0} \bar{H}^{-n}(I)$ is open and invariant under $\bar{H}$, and its complementary set is nonempty. It is therefore not very difficult to conclude that this set is invariant under $\Phi_H(G)$ (see Exercise 2.1.11). Therefore, not every orbit of $\Phi_H(G)$ is dense, and since $\Phi_H(G)$ has no finite orbit, it must preserve an exceptional minimal set. $\square$

Using the techniques of Chapter 3, Ghys and Sergiescu proved that every faithful action of G by C^2 circle diffeomorphisms is semiconjugate to the standard piecewise affine action [96]. More recently, Ghys [88]

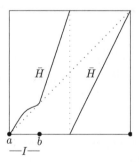

FIGURE 10

and, independently, Liousse [151] showed that the regularity hypothesis is superfluous for this claim (compare Exercise 2.2.21).

Exercise 2.1.8.   Using Theorems 1.5.1 and 2.3.2, show that every action of F by circle homeomorphisms admits a global fixed point.

Exercise 2.1.9.   Prove that if the map H is **expanding**, that is, if for every pair of distinct points x and y in the real line, one has $dist(H(x), H(y)) > dist(x, y)$, then all the orbits of the group $\Phi_H(G)$ are dense.

Exercise 2.1.10.   Prove that G admits actions by piecewise affine circle homeomorphisms having an exceptional minimal set.

Exercise 2.1.11.   Given a homeomorphism H satisfying the properties (i) and (ii) of § 1.5.2, consider the equivalence relation of the induced map $\bar{H}$ on the circle. Prove that the equivalence classes of this relation coincide with the orbits of the group $\Phi_H(G)$.

2.1.2 The case of the real line

There is no analog of Theorem 2.1.1 for general groups of homeomorphisms of the real line. In fact, it is not difficult to construct examples of subgroups of $\mathrm{Homeo}_+(\mathbb{R})$ whose action is not minimal and for which the only closed invariant subsets are the empty set and the real line itself. Nevertheless, a weak version (which in many cases is enough for applications) exists for *finitely generated* subgroups.

Proposition 2.1.12.   *Every finitely generated subgroup of* $\mathrm{Homeo}_+(\mathbb{R})$ *admits a nonempty minimal invariant closed set.*

Proof.   Let $\mathcal{G} = \{f_1, \ldots, f_k\}$ be a finite and symmetric system of generators of a subgroup Γ of $\mathrm{Homeo}_+(\mathbb{R})$ (where **symmetric** means that f^{-1} belongs to $\mathcal{G}$ for every $f \in \mathcal{G}$). If Γ admits a global fixed point, then the claim is obvious. If not, fix any point $x_0 \in \mathbb{R}$, and let x_1 be the maximum among the points $f_i(x_0)$. We claim that every orbit of Γ must intersect the interval $[x_0, x_1]$. Indeed, for every $x \in \mathbb{R}$, the supremum and the infimum of its orbit are global fixed points and therefore coincide with $+\infty$ and $-\infty$, respectively. Thus we can take x'_0 and x'_1 in this orbit such that $x'_0 < x_0 < x_1 < x'_1$. Let $f = f_{i_n} \cdots f_{i_1} \in \Gamma$ be such that $f(x'_0) = x'_1$, where each f_{i_j} belongs to $\mathcal{G}$. Let $m \in \{1, \ldots, n-1\}$ be the largest index for which $f_{i_m} \cdots f_{i_1}(x_0') < x_0$. Then $f_{i_{m+1}} f_{i_m} \cdots f_{i_1}(x_0')$ is in the orbit of x and is greater than or equal to x_0. By definition, it must be smaller than or equal to x_1.

Now let $I = [x_0, x_1]$, and on the family $\mathcal{F}$ of nonempty closed invariant subsets of $\mathbb{R}$, let us consider the order relation $\preceq$ given by $\Lambda_1 \succeq \Lambda_2$ if $\Lambda_1 \cap I \subset \Lambda_2 \cap I$. By the preceding discussion, every orbit under Γ must intersect the interval I, and so $\Lambda \cap I$ is a nonempty compact set for all $\Lambda \in \mathcal{F}$. Therefore, Zorn's lemma provides us with a maximal element for the order $\preceq$, which corresponds to the intersection with I of a nonempty minimal Γ-invariant closed subset of $\mathbb{R}$. $\square$

As in the proof of Theorem 2.1.1, one can analyze the distinct possibilities for the nonempty minimal invariant closed set Λ obtained above. To do this, notice that the boundary $\partial \Lambda$ and the set of accumulation points Λ' are also closed sets invariant under Γ. Because of the minimality of Λ, there are three possibilities:

(i) $\Lambda' = \emptyset$. In this case Λ is discrete. If it is finite, then it is made up of global fixed points. If it is infinite, then it corresponds to a sequence $(y_n)_{n \in \mathbb{Z}}$ satisfying $y_n < y_{n+1}$ for all n and without accumulation points in $\mathbb{R}$.

(ii) $\partial \Lambda = \emptyset$. In this case Λ coincides with the whole line, and hence the action is minimal.

(iii) $\partial \Lambda = \Lambda' = \Lambda$. In this case Λ is "locally" a Cantor set. Therefore, collapsing to a point the closure of each connected component of the complement of Λ, we obtain a topological line on which the original action induces (by semiconjugacy) a minimal action of Γ. However, the reader may easily construct examples showing that unlike the case of the circle, in this case the "exceptional" minimal invariant set is not necessarily unique.

2.2 Some Combinatorial Results

2.2.1 Poincaré's theory

In this section we revisit the most important dynamical invariant of circle homeomorphisms, namely the *rotation number*. We begin with the following well-known lemma about almost subadditive sequences.

Lemma 2.2.1. *Let $(a_n)_{n \in \mathbb{Z}}$ be a sequence of real numbers. Assume that there exists a constant $C \in \mathbb{R}$ such that for all m and n in $\mathbb{Z}$,*

$$|a_{m+n} - a_m - a_n| \leq C. \qquad (2.1)$$

Then there exists a unique $\rho \in \mathbb{R}$ such that the sequence $(|a_n - n\rho|)_{n\in\mathbb{Z}}$ is bounded. This number ρ is equal to the limit of the sequence (a_n/n) as n goes to $\pm\infty$ (in particular, this limit exists).

Proof. For each $n \in \mathbb{N}$, let us consider the interval $I_n = [(a_n - C)/n, (a_n + C)/n]$. We claim that I_{mn} is contained in I_n for every m and n in $\mathbb{N}$. Indeed, by (2.1) we have $|a_{mn} - ma_n| \leq (m-1)C$, from which one concludes that $a_{mn} + C \leq ma_n + mC$, and therefore,

$$\frac{a_{mn} + C}{mn} \leq \frac{a_n + C}{n}.$$

Analogously,

$$\frac{a_{mn} - C}{mn} \geq \frac{a_n - C}{n},$$

and these two inequalities together imply that $I_{mn} \subset I_n$.

Because of the preceding claim, a direct application of the finite-intersection property shows that the set $I = \cap_{n\in\mathbb{N}} I_n$ is nonempty. If ρ belongs to I, then ρ is contained in each of the intervals I_n. This allows to conclude that for every $n \in \mathbb{N}$,

$$|a_n - n\rho| \leq C, \qquad (2.2)$$

thus showing that ρ satisfies the claim of the lemma. If $\rho' \neq \rho$, then

$$|a_n - n\rho'| = |(a_n - n\rho) + n(\rho - \rho')| \geq n|\rho - \rho'| - C,$$

and therefore $|a_n - n\rho'|$ goes to infinity. Finally, dividing by n the expressions in both sides of (2.2) and then passing to the limit, one concludes that $\rho = \lim_{n\to\infty}(a_n/n)$. The case where $n < 0$ is similar, and we leave it to the reader. $\qquad \square$

We now consider the parameterization of S^1 by $[0, 1]$. Given a circle homeomorphism f, we denote by $F : \mathbb{R} \to \mathbb{R}$ any lift of f to the real line.

Proposition 2.2.2. *For each $x \in \mathbb{R}$, the limit*

$$\lim_{n\to\pm\infty} \frac{1}{n}[F^n(x) - x]$$

exists and does not depend on x.

Proof.   First notice that for every x and y in $\mathbb{R}$ and every $m \in \mathbb{Z}$,

$$||F^m(x) - x| - |F^m(y) - y|| \le 2. \tag{2.3}$$

Indeed, since $F^m(x+n) = F^m(x) + n$ for every $n \in \mathbb{Z}$, to show (2.3), we may assume that x and y belong to $[0, 1[$, and in this case inequality (2.3) is clear. Let us now fix $x \in \mathbb{R}$, and let us set $a_n = F^n(x) - x$. From

$$a_{m+n} = F^{m+n}(x) - x = \left[F^m(F^n(x)) - F^n(x)\right] + \left[F^n(x) - x\right]$$

one concludes that

$$\begin{aligned}
|a_{m+n} - a_m - a_n| &= |F^{m+n}(x) - x - (F^m(x) - x) - (F^n(x) - x)| \\
&= |F^m(F^n(x)) - F^n(x) - (F^m(x) - x)| \\
&\le 2.
\end{aligned}$$

By the preceding lemma, the expression $[F^n(x) - x]/n$ converges as n goes to infinity, and inequality (2.3) shows that the corresponding limit does not depend on x. $\square$

If we consider two different lifts of f to the real line, then the limits given by the preceding proposition coincide up to an integer number. We then define the **rotation number** of f by

$$\rho(f) = \lim_{n \to \pm\infty} \frac{1}{n}[F^n(x) - x] \quad \text{mod } 1.$$

As an example, it is easy to check that for the Euclidean rotation of angle $\theta \in [0, 1[$, one has $\rho(R_\theta) = \theta$.

Notice that for every circle homeomorphism f, every $x \in S^1$, and every $m \in \mathbb{Z}$, one has

$$\rho(f^m) = \lim_{n \to \pm\infty} \frac{1}{n}[F^{mn}(x) - x] = m \cdot \lim_{n \to \pm\infty} \frac{1}{mn}[F^{mn}(x) - x],$$

from which one concludes that $\rho(f^m) = m\rho(f)$.

If f has a **periodic point**, then $\rho(f)$ is a rational number. Indeed, if $f^p(x) = x$, then for each lift F of f, we have $F^p(x) = x + q$ for some $q \in \mathbb{Z}$ (here x denotes either a point in S^1 or one of its lifts in $\mathbb{R}$). We then obtain

$$\lim_{n \to \pm\infty} \frac{1}{pn}[F^{pn}(x) - x] = \lim_{n \to \pm\infty} \frac{1}{pn}[(x + nq) - x] = \frac{q}{p},$$

from which one concludes that

$$\rho(f) = \frac{q}{p} \quad \mod 1.$$

Conversely, the following result holds.

Proposition 2.2.3. *If $f \in \mathrm{Homeo}_+(S^1)$ has a rational rotation number, then f has a periodic point.*

Proof. Because of the equality $\rho(g^m) = m\rho(g)$, it suffices to show that every circle homeomorphism with zero rotation number has fixed points. To show this, assume that $f \in \mathrm{Homeo}_+(S^1)$ has no fixed point, and let $F : \mathbb{R} \to \mathbb{R}$ be a lift of f such that $F(0) \in\ ]0, 1[$. Notice that the function $x \mapsto F(x) - x$ has no zero on the line. Therefore, by continuity and periodicity, there exists a constant $\delta \in\ ]0, 1[$ such that for all $x \in \mathbb{R}$,

$$\delta \le F(x) - x \le 1 - \delta.$$

Letting $x = F^i(0)$ in this inequality, taking the sum from $i = 0$ to $i = n - 1$, and dividing by n, we obtain

$$\delta \le \frac{F^n(0)}{n} \le 1 - \delta.$$

If we take the limit as n goes to infinity, the last inequality gives $\delta \le \rho(f) \le 1 - \delta$. Therefore, if $f \in \mathrm{Homeo}_+(S^1)$ has no fixed point, then $\rho(f) \ne 0$. $\square$

We leave to the reader the task of showing that if the rotation number of a circle homeomorphism is rational, then all of its periodic points have the same period.

Proposition 2.2.4. *If two circle homeomorphisms are topologically conjugate, then their rotation numbers coincide.*

Proof. We need to show that $\rho(f) = \rho(gfg^{-1})$ for every f and g in $\mathrm{Homeo}_+(S^1)$. Let F and G be the lifts to the real line of f and g, respectively, such that $F(0)$ and $G(0)$ belong to $[0, 1[$. It is clear that G^{-1} is a lift of g^{-1}. To show the proposition, we need to estimate the value of the expression

$$|(GFG^{-1})^n(x) - F^n(x)| = |GF^nG^{-1}(x) - F^n(x)|.$$

It is not very difficult to check that $|G(x) - x| < 2$ and $|G^{-1}(x) - x| < 2$ for every $x \in \mathbb{R}$. Moreover, if $|x - y| < 2$, then $|F^n(x) - F^n(y)| < 3$. We thus conclude that

$$|GF^nG^{-1}(x) - F^n(x)| \leq |GF^nG^{-1}(x)$$
$$- F^nG^{-1}(x)| + |F^nG^{-1}(x) - F^n(x)| < 5.$$

If we divide by n and pass to the limit, this clearly yields $\rho(f) = \rho(gfg^{-1})$. $\square$

Remark 2.2.5. The preceding proposition may be extended to homeomorphisms that are semiconjugate. We leave the proof of this fact to the reader.

As a consequence of the preceding discussion, the dynamics of a circle homeomorphism f having rational rotation number p/q (where p and q are relatively prime integers) is completely determined by this number, the topology of the set Per(f) of periodic points of f, and the "direction" of the dynamics of f^q on each of the connected components of the complement of Per(f).

The case of an irrational rotation number is much more interesting. The fact that all the orbits of the rotation by an angle $\theta \notin \mathbb{Q}$ are dense might suggest the existence of a unique model (depending on θ) for this case. Nevertheless, it is not very difficult to construct circle homeomorphisms that have an irrational rotation number and are not topologically conjugate to the corresponding rotation. To do this, choose a point $x \in S^1$, and replace each point $R_\theta^i(x)$, $i \in \mathbb{Z}$, by an interval of length $1/2^{|i|}$. We then obtain a "larger" circle S_x^1, and the rotation R_θ induces a homeomorphism $R_{\theta,x}$ of S_x^1 by extending R_θ affinely to each interval that we added to the original circle. The map $R_{\theta,x}$ is semiconjugate to R_θ, and therefore its rotation number equals θ. However, none of the orbits of $R_{\theta,x}$ is dense; in particular, $R_{\theta,x}$ is not topologically conjugate to R_θ.

Nevertheless, the model is unique up to topological semiconjugacy.

Theorem 2.2.6. *If the rotation number $\rho(f)$ of $f \in \mathrm{Homeo}_+(S^1)$ is irrational, then f is semiconjugate to the rotation of angle $\rho(f)$. The semiconjugacy is a conjugacy if and only if all the orbits of f are dense.*

Proof. Let $F : \mathbb{R} \to \mathbb{R}$ be a lift of f to the real line such that $F(0) \in [0, 1[$. By Lemma 2.2.1 and the proof of Proposition 2.2.2, for each $x \in \mathbb{R}$, the value of

$$\varphi(x) = \sup_{n \in \mathbb{Z}} (F^n(x) - n\rho(F))$$

is finite. Moreover, the map $\varphi : \mathbb{R} \to \mathbb{R}$ satisfies the following properties:

(i) It is nondecreasing.
(ii) It satisfies $\varphi(x + 1) = \varphi(x) + 1$ for all $x \in \mathbb{R}$.
(iii) It satisfies $\varphi(F(x)) = \varphi(x) + \rho(F)$ for all $x \in \mathbb{R}$.

From these properties we see that in order to prove that φ is a semi-conjugacy, we need to show that φ is continuous. To do this, first notice that for each $x \in \varphi(\mathbb{R})$, the set $\varphi^{-1}(x)$ is either a point or a nondegenerate interval. Let us denote by $\widetilde{\text{Plan}}(F)$ the union of the interior of these intervals, and by $\widetilde{\text{Salt}}(F)$ the interior of the complement of $\varphi(\mathbb{R})$. The sets $\widetilde{\text{Plan}}(F)$ and $\widetilde{\text{Salt}}(F)$ are invariant under the integer translations on the line, and therefore they project into subsets $\text{Plan}(f)$ and $\text{Salt}(f)$ of the circle, respectively. It is easy to see that $\text{Salt}(f)$ is invariant under the rotation of angle $\rho(f)$. Since $\rho(f)$ is irrational, $\text{Salt}(f)$ must be the empty set, which implies that φ is continuous, thus inducing a semiconjugacy from f to $R_{\rho(f)}$. Finally, notice that $\text{Plan}(f)$ is invariant under f. Therefore, if the orbits under f are dense, then $\text{Plan}(f)$ is empty. In this case φ is injective, and thus it induces a conjugacy between f and $R_{\rho(f)}$. $\square$

By the preceding theorem, the combinatorics of the dynamics of a homeomorphism of irrational rotation number reduces to that of the corresponding rotation. To understand this combinatorics better, for $\theta \in [0, 1] \setminus \mathbb{Q}$, we define inductively the integers

$$q_1 = 1, \qquad q_{n+1} = \min \{q > q_n : dist(q\theta, \mathbb{N}) < dist(q_n\theta, \mathbb{N})\}.$$

It is well known (and easy to check) that the sequence $(q_n)_{n \in \mathbb{N}}$ satisfies

$$dist(q_n\theta, \mathbb{N}) = \{q_n\theta\} \text{ if and only if } dist(q_{n+1}\theta, \mathbb{N}) = 1 - \{q_{n+1}\theta\}, \quad (2.4)$$

where $\{a\} = a - [a]$ denotes the fractional part of a (see, for instance, [110]). If we project these points on the circle and keep the same notation, property (2.4) implies that either

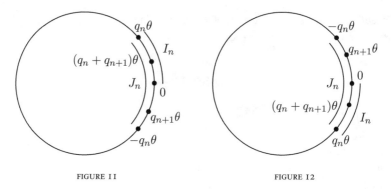

FIGURE 11   FIGURE 12

$$-q_n\theta < q_{n+1}\theta < 0 < -q_{n+1}\theta < q_n\theta < -q_n\theta$$

or

$$q_n\theta < -q_{n+1}\theta < 0 < q_{n+1}\theta < -q_n\theta < q_n\theta.$$

We denote by I_n the interval in S^1 with endpoints 0 and $q_n\theta$ that is closed at 0 and open at $q_n\theta$. The open interval with endpoints $-q_n\theta$ and $q_n\theta$ will be denoted by J_n. Notice that $(q_n + q_{n+1})\theta$ belongs to J_n (see Figures 11 and 12).

Now we claim that the intervals $R_{j\theta}(I_n)$, $j \in \{0, 1, \ldots, q_{n+1} - 1\}$, are disjoint. Indeed, if $R_{j\theta}(I_n) \cap R_{k\theta}(I_n) \neq \emptyset$ for some $0 \leq j < k < q_{n+1}$, then $R_{(k-j)\theta}(I_n) \cap I_n \neq \emptyset$, which implies that $dist((k - j)\theta, 0) < dist(q_n\theta, 0)$. However, since $k - j < q_{n+1}$, this contradicts the definition of q_{n+1}.

From the preceding discussion we deduce that the intervals $R_{j\theta}(J_n)$, $j \in \{0, 1, \ldots, q_{n+1} - 1\}$, cover the circle, and each point $x \in S^1$ is contained in at most two of them. Replacing θ by $-\theta$, we obtain two sequences of intervals I_{-n} and J_{-n} such that $J_n = I_n \cup I_{-n} = J_{-n}$. Thus we see that each point of the circle is contained in at most two intervals of the form $R_{j\theta}(J_n)$, where $j \in \{-(q_{n+1} - 1), -(q_{n+1} - 2), \ldots, 0\}$.

The preceding notation is standard and will often be used in the next sections.

Exercise 2.2.7. Prove that for every $f \in \text{Homeo}_+(S^1)$, there exists an angle $\theta \in [0, 1[$ such that the rotation number of $R_\theta \circ f$ is nonzero.

Exercise 2.2.8. Given two positive parameters $\lambda_1 > 1$ and $\lambda_2 < 1$, let us define $a = a(\lambda_1, \lambda_2)$ as the only real number such that $\lambda_1 a + \lambda_2(1 - a) = 1$.

Consider the (unique) piecewise affine homeomorphism $f = f_{\lambda_1,\lambda_2} : S^1 \to S^1$ satisfying $f(a) = 0$ and whose derivative equals λ_1 on $]0, a[$ and λ_2 on $]a, 1[$ (this example is due to Boshernitzan [21]).

(i) For $\sigma = \lambda_1/\lambda_2$, let h_σ be the homeomorphism of $[0, 1]$ defined by $h_\sigma(x) = (\sigma^x - 1)/(\sigma - 1)$. Show that $h_\sigma^{-1} \circ f_{\lambda_1,\lambda_2} \circ h_\sigma$ coincides with the rotation R_ρ, where ρ satisfies the equality $\sigma^\rho = \lambda_1$.

(ii) Conclude that $\rho(f_{\lambda_1,\lambda_2}) = \log(\lambda_1)/(\log(\lambda_1) - \log(\lambda_2))$.

Exercise 2.2.9. Let us now fix a positive real number $\sigma \neq 1$, and for each $\rho \in]0, 1[$, let us consider the circle homeomorphism $g_{\sigma,\rho} = h_\sigma \circ R_\rho \circ h_\sigma^{-1}$.

(i) Check that $g_{\sigma,\rho}$ coincides with f_{λ_1,λ_2}, where $\lambda_1 = \sigma^\rho$ and $\lambda_2 = \sigma^{\rho-1}$.

(ii) Conclude that inside the group of piecewise affine homeomorphisms of the circle, there exist continuous embeddings of the group of rotations that are not conjugate to the natural embedding by any piecewise affine homeomorphism (see [168, 169] for more on this).

2.2.2 Rotation numbers and invariant measures

According to the Bogolioubov-Krylov theorem, every circle homeomorphism f preserves a probability measure μ (see Appendix B). Notice that the value of $\mu([x, f(x)[)$ is independent of $x \in S^1$. For instance, if $y \in S^1$ is such that $y < f(x) < f(y)$, then

$$\mu([y, f(y)[) = \mu([y, f(x)[) + \mu([f(x), f(y)[)$$
$$= \mu([y, f(x)[) + \mu([x, y[) = \mu([x, f(x)[).$$

This common value will be denoted by $\rho_\mu(f)$.

Theorem 2.2.10. *The value of $\rho_\mu(f)$ coincides* (mod 1) *with the rotation number of f.*

Proof. The measure μ lifts to a σ-finite measure $\tilde\mu$ on $\mathbb{R}$. Let us fix a point $x \in S^1$, and let us consider one of its preimages in $\mathbb{R}$, which we will still denote by x. Notice that every lift F of f preserves $\tilde\mu$. Moreover, $\tilde\mu([x, x+k[) = k$ for all $k \in \mathbb{N}$. Therefore, if $F^n(x) \in [x+k, x+k+1[$, then

$$F^n(x) - x - 1 \leq k \leq \tilde\mu([x, F^n(x)[) \leq k+1 \leq F^n(x) - x + 1.$$

We thus conclude that

$$\lim_{n\to\infty} \frac{F^n(x) - x}{n} = \lim_{n\to\infty} \frac{\tilde{\mu}([x, F^n(x)[)}{n} = \lim_{n\to\infty} \frac{1}{n} \sum_{i=0}^{n-1} \tilde{\mu}([F^i(x), F^{i+1}(x)[),$$

and since for every $i \in \mathbb{N}$ one has $\tilde{\mu}([F^i(x), F^{i+1}(x)[) = \tilde{\mu}([x, F(x)[)$, this yields

$$\lim_{n\to\infty} \frac{F^n(x) - x}{n} = \tilde{\mu}([x, F(x)[).$$

Therefore, (mod 1) the equality $\rho(f) = \mu([x, f(x)[) = \rho_\mu(f)$ holds.   □

It is very useful to describe the support $supp(\mu)$ of a probability measure μ that is invariant under a circle homeomorphism f. If $\rho(f)$ is rational, then f has periodic points, and μ is supported on these points. If $\rho(f)$ is irrational, then two cases may occur: if f admits an exceptional minimal set Λ, then $supp(\mu) = \Lambda$ and μ has no atom, and if the orbits under f are dense, then the support of μ is the whole circle and μ has no atom either.

Given a probability measure μ on the circle, we will denote the group of homeomorphisms that preserve μ by Γ_μ. Notice that the rotation-number function restricted to Γ_μ is a group homomorphism into $\mathbb{T}^1$. For instance, for every f and g in Γ_μ,

$$\rho_\mu(fg) = ([x, fg(x)[) = \mu([x, g(x)[) + \mu([g(x), fg(x)[) = \rho_\mu(f) + \rho_\mu(g),$$

where the second equality holds (mod 1).

If Γ is an amenable subgroup of $\text{Homeo}_+(S^1)$, then there exists a probability measure on S^1 that is invariant under Γ (see Appendix B). As a consequence, we obtain the following proposition (an alternative proof using bounded cohomology appears in [88]).

Proposition 2.2.11. The restriction of the rotation-number function to every amenable subgroup of $\text{Homeo}_+(S^1)$ is a group homomorphism into $\mathbb{T}^1$.

Exercise 2.2.12. Prove that an irrational rotation is **uniquely ergodic**, that is, it has a *unique* invariant probability measure (namely, the Lebesgue measure). Conclude that if Γ is a subgroup of $\text{Homeo}_+(S^1)$ all of whose elements commute with a prescribed minimal circle homeomorphism, then Γ is topologically conjugate to a group of rotations.

Exercise 2.2.13. Give an example of a subgroup Γ of $\text{Homeo}_+(S^1)$ that does not preserve any probability measure on the circle, but such that the restriction of the rotation-number function to Γ is a group homomorphism into $\mathbb{T}^1$.

Remark. After reading § 2.3.2, the reader should be able to prove that for every group Γ satisfying the properties above, the image $\rho(\Gamma)$ is finite.

2.2.3 Faithful actions on the line

In this section we will show that the existence of faithful actions on the line is closely related to the possibility of endowing the corresponding group with a total order relation that is invariant under left multiplication.

Definition 2.2.14. An order relation $\preceq$ on a group Γ is *left-invariant* (resp. *right-invariant*) if for every g and h in Γ such that $g \preceq h$, one has $fg \preceq fh$ (resp. $gf \preceq hf$) for all $f \in \Gamma$. The relation is *bi-invariant* if it is invariant under multiplication on both the left and the right simultaneously. To simplify, we will use the term *ordering* for a left-invariant total order relation on a group. A group is said to be *orderable* (resp. *biorderable*) if it admits an ordering (resp. a bi-invariant ordering).

For an ordering $\preceq$ on a group Γ, we will say that $f \in \Gamma$ is *positive* if $f \succ id$. Notice that the set of these elements forms a semigroup, which is called the *positive cone* of the ordering. An element f is *between g and h* if either $g \prec f \prec h$ or $h \prec f \prec g$.

Exercise 2.2.15. Show that every orderable group is torsion free.

Exercise 2.2.16. Prove that a group Γ is orderable if and only if it contains a subsemigroup Γ_+ such that $\Gamma \setminus \{id\}$ is the disjoint union of Γ_+ and the semigroup $\Gamma_- = \{g : g^{-1} \in \Gamma_+\}$.

Exercise 2.2.17. Show that $\{(m, n) : m > 0, \text{ or } m = 0 \text{ and } n > 0\}$ corresponds to the positive cone of an ordering (called the *lexicographic ordering*) on $\mathbb{Z}^2$.

Exercise 2.2.18. Following the indications below, prove that the free group $\mathbb{F}_2$ is biorderable.

(i) Consider the (non-Abelian) ring $\mathbb{A} = \mathbb{Z}\langle X, Y \rangle$ formed by the formal power series with integer coefficients in two independent variables X and Y. Denoting by $o(k)$ the subset of $\mathbb{A}$ formed by the elements all of whose terms have degree at least k, show that $L = 1 + o(1) = \{1 + S : S \in o(1)\}$ is a subgroup (under multiplication) of $\mathbb{A}$.

(ii) If f and g are the generators of $\mathbb{F}_2$, prove that the map ϕ sending f (resp. g) to the element $1 + X$ (resp. $1 + Y$) in $\mathbb{A}$ extends in a unique way to an injective homomorphism $\phi : \mathbb{F}_2 \to L$.

(iii) Define a lexicographic-type order relation on L that is bi-invariant under multiplication by elements in L (notice that this order will not be invariant under multiplication by elements in $\mathbb{A}$). Using this order and the homomorphism ϕ, endow $\mathbb{F}_2$ with a bi-invariant ordering.

Remark. The preceding technique, due to Magnus, allows one to show that $\mathbb{F}_2$ is *residually nilpotent* (see Appendix A). Indeed, it is easy to check that $\Phi(\Gamma_i^{\text{nil}})$ is contained in $1 + o(i + 1)$ for every $i \geq 0$ (compare Exercise 2.2.25).

The following theorem gives a dynamical characterization of group orderability.

Theorem 2.2.19. *For a countable group Γ, the following are equivalent:*

(i) *Γ acts faithfully on the line by orientation-preserving homeomorphisms.*
(ii) *Γ admits an ordering.*

Proof. Suppose that Γ acts faithfully on the line by orientation-preserving homeomorphisms, and let us consider a dense sequence (x_n) in $\mathbb{R}$. Let us define $g \prec h$ if the smallest index n for which $g(x_n) \neq h(x_n)$ is such that $g(x_n) < h(x_n)$. It is not difficult to check that $\preceq$ is an ordering.

Now suppose that Γ admits an ordering $\preceq$. Choose a numbering (g_i) for Γ, set $t(g_0) = 0$, and assume that $t(g_0), \ldots, t(g_i)$ have already been defined. If g_{i+1} is greater (resp. smaller) than $g_0, \ldots, g_i$, then set $t(g_{i+1}) = \max\{t(g_0), \ldots, t(g_i)\} + 1$ (resp. $\min\{t(g_0), \ldots, t(g_i)\} - 1$). Finally, if $g_m \prec g_{i+1} \prec g_n$ for some m, n in $\{0, \ldots, i\}$, and if g_j is not between g_m and g_n for any $0 \leq j \leq i$, then define $t(g_{i+1}) = (t(g_m) + t(g_n))/2$. The group Γ naturally acts on $t(\Gamma)$ by letting $g(t(g_i)) = t(gg_i)$, and this action continuously extends to the closure of $t(\Gamma)$. Finally, this action may be extended to the whole line in such a way that the map g is affine on each interval of the complement of the closure of $t(\Gamma)$. We leave the details to the reader.   □

Remark 2.2.20. Notice that the first part of the proof does not use the countability assumption. Although this hypothesis is necessary for the second part, many properties of orderable groups involve only finitely many elements. To treat such a property, one may still use dynamical methods by considering the preceding construction for the finitely generated subgroups of the underlying group.

If we fix an ordering $\preceq$ on a countable group Γ, as well as a numbering (g_i) of it, then we will call the action constructed in the proof of Theorem 2.2.19 the *dynamical realization*. It is easy to see that if Γ is nontrivial, then this realization has no global fixed point. Another important property (which is also easy to check) is that if f is an element of Γ whose dynamical realization has two fixed points $a < b$ (which may be equal to $-\infty$ and/or $+\infty$, respectively) such that $]a, b[$ contains no fixed point of f, then there must exist some point of the form $t(g)$ inside $]a, b[$.

Exercise 2.2.21. The preceding construction is very interesting if the ordering is bi-invariant, as is shown here.

(i) Prove that for every element f in the dynamical realization of a bi-invariant ordering on a countable group, either $f(x) \leq x$ for all $x \in \mathbb{R}$, or $f(x) \geq x$ for every $x \in \mathbb{R}$. Conversely, show that every group of homeomorphisms of the real line all of whose elements satisfy this property is biorderable.

(ii) On $\mathrm{PAff}_+([0, 1])$, define an order relation $\preceq$ by letting $f \succ id$ if the right derivative of f at the rightmost point x_f such that f coincides with the identity on $[0, x_f]$ is greater than 1. Show that $\preceq$ is a bi-invariant ordering.

(iii) From (i) and (ii), conclude that Thompson's group F admits actions on the interval that are not semiconjugate to its standard piecewise affine action.

Remark. Biorderings on F were completely classified in [189].

Exercise 2.2.22. Give explicit examples of free groups of homeomorphisms of the real line to conclude that each $\mathbb{F}_n$ is orderable (compare Exercise 2.2.3 and [88, Proposition 4.5]).

For further information concerning orderable groups, we recommend [22, 142, 177]. In the opposite direction, the problem of showing that some particular classes of groups are nonorderable is also very interesting. An important result in this direction, due to Witte-Morris [253], establishes that finite-index subgroups of SL$(n, \mathbb{Z})$ are nonorderable for $n \geq 3$. (Notice that most of these groups are torsion free.)

Theorem 2.2.23. *If $n \geq 3$ and Γ is a finite-index subgroup of SL$(n, \mathbb{Z})$, then Γ is not orderable.*

Proof. Since SL$(3, \mathbb{Z})$ injects into SL$(n, \mathbb{Z})$ for every $n \geq 3$, it suffices to consider the case $n = 3$. Assume for a contradiction that $\preceq$ is an ordering on a finite-index subgroup Γ of SL$(n, \mathbb{Z})$. Notice that for large-enough $k \in \mathbb{N}$, the following elements must belong to Γ:

$$g_1 = \begin{pmatrix} 1 & k & 0 \\ 0 & 1 & 0 \\ 0 & 0 & 1 \end{pmatrix}, \qquad g_2 = \begin{pmatrix} 1 & 0 & k \\ 0 & 1 & 0 \\ 0 & 0 & 1 \end{pmatrix}, \qquad g_3 = \begin{pmatrix} 1 & 0 & 0 \\ 0 & 1 & k \\ 0 & 0 & 1 \end{pmatrix},$$

$$g_4 = \begin{pmatrix} 1 & 0 & 0 \\ k & 1 & 0 \\ 0 & 0 & 1 \end{pmatrix}, \qquad g_5 = \begin{pmatrix} 1 & 0 & 0 \\ 0 & 1 & 0 \\ k & 0 & 1 \end{pmatrix}, \qquad g_6 = \begin{pmatrix} 1 & 0 & 0 \\ 0 & 1 & 0 \\ 0 & k & 1 \end{pmatrix}.$$

It is easy to check that for each $i \in \mathbb{Z}/6\mathbb{Z}$, the following relations hold:

$$g_i g_{i+1} = g_{i+1} g_i, \qquad [g_{i-1}, g_{i+1}] = g_i^k.$$

For $g \in \Gamma$, we define $|g| = g$ if $g \succeq id$, and we let $|g| = g^{-1}$ in the other case. We also write $g \gg h$ if $g \succ h^n$ for every $n \geq 1$.

Let us now fix an index i, and let us consider the ordering $\preceq$ restricted to the subgroup of Γ generated by g_{i-1}, g_i, and g_{i+1}. One can easily check that it is possible to choose three elements a, b, and c that are positive with respect to $\preceq$, such that either $a = g_{i-1}^{\pm 1}$, $b = g_{i+1}^{\pm 1}$, $c = g_i^{\pm k}$, or $a = g_{i+1}^{\pm 1}$, $b = g_{i-1}^{\pm 1}$, $c = g_i^{\pm k}$, and such that

$$ac = ca, \qquad bc = cb, \qquad aba^{-1}b^{-1} = c^{-1}.$$

We claim that either $a \gg c$ or $b \gg c$. To show this, assume that for some $n \geq 1$, one has $c^n \succ a$ and $c^n \succ b$. Let $d_m = a^m b^m (a^{-1} c^n)^m (b^{-1} c^n)^m$. Since d_m is a product of positive elements, d_m is positive. On the other hand, it is not difficult to check that $d_m = c^{-m^2 + 2mn}$, and therefore $d_m \prec id$ for large-enough m, which is a contradiction.

The preceding claim allows us to conclude that either $|g_i| \ll |g_{i-1}|$ or $|g_i| \ll |g_{i+1}|$. If we assume that $|g_1| \ll |g_2|$, then we obtain $|g_1| \ll |g_2|$

$\ll |g_3| \ll |g_4| \ll |g_5| \ll |g_6| \ll |g_1|$, which is a contradiction. The case where $|g_1| \gg |g_2|$ is analogous. □

It follows from an important theorem due to Margulis that for $n \geq 3$, every normal subgroup of a finite-index subgroup of $SL(n, \mathbb{Z})$ is either finite or of finite index (see [158]). As a corollary, we obtain the following strong version of Witte-Morris's theorem.

Theorem 2.2.24. For $n \geq 3$, every action of a finite-index subgroup of $SL(n, \mathbb{Z})$ by homeomorphisms of the line is trivial.

Exercise 2.2.25. Prove directly that every torsion-free nilpotent group is orderable (actually, it is biorderable). More generally, prove that the same is true for every residually nilpotent group Γ for which the (Abelian) quotients $\Gamma_i^{nil}/\Gamma_{i+1}^{nil}$ are torsion free (see Appendix A).

Hint. Use Exercise A.2.

Exercise 2.2.26. Let N_n be the group of $n \times n$ upper triangular matrices with integer entries and such that each entry in the diagonal equals 1. Prove that N_n is orderable by using its natural action on $\mathbb{Z}^n$ and the lexicographic order on it.

Remark. It is easy to check that N_n is nilpotent. On the other hand, according to a classical result due to Malcev [211], every nilpotent, finitely generated, torsion-free group embeds into N_n for some n. This allows one to reobtain indirectly the first claim of Exercise 2.2.25.

2.2.4 Free actions and Hölder's theorem

The main results of this section are classical and essentially due to Hölder. Roughly, they state that free actions on the line exist only for groups admitting an order relation satisfying an Archimedean-type property. Moreover, these groups are necessarily isomorphic to subgroups of $(\mathbb{R}, +)$, and the corresponding actions are semiconjugate to actions by translations.

Definition 2.2.27. An ordering $\preceq$ on a group Γ is said to be **Archimedean** if for all g and h in Γ such that $g \neq id$, there exists $n \in \mathbb{Z}$ satisfying $g^n \succ h$.

Proposition 2.2.28.   *If Γ is a group acting freely by homeomorphisms of the real line, then Γ admits a bi-invariant Archimedean ordering.*

Proof.   Let $\preceq$ be the left-invariant order relation on Γ defined by $g \prec h$ if $g(x) < h(x)$ for some (equivalently, for all) $x \in \mathbb{R}$. This order relation is total, and using the fact that the action is free, one easily checks that it is also right-invariant and Archimedean. $\square$

The converse to the preceding proposition is a direct consequence of the following one. As we will see in Exercises 2.2.30 and 2.2.31, the hypothesis of bi-invariance can be weakened, and left-invariance is sufficient.

Proposition 2.2.29.   *Every group admitting a bi-invariant Archimedean ordering is isomorphic to a subgroup of $(\mathbb{R}, +)$.*

Proof.   Assume that a nontrivial group Γ admits a bi-invariant Archimedean ordering $\preceq$, and let us fix a positive element $f \in \Gamma$. For each $g \in \Gamma$ and each $p \in \mathbb{N}$, let us consider the unique integer $q = q(p)$ such that $f^q \preceq g^p \prec f^{q+1}$.

Claim (i). The sequence $q(p)/p$ converges to a real number as p goes to infinity.

Indeed, if $f^{q(p_1)} \preceq g^{p_1} \prec f^{q(p_1)+1}$ and $f^{q(p_2)} \preceq g^{p_2} \prec f^{q(p_2)+1}$, then

$$f^{q(p_1)+q(p_2)} \preceq g^{p_1+p_2} \prec f^{q(p_1)+q(p_2)+2}.$$

Therefore, $q(p_1) + q(p_2) \le q(p_1 + p_2) \le q(p_1) + q(p_2) + 1$. The convergence of the sequence $(q(p)/p)$ to some real number then follows from Lemma 2.2.1.

Claim (ii). If we denote by $\phi(g)$ the limit of $q(p)/p$ in Claim (i), the resulting map $\phi : \Gamma \to (\mathbb{R}, +)$ is a group homomorphism.

Indeed, let g_1 and g_2 be arbitrary elements in Γ. Let us suppose that $g_1 g_2 \preceq g_2 g_1$ (the case where $g_2 g_1 \preceq g_1 g_2$ is analogous). Since $\preceq$ is bi-invariant, if $f^{q_1} \preceq g_1^p \prec f^{q_1+1}$ and $f^{q_2} \preceq g_2^p \prec f^{q_2+1}$, then

$$f^{q_1+q_2} \preceq g_1^p g_2^p \preceq (g_1 g_2)^p \preceq g_2^p g_1^p \prec f^{q_1+q_2+2}.$$

From this, one concludes that

$$\phi(g_1) + \phi(g_2) = \lim_{p \to \infty} \frac{q_1 + q_2}{p}$$
$$\leq \phi(g_1 g_2)$$
$$\leq \lim_{p \to \infty} \frac{q_1 + q_2 + 1}{p} = \phi(g_1) + \phi(g_2),$$

and therefore $\phi(g_1 g_2) = \phi(g_1) + \phi(g_2)$.

Claim (iii). The homomorphism ϕ is one-to-one.

Notice that ϕ is order preserving in the sense that if $g_1 \preceq g_2$, then $\phi(g_1) \leq \phi(g_2)$. Moreover, $\phi(f) = 1$. Let h be an element in Γ such that $\phi(h) = 0$. Assume that $h \neq id$. Then there exists $n \in \mathbb{Z}$ such that $h^n \succeq f$. From this, one concludes that $0 = n\phi(h) = \phi(h^n) \geq \phi(f) = 1$, which is absurd. Therefore, if $\phi(h) = 0$, then $h = id$, and this concludes the proof. $\square$

If Γ is an infinite group acting freely on the line, then we can endow it with the order relation introduced in the proof of Proposition 2.2.28. This order allows us to construct an embedding ϕ from Γ into $(\mathbb{R}, +)$. If $\phi(\Gamma)$ is isomorphic to $(\mathbb{Z}, +)$, then the action of Γ is conjugate to the action by integer translations. In the other case, the group $\phi(\Gamma)$ is dense in $(\mathbb{R}, +)$. For each point x in the line, we then define

$$\varphi(x) = \sup\{\phi(h) \in \mathbb{R} : h(0) \leq x\}.$$

It is easy to see that $\varphi : \mathbb{R} \to \mathbb{R}$ is a nondecreasing map. Moreover, it satisfies the equality $\varphi(h(x)) = \varphi(x) + \phi(h)$ for all $x \in \mathbb{R}$ and all $h \in \Gamma$. Finally, φ is continuous, since otherwise the set $\mathbb{R} \setminus \varphi(\mathbb{R})$ would be a nonempty open set invariant under the translations of $\phi(\Gamma)$, which is impossible.

To summarize, if Γ is a group acting freely on the line, then its action is semiconjugate to an action by translations.

Exercise 2.2.30. If Γ admits an Archimedean ordering, then this ordering is necessarily bi-invariant. To show this claim (first remarked by Conrad [57]), follow these steps:

(i) Prove that an ordering $\preceq$ is bi-invariant if and only if its positive cone is a normal subsemigroup, that is, hgh^{-1} belongs to Γ_+ for every $g \in \Gamma_+$ and every $h \in \Gamma$ (see Exercise 2.2.16).

(ii) Let $\preceq$ be an Archimedean ordering on a group Γ. Suppose that $g \in \Gamma_+$ and $h \in \Gamma_-$ are such that $hgh^{-1} \notin \Gamma_+$, and consider the smallest positive integer n for which $h^{-1} \prec g^n$. Using the relation $hgh^{-1} \prec id$, show that $h^{-1} \prec g^{-1}h^{-1} \prec g^{n-1}$, thus contradicting the definition of n. Conclude that Γ_+ is stable under conjugacy by elements in Γ_-.

(iii) Assume now that $g \in \Gamma_+$ and $h \in \Gamma_+$ verify $hgh^{-1} \notin \Gamma_+$. In this case, $hg^{-1}h^{-1} \succ id$, and since $h^{-1} \in \Gamma_-$, by (ii) one has $h^{-1}(hg^{-1}h^{-1})h \in \Gamma_+$, that is, $g^{-1} \in \Gamma_+$, which is absurd.

Exercise 2.2.31.   As an alternative argument to that of the preceding exercise, show that if a countable group is endowed with an Archimedean ordering, then the action of the corresponding dynamical realization is free.

Let us now consider a group Γ acting freely by circle homeomorphisms. In this case, the preimage $\tilde{\Gamma}$ of Γ in $\widetilde{\text{Homeo}}_+(S^1)$ acts freely on the line. If we repeat the arguments of the proof of Proposition 2.2.29 by considering the translation $x \mapsto x + 1$ as being the positive element f, then we obtain that $\tilde{\Gamma}$ is isomorphic to a subgroup of $(\mathbb{R}, +)$, and this isomorphism projects to an isomorphism between Γ and a subgroup of $SO(2, \mathbb{R})$. We record this fact as a theorem.

Theorem 2.2.32.   *If Γ is a group acting freely by circle homeomorphisms, then Γ is isomorphic to a subgroup of $SO(2, \mathbb{R})$.*

As in the case of the real line, under the preceding hypothesis the action of Γ is semiconjugate to that of the corresponding group of rotations.

Exercise 2.2.33.   Prove that if Γ is a finitely generated subgroup of $\text{Homeo}_+(S^1)$ all of whose elements have finite order, then Γ is finite.

Remark. It is unknown whether the same is true for torsion subgroups of $\text{Homeo}_+(S^2)$.

Exercise 2.2.34.   Let Γ be a subgroup of $PSL(2, \mathbb{R})$ all of whose elements are elliptic. Prove that Γ is conjugate to a group of rotations.

Exercise 2.2.35.   Give an alternative proof of Proposition 1.1.2 using Hölder's theorem.

The preceding results show that free actions on the circle or the real line are topologically semiconjugate to the actions of groups of rotations or translations, respectively. Similarly, note that nontrivial elements of the affine group fix at most one point. In what follows, we will see that up to topological semiconjugacy and discarding a degenerate case, this property characterizes the affine group. The next result is due to Solodov.

Theorem 2.2.36. *Let Γ be a subgroup of* Homeo$_+(\mathbb{R})$ *such that every nontrivial element in Γ fixes at most one point. Suppose that there is no global fixed point for the action. Then Γ is semiconjugate to a subgroup of the affine group.*

Notice that the case we are not considering, namely, when there exists a point x_0 that is fixed by every element, the actions of Γ on $]-\infty, x_0[$ and $]x_0, \infty[$ are free and therefore are semiconjugate to actions by translations. However, the action of Γ on the line is not necessarily conjugate to the action of a subgroup of the stabilizer of some point in the affine group, since Γ may contain elements for which x_0 is a parabolic fixed point...

Proof of Theorem 2.2.36. If the action is free, then the claim of the theorem follows from Hölder's theorem. We will assume throughout that the action is not free. Under this assumption, Γ cannot be Abelian. Indeed, if Γ were Abelian, then the orbit of a fixed point x_0 of a nontrivial element $g \in \Gamma$ would be contained in the set of fixed points of g. Thus x_0 would be a global fixed point of Γ, which is a contradiction.

First step. We claim that if Γ_0 is a normal subgroup of Γ containing a nontrivial element with a fixed point, then Γ_0 has an element with an attracting fixed point.

Indeed, let $h_0 \in \Gamma_0$ be a nontrivial element such that $h_0(x_0) = x_0$ for some point $x_0 \in \mathbb{R}$. Suppose that x_0 is a parabolic fixed point of h_0. Replacing h_0 by its inverse if necessary, we may assume that $h_0(y) > y$ for $y \neq x_0$. By hypothesis, there exists an element $g \in \Gamma$ such that $x_1 = g(x_0) \neq x_0$. Replacing g by g^{-1} if necessary, we may assume that $x_1 > x_0$. Let us consider the elements $h_1 = gh_0g^{-1} \in \Gamma_0$ and $h = h_0h_1^{-1} \in \Gamma_0$. It is easy to see that $h(x_0) < x_0$ and $h(x_1) > x_1$. Thus h has a repelling fixed point in $]x_0, x_1[$, which proves the claim.

Second step. The definition of an order relation.

Given g and h in Γ, we write $g \preceq h$ if there exists $x \in \mathbb{R}$ such that $g(y) \le h(y)$ for every $y \ge x$. It is easy to check that this defines a total and bi-invariant order relation. We claim that this ordering satisfies the following weak form of the Archimedean property: if $f \in \Gamma$ has a repelling fixed point and $g \in \Gamma$, then there exists $n \in \mathbb{N}$ such that $g \preceq f^n$. Indeed, for the fixed point x_0 of f, let x_- and x_+ be such that $x_- < x_0 < x_+$. For large-enough $n \in \mathbb{N}$, we have $f^n(x_-) < g(x_-)$ and $f^n(x_+) > g(x_+)$, and therefore $g^{-1} f^n$ has a fixed point in the interval $]x_-, x_+[$. Since $g^{-1} f^n(x_+) > x_+$, this implies that $f^n(x) > g(x)$ for all $x \ge x_+$, and thus $g \preceq f^n$.

Third step. A homomorphism into the reals.

Let us fix an element $f \in \Gamma$ with a repelling fixed point. As in the proof of Hölder's theorem, for $g \in \Gamma$ such that $g \succeq id$, we define

$$\phi(g) = \lim_{p \to \infty} \left\{ \frac{q}{p} : f^q \preceq g^p \prec f^{q+1} \right\},$$

and for $g \preceq id$, we let $\phi(g^{-1}) = -\phi(g)$. The map $\phi : \Gamma \to (\mathbb{R}, +)$ is a group homomorphism satisfying $\phi(f) = 1$. Notice that if Γ_0 is a normal subgroup of Γ containing a nontrivial element having a fixed point, then by the first step of the proof there exists $h \in \Gamma_0$ with a repelling fixed point. By the second step we have $f \preceq h^n$ for large-enough $n \in \mathbb{N}$, and thus $\phi(h) \ge 1/n$. In particular, $\phi(\Gamma_0) \ne \{0\}$.

Fourth step. The action of $[\Gamma, \Gamma]$ on the line is free.

Indeed, because $[\Gamma, \Gamma]$ is normal in Γ, if Γ has a nontrivial element with a fixed point, then $\phi([\Gamma, \Gamma]) \ne 0$. Nevertheless, this is absurd, since $[\Gamma, \Gamma]$ is contained in the kernel of ϕ.

Therefore, $[\Gamma, \Gamma]$ is semiconjugate to a group of translations; that is, there exists a group homomorphism $\phi_0 : [\Gamma, \Gamma] \to (\mathbb{R}, +)$ and a continuous nondecreasing surjective map φ of the line such that $\varphi(h(x)) = \varphi(x) + \phi_0(h)$ for all $x \in \mathbb{R}$ and all $h \in [\Gamma, \Gamma]$. We claim that $\phi_0([\Gamma, \Gamma])$ is nondiscrete. If not, the conjugacies by elements in Γ would preserve the generator of $\phi_0([\Gamma, \Gamma]) \sim \mathbb{Z}$, and thus $[\Gamma, \Gamma]$ would be contained in the center of Γ. However, this is impossible, because Γ contains elements having one fixed point.

Fifth step. End of the proof.

By the preceding argument, the image $\phi_0([\Gamma, \Gamma])$ is dense in $\mathbb{R}$. It is then easy to see that the conjugacy φ of the fourth step is unique up to composition on the right with elements in the affine group. Since $[\Gamma, \Gamma]$ is normal in Γ, for each $g \in \Gamma$, the homeomorphism $\varphi g \varphi^{-1}$ belongs to the affine group, and this finishes the proof. $\quad\square$

Once again, we emphasize that neither Hölder's theorem nor Solodov's theorem can be extended in a natural way to groups acting (even minimally and smoothly) on the circle so that every nontrivial element fixes at most two points [145, 182].

2.2.5 Translation numbers and quasi-invariant measures

As in the case of actions on the circle, for a Radon measure υ on the line we may consider the group Γ_υ of the homeomorphisms that preserve it. For $g \in \Gamma_\upsilon$, we define its **translation number** with respect to υ by

$$
\tau_\upsilon(g) = \begin{cases} \upsilon([x, g(x)[) & \text{if } g(x) > x, \\ 0 & \text{if } g(x) = x, \\ -\upsilon([g(x), x[) & \text{if } g(x) < x. \end{cases}
$$

It is easy to see that this number does not depend on the choice of $x \in \mathbb{R}$.

The translation number satisfies many properties that are similar to those of the rotation number of circle homeomorphisms. For instance, for $g \in \Gamma_\upsilon$, one has

$$\tau_\upsilon(g) = 0 \text{ if and only if } g \text{ has a fixed point.} \tag{2.5}$$

Indeed, if $\text{Fix}(g) = \emptyset$, then the orbit of every point x in the line is unbounded from both sides. Let us fix $x \in \mathbb{R}$, and let us assume that $g(x) > x$ (if this is not the case, then we may replace g by g^{-1}). If $\tau_\upsilon(g) = 0$, then, letting n go to infinity in the equality $\upsilon([x, g^n(x)[) = \upsilon([g^{-n}(x), x[) = 0$, we conclude that $\upsilon(]-\infty, +\infty[) = 0$, which is absurd. Conversely, if $\text{Fix}(g)$ is nonempty, then by definition we have $\tau_\upsilon(g) = 0$.

Let us remark that a stronger property holds for elements $g \in \Gamma_\upsilon$, namely,

$$\text{if } \text{Fix}(g) \neq \emptyset, \quad \text{then } supp(\upsilon) \subset \text{Fix}(g). \tag{2.6}$$

Indeed, if $supp(\upsilon)$ is not contained in $\text{Fix}(g)$, then there is a positive υ-measure set A contained in a connected component of the complement of Fix(g) such that $A \cap g(A) = \emptyset$. At least one of the sets $\cup_{n\in\mathbb{N}} g^n(A)$ or $\cup_{n\in\mathbb{N}} g^{-n}(A)$ must be bounded and therefore of finite υ-measure. However, the υ-measure of these sets equals $\sum_{n\in\mathbb{N}} \upsilon(A) = \infty$.

Notice that the function $\tau_\upsilon : \Gamma_\upsilon \to \mathbb{R}$ is a group homomorphism from Γ_υ into $(\mathbb{R}, +)$. This property will be very important in dealing with the problem of the uniqueness (up to a scalar factor) of the invariant Radon measure.

Lemma 2.2.37. *If υ_1 and υ_2 are Radon measures that are invariant under a subgroup Γ of $\mathrm{Homeo}_+(\mathbb{R})$, then there exists $\kappa > 0$ such that the homomorphisms τ_{υ_1} and τ_{υ_2} satisfy the relation $\tau_{\upsilon_1} = \kappa \tau_{\upsilon_2}$.*

Proof. Because of property (2.5), the kernels of τ_{υ_1} and τ_{υ_2} coincide with $\Gamma_0 = \{g : \mathrm{Fix}(g) \neq \emptyset\}$. We then have two homomorphisms τ_1 and τ_2 from Γ/Γ_0 into $(\mathbb{R}, +)$. Let us fix a point x_0 in Fix(Γ_0) (the existence of such a point is ensured by (2.6)). The group Γ/Γ_0 acts freely on the orbit $\Gamma(x_0)$; hence it admits an Archimedean ordering $\preceq$, namely, the one given by $g_1\Gamma_0 \prec g_2\Gamma_0$ if $g_1(x_0) < g_2(x_0)$. Notice that both τ_{υ_1} and τ_{υ_2} preserve this order. If we fix an element $f \in \Gamma$ such that $\Gamma_0 \prec f\Gamma_0$, it is easy to see that for every $g \in \Gamma$, one has

$$\tau_i(g\Gamma_0) = \tau_i(f\Gamma_0) \cdot \lim_{p\to\infty} \left\{ \frac{q}{p} : f^q\Gamma_0 \preceq g^p\Gamma_0 \prec f^{q+1}\Gamma_0 \right\}.$$

Thus $\tau_2(f) \cdot \tau_{\upsilon_1} = \tau_1(f) \cdot \tau_{\upsilon_2}$, which concludes the proof. $\square$

If Γ preserves υ and $\tau_\upsilon(\Gamma)$ is trivial or isomorphic to $\mathbb{Z}$, one cannot expect to have uniqueness (up to a scalar factor) of the invariant measure υ. However, the case where $\tau_\upsilon(\Gamma)$ is dense in $\mathbb{R}$ is distinct.

Proposition 2.2.38. *If υ_1 and υ_2 are two Radon measures that are invariant under a subgroup Γ of $\mathrm{Homeo}_+(\mathbb{R})$ such that $\tau_{\upsilon_1}(\Gamma)$ and $\tau_{\upsilon_2}(\Gamma)$ are dense in $\mathbb{R}$, then there exists $\kappa > 0$ such that these measures satisfy the relation $\upsilon_1 = \kappa \upsilon_2$.*

Proof. By the preceding lemma, 2.2.37, after normalization we may assume that $\tau_{\upsilon_1} = \tau_{\upsilon_2}$. We will then show that $\upsilon_1 = \upsilon_2$. To do this, first notice that none of these measures has atoms. Indeed, if $\upsilon_i(\{x_0\}) > 0$, then every positive element in $\tau_{\upsilon_i}(\Gamma)$ would be greater than or equal to $\upsilon_i(\{x_0\}) > 0$, which contradicts the fact that $\tau_{\upsilon_i}(\Gamma)$ is dense. A similar argument shows that the actions of Γ on the supports of υ_1 and υ_2 are minimal.

Now we show that the supports $supp(v_1)$ and $supp(v_2)$ are actually equal. Indeed, if they were not equal, there would be a point $x \in supp(v_i) \setminus supp(v_{i+1})$ (where $i \in \mathbb{Z}/2\mathbb{Z}$). Because of the density of the orbits on the supports, we could then choose $g \in \Gamma$ such that $g(x) \neq x$ and such that the v_{i+1}-measure of the interval with endpoints x and $g(x)$ is zero. However, this would imply that $\tau_{v_{i+1}}(g) = 0$ and $\tau_{v_i}(g) \neq 0$, which is absurd.

To finish the proof of the equality of v_1 and v_2, we need to show that they give the same mass to intervals having endpoints in their common support. If $[x, y]$ is an interval of this type, we may choose $g_n \in \Gamma$ such that $g_n(x)$ converges to y. We then have

$$v_1([x, y]) = \lim_{n \to \infty} v_1([x, g_n(x)]) = \lim_{n \to \infty} \tau_{v_1}(g_n)$$

$$= \lim_{n \to \infty} \tau_{v_2}(g_n) = \lim_{n \to \infty} v_2([x, g_n(x)]) = v_2([x, y]),$$

thus concluding the proof. $\qquad\qquad\square$

The preceding discussion shows how important it is to know a priori what conditions ensure the existence of an invariant Radon measure. The following result, due to Plante [207], is an important step in this direction.

Theorem 2.2.39. *If Γ is a finitely generated virtually nilpotent subgroup of* Homeo$_+(\mathbb{R})$, *then Γ preserves a Radon measure on the line.*

We immediately state a corollary of this result that will be useful later.

Corollary 2.2.40. *If Γ is a finitely generated virtually nilpotent subgroup of* Homeo$_+(\mathbb{R})$, *then the action of the commutator subgroup $[\Gamma, \Gamma]$ has global fixed points.*

Indeed, if v is an invariant Radon measure, then the translation number of every element of $[\Gamma, \Gamma]$ with respect to v is zero. The claim of the corollary then follows from (2.5).

The original proof by Plante of Theorem 2.2.39 involves very interesting ideas related to growth of groups (see § 4.2.2), the notion of a pseudogroup (see § 3.2), and group amenability (see Appendix B), which apply in more general situations. Nevertheless, as an application of the methods from § 2.1.1, we will give a more direct proof based on the fact that nilpotent groups do not contain free semigroups.

Definition 2.2.41. Two elements f and g in a group Γ generate a *free semigroup* if the elements of the form $f^n g^{m_r} f^{n_r} \cdots g^{n_1} f^{n_1} g^m$, where n_j and m_j are positive integers, $m \geq 0$, and $n \geq 0$, are two-by-two different for different choices of the exponents.

Exercise 2.2.42. Prove that virtually nilpotent groups do not contain free semigroups on two generators.

Hint. For nilpotent groups, use induction on the nilpotence degree.

To find free semigroups inside groups acting on the line, the following notion is quite appropriate.

Definition 2.2.43. Two orientation-preserving homeomorphisms of the real line are *crossed* on an interval $]a, b[$ if one of them fixes a and b but no other point in $[a, b]$, while the other one sends either a or b into $]a, b[$. Here we allow the case $a = -\infty$ or $b = +\infty$.

In foliation theory the notion of crossed elements corresponds to that of *resilient leaves* (*feuilles ressort* in French terminology); see, for instance, [46]. However, the latter is more general, since it also applies to *pseudogroups* of homeomorphisms of one-dimensional manifolds (cf. Definition 3.2.1). In addition, crossed elements are dynamically relevant since they somewhat correspond to one-dimensional versions of *Smale horseshoes* [202].

The next elementary criterion showing that certain semigroups are free is well known.

Lemma 2.2.44. Every subgroup of $\mathrm{Homeo}_+(\mathbb{R})$ having crossed elements contains a free semigroup on two generators.

Proof. Suppose that for two elements f and g in $\mathrm{Homeo}_+(\mathbb{R})$ there exists an interval $[a, b]$ such that $\mathrm{Fix}(f) \cap [a, b] = \{a, b\}$ and $g(a) \in]a, b[$ (the case where $g(b) \in]a, b[$ is similar). Replacing f by its inverse if necessary, we may assume that $f(x) < x$ for every $x \in]a, b[$. Let $c = g(a) \in]a, b[$, and let d' be a point in $]c, b[$. Since $g f^k(a) = c$ for every $k \in \mathbb{N}$, and since $g f^k(d')$ converges to c as k goes to infinity, for large-enough $k \in \mathbb{N}$, the map $g f^k$ has a fixed point in $]a, d'[$. Let $n \in \mathbb{N}$ be such an integer, and let $d > c$ be the infimum of the set of fixed points of $g f^n$ in $]a, b[$. Let $m \in \mathbb{N}$ be large enough that $f^m(d) < c$. Then the positive Ping-Pong lemma applied to the restrictions of f^m and gf^n to $[a, b]$ shows that the semigroup generated by these elements is free (see Exercise 2.3.12).   □

According to Exercise 2.2.5 and Lemma 2.2.44, the following result corresponds to a generalization of Theorem 2.2.39 (compare [14, 15, 231]).

Proposition 2.2.45. *Let* Γ *be a finitely generated group of orientation-preserving homeomorphisms of the real line. If* Γ *has no crossed elements, then* Γ *preserves a Radon measure on* $\mathbb{R}$.

Proof. If Γ has a global fixed point, then the claim is obvious: the Dirac measure with mass on any of these points is invariant under the action. Assume throughout that there is no global fixed point. According to Proposition 2.1.12, Γ preserves a nonempty minimal invariant closed set Λ. Moreover, by the comments after the proof of that proposition, there are three possibilities.

Case (i). $\Lambda' = \emptyset$.

In this case, Λ coincides with the set of points of an increasing sequence $(y_n)_{n \in \mathbb{Z}}$ without accumulation points in $\mathbb{R}$. One then easily checks that the Radon measure $\sum_{n \in \mathbb{Z}} \delta_{y_n}$ is invariant under Γ.

Case (ii). $\partial \Lambda = \emptyset$.

In this case, the action of Γ is minimal. We claim that this action is also free. Indeed, if it is not, then there exist an interval in $\mathbb{R}$ of the form $[u, v[$ or $]u, v]$ and an element $g \in \Gamma$ fixing $]u, v[$ and with no fixed point inside. Since the action is minimal, there must be some $h \in \Gamma$ sending u or v inside $]u, v[$; however, this implies that g and h are crossed on $[u, v]$, thus contradicting our assumption. Now because the action of Γ is free, Hölder's theorem implies that Γ is topologically conjugate to a (dense) group of translations. Pulling back the Lebesgue measure by this conjugacy, we obtain an invariant Radon measure for the action of Γ.

Case (iii). $\partial \Lambda = \Lambda' = \Lambda$.

Collapsing to a point the closure of each connected component of the complement of the "local Cantor set" Λ, we obtain a topological line with an action of Γ induced by semiconjugacy. As in the second case, one easily checks that the induced action is free and hence preserves a Radon measure. Pulling back this measure by the semiconjugacy, one obtains a Radon measure on $\mathbb{R}$ that is invariant under the original action.   □

Exercise 2.2.46. Let Γ be a nonnecessarily finitely generated group of homeomorphisms of the line without crossed elements.

   (i) Prove directly (i.e., without using Proposition 2.2.45) that the set Γ_0 formed by the elements of Γ having fixed points is a normal subgroup of Γ.
  (ii) Prove that the group Γ/Γ_0 admits an Archimedean ordering.
 (iii) Using (i) and (ii), give an alternative proof of Proposition 2.2.45.

Exercise 2.2.47. Let $\mathcal{G} = \{f_1, \ldots, f_k\}$ be a system of generators of a group Γ of homeomorphisms of the line having no global fixed point and without crossed elements. Show that at least one of these generators does not have fixed points.

Hint. Suppose for a contradiction that all the f_i's have fixed points, and let $x_1 \in \mathbb{R}$ be any fixed point of f_1. If f_2 fixes x_1, then the point $x_2 = x_1$ is fixed by both f_1 and f_2. If not, choose a fixed point $x_2 \in \mathbb{R}$ of f_2 such that f_2 does not fix any point between x_1 and x_2. Show that x_2 is still fixed by f_1. Continuing in this way, show that there is a point that is simultaneously fixed by all the f_i's, thus contradicting the hypothesis.

The reader will easily check that the finite-generation hypothesis is necessary for Proposition 2.2.45 (see Exercise 2.2.50). However, throughout the proof this hypothesis was used only to ensure the existence of a nonempty minimal invariant closed set, which in its turn easily follows from the existence of an element without fixed points. Therefore, if such a condition is assumed a priori, then the theorem still holds (compare Exercise 2.2.57). We record this fact in the case of Abelian groups as a proposition.

Proposition 2.2.48. *Every Abelian subgroup of* $\mathrm{Homeo}_+(\mathbb{R})$ *having elements without fixed points preserves a Radon measure on the line.*

Exercise 2.2.49. In an alternative way, prove the preceding proposition above by considering the action induced on the topological circle obtained as the quotient of the line by the action of an element without fixed points.

Exercise 2.2.50. Give an example of a countable, Abelian, infinitely generated subgroup of $\mathrm{Homeo}_+(\mathbb{R})$ for which there is no invariant Radon measure on the line (see [206] in case of problems with this).

Remark 2.2.51. If a finitely generated group acts on the real line without global fixed points and preserves a Radon measure, then the corresponding translation-number function provides us with a nontrivial homomorphism into $(\mathbb{R}, +)$. However, there is a large variety of finitely generated orderable groups for which there is no such homomorphism (compare § 2.2.6). A concrete example is the preimage $\tilde{G}$ in $\widetilde{\mathrm{Homeo}}_+(\mathbb{R})$ of Thompson's group G. Indeed, although $\tilde{G}$ is not simple, it is a **perfect** group; that is, it coincides with its first derived group [44]. Another (historically important) example will be discussed in § 5.1. For these groups, all nontrivial actions on the line must have crossed elements.

Theorem 2.2.39 concerns nilpotent groups and leads naturally to the case of solvable groups. To begin with, notice that if $t \neq 0$ and $\kappa \neq 1$, then the subgroup $\mathrm{Aff}_+(\mathbb{R})$ generated by $f(x) = x + t$ and $g(x) = \kappa x$ does not preserve any Radon measure on the line. However, since the elements in the affine group preserve the Lebesgue measure up to a factor, if one bears in mind Proposition 1.2.2, it is natural to look for conditions that ensure that a solvable group of $\mathrm{Homeo}_+(\mathbb{R})$ leaves a Radon measure quasi-invariant on the line. Once again, we will consider only finitely generated groups, since there exist non–finitely generated Abelian groups of homeomorphisms of the line that do not admit any nontrivial quasi-invariant Radon measure (compare Exercise 2.2.50).

Solvable groups are constructed starting from Abelian groups by successive extensions. We therefore begin with an elementary remark: if Γ_0 is a normal subgroup of a subgroup Γ of $\mathrm{Homeo}_+(\mathbb{R})$ and preserves a Radon measure υ, then for each $g \in \Gamma$, the measure $g(\upsilon)$ is also invariant under Γ_0. Indeed, for every $h \in \Gamma_0$, we have

$$(ghg^{-1})(g(\upsilon)) = g(h(\upsilon)) = g(\upsilon),$$

where $g(\upsilon)(A) = g_*(\upsilon)(A) = \upsilon(g^{-1}(A))$ for every Borel set $A \subset \mathbb{R}$. By Lemma 2.2.37, there exists a constant $\kappa(g)$ such that $\tau_{g(\upsilon)} = \kappa(g)\tau_\upsilon$.

Exercise 2.2.52.   Show that $\kappa(g)$ does not depend on the Γ_0-invariant Radon measure v.

The next three lemmas deal with the problem of showing the existence of a quasi-invariant measure for a group if we start from the quasi-invariance of some measure for a normal subgroup.

Lemma 2.2.53.   *Let Γ_0 be a normal subgroup of a subgroup Γ of $\mathrm{Homeo}_+(\mathbb{R})$, and let v be a Radon measure invariant under Γ_0. Suppose that $\tau_v(\Gamma_0) \neq \{0\}$. If $\kappa(\Gamma) = \{1\}$ and Γ/Γ_0 is amenable, then there exists a Radon measure that is invariant under Γ.*

Proof.   If $\tau_v(\Gamma_0)$ is dense in $\mathbb{R}$, then Proposition 2.2.38 shows that v is invariant under Γ. Assume throughout that $\tau_v(\Gamma_0)$ is cyclic, and let us normalize v so that $\tau_v(\Gamma_0)$ equals $\mathbb{Z}$. Let $h_0 \in \Gamma_0$ be such that $\tau_v(h_0) = 1$, and let Γ_0^* be the kernel of τ_v. The set $\mathrm{Fix}(\Gamma_0)$ is nonempty (it contains the support of v) and is invariant under Γ_0; hence it is unbounded in both directions. Moreover, Γ_0^* coincides with $\{h \in \Gamma_0 : \mathrm{Fix}(h) \neq \emptyset\}$, which shows that Γ_0^* is normal in Γ (and not only in Γ_0).

We will now use a similar idea to that of the proof of Proposition 2.2.48. We first claim that Γ_0/Γ_0^* is contained in the center of Γ/Γ_0^*. To prove this, it suffices to show that for every $g \in \Gamma$ and every $x \in \mathrm{Fix}(\Gamma_0^*)$, one has $g^{-1}h_0 g(x) = h_0(x)$. Consider a measure v_1 giving mass 1 to each point in the set $\{h_0^n(x), n \in \mathbb{Z}\}$. Notice that v_1 is invariant under Γ_0. From the hypothesis $\kappa(\Gamma) = 1$ we conclude that

$$v_1([x, g^{-1}h_0 g(x)[) = \tau_{v_1}(g^{-1}h_0 g)$$
$$= \tau_{g(v_1)}(h_0) = \kappa(g)v_1([x, h_0(x)[) = v_1([x, h_0(x)[),$$

which implies that $g^{-1}h_0 g(x) \leq h_0(x)$. On the other hand, if we consider the measure v_2 that gives mass 1 to each point in the set $g^{-1}h_0^n g(x)$, and if we replace h_0 by $g^{-1}h_0 g$, the same argument shows that $h_0(x) \leq g^{-1}h_0 g(x)$.

Taking $x_0 \in \mathrm{Fix}(\Gamma_0^*)$, we notice that the group $(\Gamma/\Gamma_0^*)/(\Gamma_0/\Gamma_0^*) = \Gamma/\Gamma_0$ acts on the quotient space $\mathrm{Fix}(\Gamma_0^*)/\langle h_0 \rangle(x_0)$, which is compact. By the amenability hypothesis, Γ/Γ_0 preserves a probability measure on this space, which lifts to a Γ-invariant Radon measure on $\mathrm{Fix}(\Gamma_0^*)$. $\square$

Lemma 2.2.54.   *Let Γ_0 be a normal subgroup of a subgroup Γ of $\mathrm{Homeo}_+(\mathbb{R})$, and let v be a Radon measure invariant under Γ_0. Suppose that $\tau_v(\Gamma_0) \neq \{0\}$. If $\kappa(\Gamma) \neq \{1\}$, then v is quasi-invariant under Γ.*

Proof.   Let $h \in \Gamma_0$ and $g \in \Gamma$ be such that $\tau_\upsilon(h) \neq 0$ and $\kappa(g) \neq 1$. Since the equality

$$\tau_\upsilon(g^{-n}h^m g^n) = \tau_{g^n(\upsilon)}(h^m) = m\kappa(g)^n \tau_\upsilon(h)$$

holds for every m and n in $\mathbb{Z}$, the image $\tau_\upsilon(\Gamma)$ must be dense in $\mathbb{R}$. Proposition 2.2.38 then shows that υ is quasi-invariant under Γ. $\square$

Lemma 2.2.55.   *Let Γ_0 be a normal subgroup of a subgroup Γ of* Homeo$_+(\mathbb{R})$, *and let υ be a Radon measure quasi-invariant under Γ_0. Let Γ_0^* be the subgroup of the elements $h \in \Gamma_0$ such that $h(\upsilon) = \upsilon$. If $\tau_\upsilon(\Gamma_0^*) \neq \{0\}$ and $\kappa(\Gamma_0) \neq \{1\}$, then υ is quasi-invariant under Γ.*

Proof.   Since $\tau_\upsilon(\Gamma_0^*) \neq \{0\}$ and $\kappa(\Gamma_0) \neq \{1\}$, it is easy to see that an element $h \in \Gamma_0$ belongs to Γ_0^* if and only if either Fix$(h) = \emptyset$ or Fix(h) is unbounded in both directions. Because both conditions are stable under conjugacy, one concludes that Γ_0^* is normal in Γ. It is also clear that $\kappa(\Gamma) \neq \{1\}$. The claim then follows from Lemma 2.2.54. $\square$

We are now ready to deal with the problem of the existence of a quasi-invariant Radon measure for a large family of solvable subgroups of Homeo$_+(\mathbb{R})$.

Theorem 2.2.56.   *Let Γ be a solvable subgroup of* Homeo$_+(\mathbb{R})$. *Suppose that Γ admits a chain of subgroups $\{id\} = \Gamma_n \lhd \Gamma_{n-1} \lhd \ldots \lhd \Gamma_0 = \Gamma$ such that each Γ_i is finitely generated and each quotient Γ_{i-1}/Γ_i is Abelian. Then there exists a Radon measure that is quasi-invariant under Γ.*

Proof.   We will assume that Γ does not preserve any Radon measure. We then have Fix$(\Gamma) = \emptyset$, since otherwise the Dirac delta with mass on a global fixed point would be an invariant measure. Let $j > 0$ be the smallest index for which Fix$(\Gamma_j) \neq \emptyset$. The Abelian and finitely generated group Γ_{j-1}/Γ_j acts on the closed unbounded set Fix(Γ_j) by homeomorphisms that preserve the order. We leave to the reader the task of showing the existence of an invariant measure for this action by using (a slight modification of) Theorem 2.2.39. This measure naturally induces a Radon measure υ on the line that is invariant under Γ_{j-1} and whose support is contained in Fix(Γ_j). Notice that $\tau_\upsilon(\Gamma_{j-1}) \neq \{0\}$, since Fix$(\Gamma_{j-1}) = \emptyset$.

Let $k > 0$ be the smallest index for which Γ_k preserves a Radon measure. Slightly abusing of notation, let us still denote this measure by υ. As before, it is easy to see that $\tau_\upsilon(\Gamma_k) \neq \{0\}$. According to Lemma 2.2.53, we have $\kappa(\Gamma_{k-1}) \neq \{1\}$, and by Lemma 2.2.54 this implies that υ is quasi-invariant under Γ_{k-1}. Lemma 2.2.55 then allows us to prove by induction that υ is quasi-invariant under Γ_{k-2}, Γ_{k-3}, and so on. Thus we conclude that υ is quasi-invariant under Γ. $\square$

The hypothesis of finite generation for each Γ_i is satisfied by an important family of groups, namely, that of *polycyclic groups* (see Appendix A). In addition, it may be weakened, as is shown in the following exercise.

Exercise 2.2.57. The dynamics of a subgroup Γ of $\mathrm{Homeo}_+(\mathbb{R})$ is said to be *boundedly generated* if there exists a system of generators $\mathcal{G}$ of Γ and a point $x_0 \in \Gamma$ such that $\{h(x_0) : h \in \mathcal{G}\}$ is a bounded subset of the line. Prove that Theorems 2.2.39 and 2.2.56, as well as Proposition 2.2.45, are true if the hypothesis of finite generation is replaced by a hypothesis of boundedly generated dynamics.

2.2.6 An application to amenable, orderable groups

In this section we will use some previously developed ideas to give a "dynamical proof" of the following algebraic result due to Witte-Morris [252] (we recommend reading Appendix B for the concept of amenability; see also Remark 2.2.51, the example after Exercise 5.1.5, and Exercise A.3).

Theorem 2.2.58. *Every orderable, finitely generated, infinite, amenable group admits a nontrivial homomorphism into $(\mathbb{R}, +)$.*

This theorem settles an old problem in the theory of left-orderable groups.[1] We will follow essentially the same (brilliant) idea of Witte-Morris, but unlike [252] we will avoid the use of the algebraic theory of $\mathcal{C}$-orderable groups.

The first ingredient of the proof goes back to an idea independently due to Ghys and Sikora (see, for instance, [230]), which consists in the introduction of a *space of orderings* associated to an orderable group and of a natural action of the group on it. More precisely, for a finitely generated

---

1. A conjecture leading to Theorem 2.2.58 appears in the work by Linnell [150], but quite remarkably, it is already suggested (in a different form) in an old seminal article by Thurston to be discussed in § 5.1 (see [240, page 348]).

orderable group Γ, let us denote by $\mathcal{O}(\Gamma)$ the set of all orderings on Γ. If we fix a finite system $\mathcal{G}$ of generators for Γ, then we may define the distance between $\leq$ and $\preceq$ in $\mathcal{O}(\Gamma)$ by setting $dist(\leq, \preceq) = e^{-n}$, where n is the maximum nonnegative integer such that the orderings $\leq$ and $\preceq$ coincide on the ball $B_{\mathcal{G}}(n)$ of radius n in Γ (see Appendix B for the notion of a *ball* inside a group). In other words, n is the largest nonnegative integer such that for all g and h in $B_{\mathcal{G}}(n)$, one has $g \leq h$ if and only if $g \preceq h$. If we also let $dist(\preceq, \preceq) = 0$ for every ordering $\preceq$, then it is easy to check that the function $dist$ thus defined is a distance on $\mathcal{O}(\Gamma)$ (which depends on $\mathcal{G}$). Actually, the resulting metric space is ultrametric and compact.

Remark 2.2.59. The structure of the space of orderings of an orderable group is interesting in itself. A quite elegant result by Tararin [142] completely describes all orderable groups admitting only finitely many orderings. If a group admits infinitely many orderings, then it necessarily admits uncountably many [147, 177, 190]. For higher-rank torsion-free Abelian groups [230], for non-Abelian torsion-free nilpotent groups [177], and for non-Abelian free groups [164, 177], the spaces of orderings are known to be homeomorphic to the Cantor set. However, there exist relevant examples of groups whose spaces of orderings are infinite but contain isolated points [56, 69, 176, 177]. According to Linnell [148], one of the reasons for the interest in all this concerns the structure of the semigroup formed by the positive elements, as stated in the following exercise.

Exercise 2.2.60. Prove that if $\preceq$ is a nonisolated point in the space of orderings of a finitely generated group, then its positive cone is not finitely generated as a semigroup.

The group Γ acts (continuously) on $\mathcal{O}(\Gamma)$ by right multiplication: given an ordering $\preceq$ with positive cone Γ_+ and an element $f \in \Gamma$, the image of $\preceq$ under f is the ordering $\preceq_f$ whose positive cone is the conjugate $f\Gamma_+ f^{-1}$ of Γ_+ by f. In other words, one has $g \preceq_f h$ if and only if $fgf^{-1} \preceq fhf^{-1}$, which is equivalent to $gf^{-1} \preceq hf^{-1}$.

We will say that an ordering $\preceq$ is ***right-recurrent*** if for every pair of elements f and h in Γ such that $f \succ id$, there exists $n \in \mathbb{N}$ satisfying $fh^n \succ h^n$. For instance, every bi-invariant ordering is right-recurrent. The following corresponds to the main step in the proof of Witte-Morris's theorem.

Proposition 2.2.61. *If Γ is a finitely generated, amenable, orderable group, then Γ admits a right-recurrent ordering.*

To prove this proposition, we need the following weak form of the Poincaré recurrence theorem.

Theorem 2.2.62. *If T is a measurable map that preserves a probability measure μ on a space M, then for every measurable subset A of M and μ-almost every point $x \in A$, there exists $n \in \mathbb{N}$ such that $T^n(x)$ belongs to A.*

Proof. The set of points in A that do not come back to A under iterates of T is $B = A \setminus \cup_{n \in \mathbb{N}} T^{-n}(A)$. One easily checks that the sets $T^{-i}(B)$, with $i > 0$, are two-by-two disjoint. Since T preserves μ, these sets have the same measure, and since μ has total mass 1, the only possibility is that this measure equals zero. Therefore, $\mu(B) = 0$, that is, μ-almost every point in A comes back to A under some iterate of T.   □

Exercise 2.2.63. Prove that under the hypothesis of the preceding theorem, μ-almost every point in A comes back to A under *infinitely many* iterates of T.

Proof of Proposition 2.2.61. By definition, if a finitely generated orderable group Γ is amenable, then its action on $\mathcal{O}(\Gamma)$ preserves a probability measure μ. We will show that μ-almost every point in $\mathcal{O}(\Gamma)$ is right-recurrent. To do this, for each $g \in \Gamma$, let us consider the subset A_g of $\mathcal{O}(\Gamma)$ formed by all the orderings $\preceq$ on Γ such that $g \succ id$. By the Poincaré recurrence theorem, for each $f \in \Gamma$, the set $B_g(f) = A_g \setminus \cup_{n \in \mathbb{N}} f^n(A_g)$ has null μ-measure. Therefore, the measure of $B_g = \cup_{f \in \Gamma} B_g(f)$ is also zero, as well as the measure of $B = \cup_{g \in \Gamma} B_g$. Let us consider an arbitrary element $\preceq$ in the (μ-full measure) set $A = \mathcal{O}(\Gamma) \setminus B$. Given $g \succ id$ and $f \in \Gamma$, from the inclusion $B_g(f) \subset B$ we deduce that $\preceq$ does not belong to $B_g(f)$, and thus there exists $n \in \mathbb{N}$ such that $\preceq$ belongs to $f^n(A_g)$. In other words, one has $g \succ_{f^{-n}} id$, that is, $gf^n \succ f^n$. Since $g \succ id$ and $f \in \Gamma$ were arbitrary, this shows the right-recurrence of $\preceq$.   □

The right-recurrence of an ordering has dynamical consequences, as is shown below.

Proposition 2.2.64. *Let Γ be a countable group admitting a right-recurrent ordering $\preceq$. If $(g_n)_{n \geq 0}$ is any numbering of Γ, then the dynamical realization of Γ associated to $\preceq$ and this numbering is a subgroup of $\mathrm{Homeo}_+(\mathbb{R})$ without crossed elements.*

Proof. The claim is obvious if Γ is trivial. Therefore, in what follows, we will assume that Γ is infinite. Let us suppose that there exist f and g in Γ and an interval $[a, b]$ such that (for the dynamical realizations one has) $\text{Fix}(f) \cap [a, b] = \{a, b\}$ and $g(a) \in \,]a, b[$ (the case where $g(b)$ belongs to $]a, b[$ is analogous). Replacing f by its inverse if necessary, we may assume that $f(x) < x$ for all $x \in \,]a, b[$. As we have already observed after Theorem 2.2.19, there must exist some $g_i \in \Gamma$ such that $t(g_i)$ belongs to the interval $]a, b[$. Replacing g_i by $f^n g_i$ for a large $n > 0$ if necessary, we may assume that $t(g_i)$ is actually contained in $]a, c[$, where $c = g(a)$. Now conjugating f and g by the element g_i (and replacing the points a, b, and c by their images under g_i), we may suppose that $t(id)$ belongs to $]a, c[$.

Choose a point $d \in \,]c, b[$. Since $gf^n(a) = c$ for all $n \in \mathbb{N}$, and since $gf^n(d)$ converges to $c < d$ when n goes to infinity, if $n \in \mathbb{N}$ is large enough, then the map $h_n = gf^n$ satisfies $h_n(a) > a$, $h_n(d) < d$, $\text{Fix}(h_n) \cap \,]a, d[\,\subset\, [c_n, c'_n] \subset \,]c, h_n(d)[$, and $\{c_n, c'_n\} \subset \text{Fix}(h_n)$, for some sequences of points c_n and c'_n converging to c on the right (see Figure 13). Notice that the element h_n is positive, since from $h_n(t(id)) > h_n(a) = c > t(id)$ one deduces that $t(h_n) > t(id)$, and by the construction of the dynamical realization this implies that $h_n \succ id$. Let us fix $m > n$ large enough that the preceding properties hold for h_m and h_n, and such that $[c_m, c'_m] \subset \,]c, c_n[$. Let us fix $k \in \mathbb{N}$ large enough that $h_n^k(a) > h_m(c_n)$, and let us define $h = h_m^k$. For each $i \in \mathbb{N}$, one has $h^i(t(id)) \in \,]h_m(c_n), c_n[$, and thus $h_m h^i(t(id)) < h_m(c_n) < h(a) < h(t(id))$. Therefore, $h_m h^i \prec h \prec h^i$ for all $i \in \mathbb{N}$. However, this contradicts the hypothesis of right-recurrence for $\preceq$. $\square$

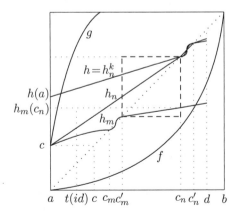

FIGURE 13

In the preceding proof, we used a property that is actually weaker than right-recurrence, namely, for all positive elements f and h in Γ, one has $fh^n \succ h$ for some $n \in \mathbb{N}$. This is called the ***Conrad property***, and the groups that admit an ordering satisfying it are said to be *C*-***orderable***. A very rich literature exists on this property, mostly from an algebraic viewpoint (see, for instance, [57, 142], as well as Remark 5.1.8). The preceding proof, together with Exercise 2.2.65, shows that it has a natural dynamical counterpart: roughly, the Conrad property is the algebraic counterpart to the condition of nonexistence of crossed elements for the corresponding action on the real line [177, 190].

Exercise 2.2.65. Let (x_n) be a dense sequence of points in the line, and let Γ be a subgroup of $\text{Homeo}_+(\mathbb{R})$. Prove that if Γ has no crossed elements, then the order relation induced from (x_n), as in the proof of Theorem 2.2.19, satisfies the Conrad condition.

Exercise 2.2.66. Let $\preceq$ be an ordering on a group Γ.

(i) Show that if $\preceq$ has the Conrad property, then $fh^2 \succ h$ for all positive elements f and h in Γ.

Hint. Following [177], suppose that the opposite inequality holds and show that for the positive elements f and $g = fh$ in Γ, one has $fg^n \prec g$ for every $n \in \mathbb{N}$.

(ii) More generally, prove that for every positive integer n, one has $fh^n \succ h^{n-1}$.

(iii) Give examples of right-recurrent orderings $\preceq$ such that $fh^2 \prec h^2$ for some $f \succ id$ and $h \succ id$.

Exercise 2.2.67. Give examples of *C*-orderable groups that do not admit right-recurrent orderings (see [252, Example 4.5] in case of problems with this).

Exercise 2.2.68. Show that the Klein group $\Gamma = \langle f, g : fgf^{-1} = g^{-1} \rangle$ is not biorderable, although it admits right-recurrent orderings.

Proof of Theorem 2.2.58. If Γ is amenable, finitely generated, and orderable, then Proposition 2.2.61 provides us with a right-recurrent ordering on Γ. By Proposition 2.2.64, the dynamical realization associated to this ordering and any numbering $(g_n)_{n \geq 0}$ of Γ is a subgroup of $\text{Homeo}_+(\mathbb{R})$ (isomorphic to Γ and) without crossed elements. By Proposition 2.2.45, Γ preserves a Radon measure υ on the line. Let us now recall that if Γ is nontrivial, then its dynamical realizations have no global fixed point. Therefore, the translation-number function with respect to υ is a nontrivial homomorphism into $(\mathbb{R}, +)$ (see (2.5) and (2.6)). $\square$

Exercise 2.2.69. Give a faithful action of the Baumslag-Solitar group $\Gamma = \langle f, g : fgf^{-1} = g^2 \rangle$ on the line without crossed elements (see [218] for more on this).

Remark. Notice that Γ embeds in the affine group by identifying f with $x \mapsto 2x$ and g with $x \mapsto x + 1$.

Exercise 2.2.70. Show that finitely generated subgroups of the group of piecewise affine homeomorphisms of the interval admit nontrivial homomorphisms into $(\mathbb{R}, +)$ (see also Exercise 2.2.21).

Remark. Recall that the group $\text{PAff}_+([0, 1])$ does not contain free subgroups on two generators (cf. Theorem 1.5.1). Since this is also the case for countable amenable groups, this leads naturally to the following question: do finitely generated subgroups of $\text{Homeo}_+([0, 1])$ without free subgroups on two generators admit nontrivial homomorphisms into $(\mathbb{R}, +)$? See [149] for an interesting result pointing in the positive direction.

2.3 Invariant Measures and Free Groups

2.3.1 A weak version of the Tits alternative

A celebrated theorem by Tits establishes that every finitely generated subgroup of $GL(n, \mathbb{R})$ either contains a free subgroup on two generators or is virtually solvable (see [25, 27] for modern versions of this result). This dichotomy, known as the *Tits alternative*, does not hold for groups of circle homeomorphisms, as is shown by the following exercise.

Exercise 2.3.1. Show that Thompson's group F is not virtually solvable (however, recall that by Theorem 1.5.1, F does not contain free subgroups on two generators).

Nevertheless, a weak version of the alternative may be established. The following result was conjectured by Ghys and proved by Margulis in [157]. We will develop the proof proposed some years later by Ghys himself in [88]. The idea of this proof will be pursued in a probabilistic setting in § 2.3.2.

Theorem 2.3.2. *If Γ is a subgroup of* $\text{Homeo}_+(S^1)$*, then either there exists a probability measure on* S^1 *that is invariant under Γ, or Γ contains a free subgroup on two generators.*

Since virtually solvable groups are amenable (cf. Exercise A.7), every action of such a group on the circle admits an invariant probability measure. Therefore, Theorem 2.3.2 may be considered a weak form of the Tits alternative. However, the weaker alternative is not a dichotomy. Indeed, according to § 2.2.3, the free group on two generators admits faithful actions on the interval, which induce actions on S^1 with a global fixed point, and hence with an invariant probability measure.

To prove Margulis's theorem, we begin by noticing that by Theorem 2.1.1 there are three distinct cases according to the type of minimal invariant closed set. If Γ has a finite orbit, there is obviously an invariant probability measure, namely, the mean of the Dirac measures with mass on the points of the orbit. We claim that the case where there exists an exceptional minimal set Λ reduces to the case where all the orbits are dense. Indeed, the associated action on the topological circle S^1_Λ is minimal. If there exists an invariant probability measure for this action, then it cannot have atoms, and its support must be total. Hence it induces an invariant measure on S^1 whose support is Λ. Moreover, if two elements in Γ project into homeomorphisms of S^1_Λ generating a free group, then these elements generate a free subgroup of Γ.

By the preceding discussion, to prove Margulis's theorem, we need to consider only the case where all the orbits are dense (i.e., the minimal case).

Definition 2.3.3. An action of a group Γ by circle homeomorphisms is said to be **equicontinuous** if for every $\delta > 0$, there exists $\varepsilon > 0$ such that if $dist(x, y) < \varepsilon$, then $dist(g(x), g(y)) < \delta$ for all $g \in \Gamma$. It is said to be **expansive** if for each $x \in S^1$, there exists an open interval I containing x and a sequence (g_n) of elements in Γ such that the length of the intervals $g_n(I)$ converges to zero.

Lemma 2.3.4. *Every minimal action by circle homeomorphisms is either equicontinuous or expansive.*

Proof. If the action is not equicontinuous, then there exists $\delta > 0$ such that for all $n \in \mathbb{N}$, there exist x_n and y_n in S^1 and g_n in the acting group Γ such that $]x_n, y_n[|$ goes to zero and $dist(g_n(x_n), g_n(y_n)) \geq \delta$. Passing to a subsequence if necessary, we may assume that $g_n(x_n)$ (resp. $g_n(y_n)$) converges to some $a \in S^1$ (resp. $b \in S^1$). Notice that $dist(a, b) \geq \delta$. By the compactness of S^1, and since the action is minimal, there exist $h_1, \ldots, h_k$ in Γ such that $S^1 = \cup_{i=1}^{k} h_i(]a, b[)$. Let $\varepsilon > 0$ be the Lebesgue number of this covering. If $dist(x, y) < \varepsilon$, then x and y belong to $h_j(]a, b[)$ for some $j \in \{1, \ldots, k\}$. The points $g_n^{-1} h_j^{-1}(x)$ and $g_n^{-1} h_j^{-1}(y)$ then belong to $]x_n, y_n[$ for n large enough, and hence the distance between them tends to zero. This shows that the ε-neighborhood of every point in the circle is "contractable" to a point by elements in Γ, and therefore the action is expansive. $\square$

Exercise 2.3.5. Let Γ be a group of circle homeomorphisms acting minimally. Show that the action of Γ is expansive if and only if there exists $\varepsilon > 0$ such that for all $x \neq y$ in S^1, there exists $g \in \Gamma$ satisfying $dist(g(x), g(y)) \geq \varepsilon$.

If the action of a group Γ by circle homeomorphisms is equicontinuous, then, by the Ascoli-Arzela's theorem, the closure of Γ in $\mathrm{Homeo}_+(S^1)$ is compact. Therefore, by Proposition 1.1.2, there exists a probability measure on S^1 that is invariant under Γ and allows conjugating Γ to a group of rotations. Hence to complete the proof of Margulis's theorem, it suffices to show the following proposition.

Proposition 2.3.6. *If the action of a group Γ by circle homeomorphisms is minimal and expansive, then Γ contains a free subgroup on two generators.*

Margulis's argument consists in proving directly an equivalent claim to this proposition by using the Klein ping-pong lemma. The proof by Ghys also uses this idea, but it is based on a preliminary study of the "maximal domain of contraction" that not only allows proving the proposition but provides useful additional information.

Definition 2.3.7. If the action of a group Γ on the circle is minimal and expansive, then we will say that it is ***strongly expansive*** if for every open interval I whose closure is not the whole circle, there exists a sequence of elements $g_n \in \Gamma$ such that the length $|g_n(I)|$ converges to zero.

Not every minimal expansive action is strongly expansive; consider, for instance, the case of subgroups of $PSL_k(2, \mathbb{R})$ acting minimally, with $k \geq 2$. In what follows, we will see that in the minimal expansive case, these "finite coverings" are the only obstructions to strong expansiveness.

Lemma 2.3.8. *If the action of a subgroup Γ of $\mathrm{Homeo}_+(S^1)$ is minimal and expansive, then there exists a finite-order homeomorphism $R : S^1 \to S^1$ commuting with all the elements in Γ such that the action of Γ on the quotient circle $S^1/ \sim$ obtained as the space of orbits of R is strongly expansive.*

Proof. For each $x \in S^1$, let us define $R(x) \in S^1$ as the "supremum" of the points y for which the interval $]x, y[$ is contractable, that is, there exists a sequence (g_n) of elements in Γ such that the length $|]g_n(x), g_n(y)[|$ converges to zero. The fact that R commutes with all the elements in Γ follows immediately from the definition. Moreover, R is a monotonic function "of degree 1". We claim that R is a finite-order homeomorphism.

To show that R is "strictly increasing" we argue by contradiction, and we consider the (nonempty) set $\mathrm{Plan}(R)$ formed by the union of the interiors of the intervals on which R is constant. Since R centralizes Γ, this set is invariant under the action. Hence by minimality, $\mathrm{Plan}(R)$ is the whole circle, and this implies that R is a constant map, which is impossible. To show that R is continuous, one may make a similar argument using the union $\mathrm{Salt}(R)$ of the interior of the intervals that are avoided by the image of R (compare the proof of Theorem 2.2.6).

The rotation number of R cannot be irrational. Indeed, if it were irrational and R admitted an exceptional minimal set, then, since R centralizes Γ, this set would be invariant under Γ, which contradicts the minimality of the action. If $\rho(R)$ were irrational and the orbits of R were dense, then the unique invariant probability measure invariant under R would be invariant under Γ, and hence Γ would be topologically conjugate to a group of rotations (see Exercise 2.2.12), which contradicts the expansiveness of the action.

The homeomorphism R has finite order. Indeed, since its rotation number is rational, it admits periodic points. The set $\mathrm{Per}(R)$ of these points is closed, nonempty, and invariant under Γ (once again, the last property follows from the fact that R centralizes Γ). From the minimality of the action of Γ, one concludes that $\mathrm{Per}(R)$ coincides with the whole circle. To finish the proof of the lemma, notice that by the definition of R, the action of Γ on $\mathrm{S}^1/\sim$ is strongly expansive. $\qquad\square$

In the context of the preceding lemma, the order of the homeomorphism R will be called the **degree** of Γ and will be denoted by $d(\Gamma)$. Notice that $d(\Gamma) = 1$ if and only if R is the identity, that is, if the original action of Γ is (minimal and) strongly expansive.

Let us come back to the proof of Proposition 2.3.6. As we already mentioned, the strategy consists in finding two elements whose action on S^1, up to a "finite covering", is similar to the action of a Schottky subgroup of $\mathrm{PSL}(2,\mathbb{R})$. Indeed, such a pair of elements generates a free group, as the next Klein's ping-pong lemma shows.

Lemma 2.3.9. *Let* M *be a set and* Γ *a group of bijections of* M. *Assume that there exist nonempty disjoint subsets* A_0 *and* A_1 *of* M, *and elements* g_0 *and* g_1 *in* Γ, *such that* $g_0^n(A_1) \subset A_0$ *and* $g_1^n(A_0) \subset A_1$ *for all* $n \in \mathbb{Z} \setminus \{0\}$. *Then* g_0 *and* g_1 *generate a free group.*

Proof. We need to show that every nontrivial word $W = g_0^{i_1} g_1^{j_1} \cdots g_0^{i_k} g_1^{j_k}$ that is **reduced** (i.e., such that the exponents are nonzero, possibly with the exception of the first and/or the last ones) represents a nontrivial element in Γ. To do this, it suffices to notice that if $i_1 = 0$ and $j_k \neq 0$ (resp. if $i_1 \neq 0$ and $j_k = 0$), then (the element represented by) W sends A_0 into A_1 (resp. A_1 into A_0), and hence it cannot equal the identity. Moreover, if i_1 and j_k are both nonzero, then by conjugating g appropriately, one obtains a word to which this argument may be applied. $\qquad\square$

If the action of a group Γ on the circle is minimal and expansive, then we consider the associated action on the quotient circle $\mathrm{S}^1/\sim$. This action is minimal and strongly expansive, and hence there exist two sequences of intervals L_n and L_n' in $\mathrm{S}^1/\sim$ converging to points a and b, respectively, in such a way that for each $n \in \mathbb{Z}$, there exists $f_n \in \Gamma$ such that $f_n(\mathrm{S}^1 \setminus L_n) = L_n'$. We can assume that $a \neq b$, since otherwise we may replace f_n by gf_n for some element $g \in \Gamma$ that does not fix a' (such an element exists by minimality). We claim that there exists $f \in \Gamma$ such that $a' = f(a)$ and $b' = f(b)$ are different from both a and b. Although this could be left as an exercise, it also follows from a very interesting (and, surprisingly, not well-known) lemma due to Newmann.[2]

Lemma 2.3.10. *No group can be written as a finite union of left classes with respect to infinite-index subgroups.*

Proof. Suppose that $\Gamma = S_1[g_1] \cup \ldots \cup S_k[g_k]$ is a decomposition of a group into left classes $[g_i]$, where each set S_i is finite and the classes $[g_i]$ are taken with respect to different subgroups. We will show by induction on k that a certain class $[g_i]$ has finite index in Γ.

If $k = 1$, there is nothing to prove. Suppose that the claim holds for $k \leq n$, and let us consider a decomposition $\Gamma = S_1[g_1] \cup \ldots \cup S_{n+1}[g_{n+1}]$, as above. If $S_1[g_1] = \Gamma$, then $[g_1]$ has finite index in Γ. In the other case, there exists $g \in \Gamma$ such that $g[g_1] \cap S_1[g_1] = \emptyset$. We then have

$$g[g_1] \subset S_2[g_2] \cup \ldots \cup S_{n+1}[g_{n+1}],$$

and thus

$$S_1[g_1] \subset T_2[g_2] \cup \ldots \cup T_{n+1}[g_{n+1}],$$

where $T_i = \{h_1 g^{-1} h_i : h_1 \in S_1, h_i \in S_i\}$ is finite. Hence

$$\Gamma = (T_2 \cup S_2)[g_2] \cup \ldots \cup (T_{n+1} \cup S_{n+1})[g_{n+1}],$$

and by the induction hypothesis this implies the existence of an index $j \in \{2, \ldots, n+1\}$ such that $[g_j]$ has finite index in Γ. $\square$

The preceding lemma allows us to choose $f \in \Gamma$ and $n \in \mathbb{N}$ such that the intervals L_n, L'_n, $f(L_n)$, and $f(L'_n)$, are disjoint. Indeed, Γ cannot be written as the union $\Gamma_{a,a} \cup \Gamma_{a,b} \cup \Gamma_{b,a} \cup \Gamma_{b,b}$, where $\Gamma_{c,c'}$ denotes the— perhaps empty—left class of the elements in Γ sending c into c'. Therefore, there must exist $f \in \Gamma$ such that $\{f(a), f(b)\} \cap \{a, b\} = \emptyset$, and hence the claimed property holds for large-enough $n \in \mathbb{N}$. To complete the proof of Proposition 2.3.6, we fix such an n, and we let $g_0 = f_n$ and $g_1 = f g_0 f^{-1}$. If we denote by J_0, I_0, J_1, and I_1 the preimages under R of L_n, L'_n, $f(L_n)$, and $f(L'_n)$, respectively, then one easily checks that for all $k \in \mathbb{Z} \setminus \{0\}$, one has

2. It is interesting to remark that some of the arguments in the proof of the Tits alternative for linear groups where the properties of Zariski's topology are strongly used may be simplified by using Newmann's lemma.

$g_0^k(I_1 \cup J_1) \subset I_0 \cup J_0$ and $g_1^k(I_0 \cup J_0) \subset I_1 \cup J_1$. Thus one may apply Lemma 2.3.9 to conclude that g_0 and g_1 generate a free group.

Exercise 2.3.11. Show that a group Γ of circle homeomorphisms preserves a probability measure on S^1 if and only if for every pair of elements in Γ, there is a common invariant probability measure.

Exercise 2.3.12. Prove the following "positive version" of the ping-pong lemma: if g_0 and g_1 are two bijections of a set M for which there exist disjoint nonempty subsets A_0 and A_1 such that $g_0^n(A_1) \subset A_0$ and $g_1^n(A_0) \subset A_1$ for every *positive* integer n, then g_0 and g_1 generate a free semigroup (cf. Definition 2.2.41).

Exercise 2.3.13. Prove that Thompson's group F satisfies no nontrivial *law*; that is, for every nontrivial word W in two letters, there exist f and g in F such that $W(f, g)$ is different from the identity. (This result holds for most groups of piecewise homeomorphisms of the interval, as was first shown in [34].)

Hint. Let $W = g_0^{i_1} g_1^{j_1} \cdots g_0^{i_k} g_1^{j_k}$ be a nontrivial reduced word, with $i_1 \neq 0$ and $j_k \neq 0$. Take two elements f and g having at least $2k$ fixed points $x_1, \ldots, x_{2k}$ and $y_1, \ldots, y_{2k}$, respectively, such that $x_1 < y_1 < x_2 < y_2 < \ldots < x_{2k} < y_{2k}$. Use a similar argument to that of the proof of the ping-pong lemma to show that if p belongs to the "middle open interval" determined by these points, then $W(f^n, g^n)(p)$ is different from p for large-enough n.

2.3.2 A probabilistic viewpoint

Let Γ be a countable group of circle homeomorphisms and p a probability measure on Γ that is **nondegenerate** (in the sense that its support generates Γ as a semigroup). Let us consider the **diffusion operator** defined on the space of continuous functions by

$$D\psi(x) = \int_\Gamma \psi(g(x)) \, dp(g). \tag{2.7}$$

Let $\nu \mapsto p * \nu$ be the dual action of this operator on the space of probability measures on the circle (this dual action is also called **convolution**). Such a measure is said to be **stationary** (with respect to p) if $p * \mu = \mu$, that is, if for every continuous function $\psi : S^1 \to \mathbb{R}$, one has

$$\int_{S^1} \psi(x)\,d\mu(x) = \int_\Gamma \int_{S^1} \psi(g(x))\,d\mu(x)\,dp(g). \qquad (2.8)$$

The existence of at least one stationary measure is ensured by the Kakutani fixed-point theorem [258]; clearly, it may also be deduced from a simple argument using Birkhoff sums (i.e., using the technique of Bogolioubov and Krylov: see Appendix B).

Lemma 2.3.14. *If the orbits under Γ are dense, then μ has total support and no atoms. If Γ admits a minimal invariant Cantor set, then this set coincides with the support of μ, and μ has no atoms either.*

Proof. Let us first show that if μ has atoms, then Γ admits finite orbits (concerning this case, see Exercise 2.3.2). Indeed, if x is a point with maximal positive mass, then from the equality

$$\mu(x) = \int_\Gamma \mu(g^{-1}(x))\,dp(g)$$

one concludes that $\mu(g^{-1}(x)) = \mu(x)$ for every g in the support of p. Since p is nondegenerate, by repeating this argument, one concludes that the equality $\mu(g^{-1}(x)) = \mu(x)$ actually holds for every element $g \in \Gamma$. Since the total mass of μ is finite, the only possibility is that the orbit of x is finite.

If the action of Γ is minimal, then because the support of μ is a closed invariant set, it must be the whole circle. If Γ admits an exceptional minimal set Λ, then because this set is unique, it must be contained in the support of μ. Therefore, to prove that Λ and $supp(\mu)$ coincide, we need to verify that $\mu(I) = 0$ for every connected component of $S^1 \setminus \Lambda$. If this were not the case, then, choosing such a component I with maximal measure, one would conclude—by an argument similar to that of the case of finite orbits—that the orbit of I is finite. However, this contradicts the fact that the orbits of the endpoints of I are dense in Λ. $\qquad \square$

The existence of stationary measures allows establishing a nontrivial result of regularity after conjugacy for general group actions on the circle [62, 65]. We point out that according to [113], such a result is no longer true in dimension greater than 1 (see, however, Exercise 2.3.17).

Proposition 2.3.15. *Every countable subgroup of $\mathrm{Homeo}_+(S^1)$ (resp. $\mathrm{Homeo}_+([0,1])$) is topologically conjugate to a group of bi-Lipschitz homeomorphisms.*

Proof. Let us first consider the case of a countable subgroup Γ of $\text{Homeo}_+(S^1)$ whose orbits are dense. Endow this group with a non-degenerate **symmetric** probability measure p, where *symmetric* means that $p(g) = p(g^{-1})$ for every $g \in \Gamma$. Let us consider an associated stationary measure μ on S^1. For each interval $I \subset S^1$ and each element $g \in supp(p)$, one has

$$\mu(I) = \sum_{h \in supp(p)} \mu(h^{-1}(I))\, p(h) \geq \mu(g(I))\, p(g^{-1}),$$

and hence

$$\mu(g(I)) \leq \frac{1}{p(g)}\, \mu(I). \tag{2.9}$$

Let us now take a circle homeomorphism φ sending μ into the Lebesgue measure. If J is an arbitrary interval in S^1, then by (2.9), for every $g \in supp(p)$, one has

$$|\varphi \circ g \circ \varphi^{-1}(J)| = \mu(g \circ \varphi^{-1}(J)) \leq \frac{1}{p(g)}\, \mu(\varphi^{-1}(J)) = \frac{1}{p(g)}\, |J|.$$

Therefore, for every $g \in supp(p)$, the homeomorphism $\varphi \circ g \circ \varphi^{-1}$ is Lipschitz with constant $1/p(g)$. Since p is nondegenerate, this implies the proposition in the minimal case.

If Γ is an arbitrary countable subgroup of $\text{Homeo}_+(S^1)$, then, if we add an irrational rotation and consider the generated group, the problem reduces to that of dense orbits. By the preceding argument, the new group—and hence the original one—is topologically conjugate to a group of bi-Lipschitz homeomorphisms. Finally, if we identify the endpoints of the interval $[0, 1]$, each subgroup Γ of $\text{Homeo}_+([0, 1])$ induces a group of circle homeomorphisms with a marked global fixed point. Therefore, if Γ is countable, then this new group is conjugate by an element φ of $\text{Homeo}_+(S^1)$ to a group of bi-Lipschitz circle homeomorphisms. To obtain a genuine conjugacy inside $\text{Homeo}_+([0, 1])$, it suffices to compose φ with a rotation in such a way that the marked point of the circle is sent to itself. $\square$

Exercise 2.3.16. Using the argument of the preceding proof, show that for each $\varepsilon > 0$, every homeomorphism of the circle or the interval is topologically conjugate to a Lipschitz homeomorphism whose derivative (well defined at almost every point) is less than or equal to $1 + \varepsilon$ (compare Exercise 4.2.22).

Remark. If one uses the well-known inequality $h_{top}(T) \leq d \log(Lip(T))$ for the topological entropy of Lipschitz maps T on d-dimensional compact manifolds, the preceding claim gives a short and conceptual proof of the fact that the topological entropy of any homeomorphism of the circle or the interval is zero (see [249]).

Exercise 2.3.17. Prove that every countable group of homeomorphisms of a compact manifold is topologically conjugate to a group of absolutely continuous homeomorphisms.

Hint. Use the classical Oxtoby-Ulam theorem, which establishes that every probability measure of total support and without atoms on a compact manifold is the image of the Lebesgue measure under a certain homeomorphism (see [99] for a concise presentation of this result). Use also the fact that every compact manifold supports minimal countable group actions (see [78] for minimal, *finitely generated* group actions).

The preceding definitions extend to any action on a measurable space M of a countable group Γ provided with a probability measure p. For example, if one looks at the action (by left translations) of Γ on itself, the convolution operator may be iterated; the nth convolution of p with itself will be denoted by p^{*n}.

In the general case, we will denote by Ω the space of the sequences $(g_1, g_2, \ldots) \in \Gamma^{\mathbb{N}}$ endowed with the product measure $\mathbb{P} = p^{\mathbb{N}}$. If σ is the (one-side) *shift* on Ω, that is, $\sigma(g_1, g_2, \ldots) = (g_2, g_3, \ldots)$, then one easily checks that a probability measure μ on M is stationary with respect to p if and only if the measure $\mathbb{P} \times \mu$ is invariant under the *skew-product* map $T : \Omega \times M \to \Omega \times M$ given by

$$T(\omega, x) = (\sigma(\omega), h_1(\omega)(x)) = (\sigma(\omega), g_1(x)), \quad \omega = (g_1, g_2, \ldots).$$

Exercise 2.3.18. A continuous function $\psi : M \to \mathbb{R}$ is said to be *harmonic* if it is invariant under the diffusion, that is, $D(\psi) = \psi$. Prove the following version of the *maximum principle*: if ψ is harmonic, then the set of points at which ψ attains its maximum is invariant under Γ. Prove that the same holds for *superharmonic* functions, that is, for functions ψ satisfying $D\psi \geq \psi$.

We now concentrate on the case where M is a compact metric space. Following the seminal work of Furstenberg [83], in order to study the evolution of the random compositions, we consider the **inverse process** given by $\bar{h}_n(\omega) = g_1 \cdots g_n$. The main reason for doing this lies in the following observation: if ψ is a continuous function defined on M and μ is a stationary measure, then the sequence of random variables

$$\xi_n(\omega) = \int_M \psi \, d(g_1 \cdots g_n(\mu)) = \int_M \psi \, d(\bar{h}_n(\omega)(\mu))$$

is a martingale [80]. Indeed, for every $g_1, \ldots, g_n$ in Γ, an equality of type (2.8) applied to the function $x \mapsto \psi(g_1 \cdots g_n(x))$ yields

$$\int_\Gamma \int_M \psi d(g_1 \cdots g_n g(\mu)) dp(g) = \int_M \psi d(g_1 \cdots g_n(\mu)),$$

that is, $\mathbb{E}(\xi_{n+1}|g_{n+1}) = \xi_n$. By the martingale convergence theorem, the sequence $(\xi_n(\omega))$ converges for almost every $\omega \in \Omega$. Let (ψ_k) be a dense sequence in the space of continuous functions on M. By the compactness of the space of probability measures on M, for a total probability subset $\Omega_0 \subset \Omega$, the sequence $g_1 g_2 \cdots g_n(\mu) = \bar{h}_n(\omega)(\mu)$ converges (in the weak* topology) to a probability $\omega(\mu)$. Moreover, the map $\omega \mapsto \omega(\mu)$ is well defined at almost every ω and is measurable (see [158, page 199] for more details on this).

Proposition 2.3.19. *Let Γ be a countable subgroup of* $\mathrm{Homeo}_+(S^1)$ *whose action is minimal. If the property of strong expansiveness is satisfied, then for almost every sequence $\omega \in \Omega_0$, the measure $\omega(\mu)$ is a Dirac measure.*

Proof. We will show that for every $\varepsilon \in \,]0, 1]$, there exists a total probability subset Ω_ε of Ω_0 such that for every $\omega \in \Omega_\varepsilon$, there exists an interval I of length $|I| \leq \varepsilon$ such that $\omega(\mu)(I) \geq 1 - \varepsilon$. This allows us to conclude that for all ω contained in the total probability subset $\Omega^* = \cap_{n \in \mathbb{N}} \Omega_{1/n}$, the measure $\omega(\mu)$ is a Dirac measure.

Let us fix $\varepsilon > 0$. For each $n \in \mathbb{N}$, let us denote by $\Omega^{n,\varepsilon}$ the set of the sequences $\omega \in \Omega_0$ such that for all $m \geq 0$ and every interval I in S^1 of length $|I| \leq \varepsilon$, one has $\bar{h}_{n+m}(\omega)(\mu)(I) < 1 - \varepsilon$. We need to show that the probability of $\Omega^{n,\varepsilon}$ is zero. We will begin by exhibiting a finite subset $\mathcal{G}_\varepsilon$ of $supp(p)$, as well as an integer $l \in \mathbb{N}$, such that for all $r \in \mathbb{N}$ and all $(g_1, \cdots, g_r) \in \Gamma^r$, there exist an interval I of length $|I| \leq \varepsilon$, an integer $\ell \leq 2l$, and elements $f_1, \ldots, f_\ell$ in $\mathcal{G}_\varepsilon$ satisfying

$$g_1 \cdots g_r f_1 \cdots f_\ell(\mu)(I) \geq 1 - \varepsilon. \tag{2.10}$$

To do this, let us fix two different points a and b in S^1, as well as an integer $q > 1/\varepsilon$, and let us take q different points $a_1, \ldots, a_q$ in the orbit of a under Γ. For each $i \in \{1, \ldots, q\}$, let us choose $h_i \in \Gamma$ and an open interval U_i containing a_i in such a way that the U_i's are two-by-two disjoint, $h_i(a) = a_i$, and $h_i(U) = U_i$ for some neighborhood U of a not containing b. Let us now consider a neighborhood V of b disjoint from U and such that $\mu(S^1 \setminus V) \geq 1 - \varepsilon$. By minimality and strong expansiveness, there exists $h \in \Gamma$ such that $h(S^1 \setminus V) \subset U$. Now each element in $\{h_1, \ldots, h_q, h\}$ may be written as a product of elements in the support of μ. This may be done in many different ways, but if we fix once and for all a particular choice for h and each h_i, then the set $\mathcal{G}_\varepsilon$ of the elements in $supp(p)$ that are used is finite. Let l be the maximal number of factors appearing in these choices. To check (2.10), notice that for $g = g_1 \cdots g_r$, the intervals $g(U_i)$ are two-by-two disjoint, and hence the length of at least one of them is bounded from above by ε. For such an interval $I = g(U_i)$, we have

$$g_1 \cdots g_r h_i h(\mu)(I) = \mu(h^{-1}(U)) \geq \mu(S^1 \setminus V) \geq 1 - \varepsilon,$$

which concludes the proof of (2.10).

Now let $\rho = \min\{p(f) : f \in \mathcal{G}_\varepsilon\}$, and let us define $\Omega_{n,m}^\varepsilon$ as the set of $\omega \in \Omega_0$ such that for every interval I of length $|I| \leq \varepsilon$ and all $k \leq m$, one has $\bar{h}_{n+k}(\omega)(\mu)(I) < 1 - \varepsilon$. By (2.10) we have

$$\mathbb{P}(\Omega_{n,2lt}^\varepsilon) \leq (1 - \rho^{2l})^t.$$

If we pass to the limit as t goes to infinity, this allows us to conclude that $\mathbb{P}(\Omega^{n,\varepsilon}) = 0$, which finishes the proof. $\qquad\square$

If we use a well-known argument in the theory of random walks on groups, Proposition 2.3.19 allows us to prove a general uniqueness result for the stationary measure. Let us point out that according to [63], this result still holds in the much more general context of codimension-one foliations (the notion of stationary measure for this case is that of Garnett; see [45, 85]).

Theorem 2.3.20. *Let Γ be a countable group of circle homeomorphisms endowed with a nondegenerate probability measure p. If Γ does not preserve any probability measure on S^1, then the stationary measure with respect to p is unique.*

Proof. First suppose that the action of Γ is minimal and strongly expansive, and let μ be a probability measure on S^1 that is stationary with respect to p. For each $\omega \in \Omega$ such that $\lim_{n \to \infty} \bar{h}_n(\omega)(\mu)$ exists and is a delta measure, let us denote by $\varsigma_\mu(\omega)$ the atom of the measure $\omega(\mu)$, that is, the point in S^1 for which $\omega(\mu) = \delta_{\varsigma_\mu(\omega)}$. The map $\varsigma_\mu : \Omega \to S^1$ is well defined at almost every sequence and measurable. We claim that the measures μ and $\varsigma_\mu(\mathbb{P})$ coincide. Indeed, since μ is stationary,

$$\mu = p^{*n} * \mu = \sum_{g \in \Gamma} p^{*n}(g)\, g(\mu) = \int_\Omega \bar{h}_n(\omega)(\mu)\, d\mathbb{P}(\omega).$$

Therefore, passing to the limit as n goes to infinity, we obtain

$$\mu = \int_\Omega \lim_{n \to \infty} h_n(\omega)(\mu)\, d\mathbb{P}(\omega) = \int_\Omega \delta_{\varsigma_\mu(\omega)}\, d\mathbb{P}(\omega) = \varsigma_\mu(\mathbb{P}).$$

Now consider two stationary probabilities μ_1 and μ_2. The measure $\mu = (\mu_1 + \mu_2)/2$ is also a stationary probability, and the function ς_μ satisfies, for $\mathbb{P}$-almost every $\omega \in \Omega$,

$$\frac{\delta_{\varsigma_{\mu_1}(\omega)} + \delta_{\varsigma_{\mu_2}(\omega)}}{2} = \delta_{\varsigma_\mu(\omega)}.$$

Clearly, this is impossible unless ς_{μ_1} and ς_{μ_2} coincide almost surely. Thus, by the claim of the first part of the proof,

$$\mu_1 = \varsigma_{\mu_1}(\mathbb{P}) = \varsigma_{\mu_2}(\mathbb{P}) = \mu_2.$$

Suppose now that the action of Γ is minimal and expansive, but not strongly expansive. Let μ be a stationary probability with respect to p. By Lemma 2.3.8, there exists a finite-order homeomorphism $R : S^1 \to S^1$ that commutes with all the elements of Γ and such that the action induced on the topological circle $S^1/\sim$ obtained as the space of orbits under R is minimal and strongly expansive. For each $x \in S^1$, let us set $\psi(x) = \mu([x, R(x)[)$. Since μ has no atom, ψ is a continuous function, and since R centralizes Γ, it is harmonic. Therefore, the set of points at which ψ attains its maximum value is invariant under Γ (see Exercise 2.3.18). Since the orbits under Γ are dense, ψ is constant; in other words, μ is invariant under R. On the other hand, μ projects into a stationary probability measure for the action of Γ on $S^1/\sim$. By the first part of the proof, this projected measure is unique, and together with the R-invariance of μ, this proves that μ is unique as well.

If Γ admits an exceptional minimal set, then this set coincides with the support of μ. By collapsing to a point each connected component in its complementary set, one obtains a minimal action. If this action is expansive, then the preceding arguments give the uniqueness of the stationary measure. To complete the proof, it suffices to notice that in all the cases that have not been considered, Γ preserves a probability measure of the circle. $\square$

Let us define the ***contraction coefficient*** contr(h) of a circle homeomorphism h as the infimum of the numbers $\varepsilon > 0$ such that there exist closed intervals I and J in S^1 of length less than or equal to ε such that $h(\overline{S^1 \setminus I}) = J$. This notion allows us to give a "topological version" of Proposition 2.3.19 for the composition in the "natural order".

Proposition 2.3.21. *Under the hypothesis of Proposition 2.3.19, for almost every sequence* $\omega = (g_1, g_2, \ldots) \in \Omega$*, the contraction coefficient of* $h_n(\omega) = g_n \cdots g_1$ *converges to zero as* n *goes to infinity.*

Proof.   Since μ has total support and no atoms, there exists a circle homeomorphism φ sending μ into the Lebesgue measure. Hence, since the claim to be proved is invariant under topological conjugacy, we may assume that μ coincides with the Lebesgue measure.

Let $\bar{p}$ be the probability on Γ defined by $\bar{p}(g) = p(g^{-1})$, and let $\bar{\Omega}$ be the probability space $\Gamma^{\mathbb{N}}$ endowed with the measure $\bar{p}^{\mathbb{N}}$. On this space let us consider the inverse process $\bar{h}_n(\bar{\omega}) = g_1 \cdots g_n$, where $\bar{\omega} = (g_1, g_2, \ldots)$. By Proposition 2.3.19, for almost every $\bar{\omega} \in \bar{\Omega}$ and all $\varepsilon > 0$, there exists a positive integer $n(\varepsilon, \bar{\omega})$ such that if $n \geq n(\varepsilon, \bar{\omega})$, then there exists a closed interval I such that $\mu(I) \leq \varepsilon$ and $\bar{h}_n(\bar{\omega})(\mu)(I) \geq 1 - \varepsilon$. If we denote by J the closure of $S^1 \setminus g_n^{-1} \cdots g_1^{-1}(I)$, then one easily checks that $|I| = \mu(I) \leq \varepsilon$,

$$|J| = 1 - |g_n^{-1} \cdots g_1^{-1}(I)| = 1 - \mu(\bar{h}_n(\bar{\omega})^{-1}(I)) = 1 - \bar{h}_n(\bar{\omega})(\mu)(I) \leq \varepsilon,$$

and $g_n^{-1} \cdots g_1^{-1}(\overline{S^1 \setminus I}) = J$. Therefore, contr$(g_n^{-1} \cdots g_1^{-1}) \leq \varepsilon$ for all $n \geq n(\varepsilon, \bar{\omega})$. The proof is then completed by noticing that the map $(g_1, g_2, \ldots) \mapsto (g_1^{-1}, g_2^{-1}, \ldots)$ identifies the spaces $(\bar{\Omega}, \bar{p}^{\mathbb{N}})$ and $(\Omega, p^{\mathbb{N}})$. $\square$

The contraction coefficient is always realized, in the sense that for every circle homeomorphism h, there exist intervals I and J such that $\max\{|I|, |J|\} = \text{contr}(h)$ and $h(\overline{S^1 \setminus I}) = J$ (however, these intervals are not necessarily unique). As a consequence, and according to the preceding proof, for almost every $\omega \in \Omega$, we may choose two sequences of closed intervals $I_n(\omega)$ and $J_n(\omega)$ whose lengths go to zero and such that $h_n(\omega)(\overline{S^1 \setminus I_n(\omega)}) = J_n(\omega)$ for all $n \in \mathbb{N}$. For any of these choices, the intervals $I_n(\omega)$ converge to the point $\varsigma_\mu(\omega)$.

Exercise 2.3.22. Let Γ be a group of circle homeomorphisms acting minimally.

(i) Show that if the Γ-action is strongly expansive, then the centralizer of Γ in $\mathrm{Homeo}_+(S^1)$ is trivial.
(ii) More generally, show that if the Γ-action is expansive, then the only circle homeomorphisms commuting with all the elements of Γ are the powers of the map R from Lemma 2.3.8.

Exercise 2.3.23. Using the preceding exercise, prove that if a product of groups $\Gamma = \Gamma_1 \times \Gamma_2$ acts on the circle, then at least one of the factors preserves a probability measure on S^1.

Exercise 2.3.24. The results in this section were obtained by the author in collaboration with Deroin and Kleptsyn in [65] (partial results appear in [139]). Nevertheless, we must point out the existence of a prior work on this topic, namely, the article [3] by Antonov, where almost equivalent results are stated (and proved) in purely probabilistic language. The following items contain the essence of [3].

(i) Given a probability measure p on a (nonnecessarily finite) system of generators of a countable subgroup Γ of $\mathrm{Homeo}_+(S^1)$ without finite orbits, consider the probability $\bar{p}$ on Γ defined by $\bar{p}(g) = p(g^{-1})$ (compare Proposition 2.3.21). If $\bar{\mu}$ is a measure on S^1 that is stationary with respect to $\bar{p}$, show that for every x and y in S^1, the sequence of random variables

$$\xi_n^{x,y}(\omega) = \bar{\mu}([g_n \cdots g_1(x), g_n \cdots g_1(y)])$$

is a martingale. In particular, this sequence converges almost surely. We want to show that if there is no nontrivial circle homeomorphism centralizing Γ, then the corresponding limit equals 0 or 1.

(ii) Let ν be a stationary measure for the diagonal action of Γ on the torus $S^1 \times S^1$. Show that the distribution of $\xi_n^{x,y}$ with respect to $\mathbb{P} \times \nu$ coincides with that of $\xi_{n+1}^{x,y}$.
(iii) Using the relation (which holds for every square integrable martingale)

$$\mathbb{E}(\xi_{n+1}^2) = \mathbb{E}(\xi_n^2) + \mathbb{E}((\xi_{n+1} - \xi_n)^2),$$

conclude that for ν-almost every point $(x, y) \in S^1 \times S^1$, one has $\xi_{n+1}^{x,y} = \xi_n^{x,y}$.

(iv) Defining $\psi((x, y)) = \bar{\mu}([x, y])$, conclude from (iii) that for ν-almost every point (x, y) in $S^1 \times S^1$ and all $g \in \Gamma$, one has $\psi((x, y)) = \psi((g(x), g(y)))$.

(v) Using (iv), prove that the support of the measure ν is contained in the diagonal.

Hint. Without loss of generality, we can reduce the general case to that where all the orbits of Γ on S^1 are dense. Suppose that $\alpha \in]0, 1[$ is such that the set

$$X_{\alpha,\varepsilon} = \{(x, y) \in S^1 \times S^1 : \psi((x, y)) \in [\alpha - \varepsilon, \alpha + \varepsilon]\}$$

has positive ν-measure for every $\varepsilon > 0$. Conclude from (iv) that the normalized restriction of ν to $X_{\alpha,\varepsilon}$ is a stationary probability. If we let ε go to 0 and pass to a weak limit, this gives a stationary probability $\bar{\nu}$ on the torus concentrated on the set

$$\{(x, y) \in S^1 \times S^1 : \psi([x, y]) = \alpha\}.$$

The density of the orbits then allows defining a unique homeomorphism R of S^1 that satisfies $\psi((x, R(x))) = \alpha$. Show that R commutes with all the elements of Γ, thus contradicting the hypothesis at the end of item (i).

(vi) Fixing x and y in S^1, consider the Dirac measure $\delta_{(x,y)}$ with mass on the point $(x, y) \in S^1 \times S^1$. Show that every measure in the adherence of the set of measures

$$\nu_n = \frac{1}{n} \sum_{j=0}^{n-1} p^{*j} * \delta_{(x,y)}$$

is a stationary probability concentrated on the diagonal. Finally, using the fact that $\xi_n^{x,y}$ converges almost surely, conclude that the limit of $\xi_n^{x,y}$ is equal to 0 or 1.

Exercise 2.3.25.   Prove that if Γ is a subgroup of $\text{Homeo}_+([0, 1])$ without global fixed point in the interior and p is a symmetric nondegenerate probability measure on Γ, then every probability measure on $[0, 1]$ that is stationary (with respect to p) is supported at the endpoints of $[0, 1]$.

Hint. Let μ be a stationary measure supported on $]0, 1[$. Show first that μ has no atom. Then, by collapsing the connected components of the complement of the support and reparameterizing the interval, reduce the general case to that where μ coincides with the Lebesgue measure. One then has,

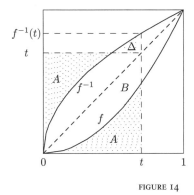

$$A = \int_0^t f(s)\,ds$$
$$B = \int_0^t f^{-1}(s)\,ds$$
$$\Downarrow$$
$$A + B = t^2 + \Delta$$

FIGURE 14

for all $s \in \,]0, 1[$,

$$s = \mu([0, s]) = \int_\Gamma \mu(g^{-1}([0, s]))\,dp(g)$$
$$= \int_\Gamma \frac{\mu(g([0, s])) + \mu(g^{-1}([0, s]))}{2}\,dp(g),$$

which gives

$$s = \int_\Gamma \frac{g(s) + g^{-1}(s)}{2}\,dp(g).$$

Hence, by integrating between 0 and an arbitrary point $t \in \,]0, 1[$,

$$t^2 = \int_\Gamma \int_0^t (g(s) + g^{-1}(s))\,ds\,dp(g). \tag{2.11}$$

Now, by looking at Figure 14, conclude that for every homeomorphism f of the interval and every $t \in [0, 1]$, one has

$$\int_0^t (f(s) + f^{-1}(s))\,ds \geq t^2,$$

where the equality holds if and only if $f(t) = t$. Conclude that t is a global fixed point for the action, thus contradicting the hypothesis (see [65] for an alternative and more conceptual proof using Garnett's ergodic theorem).

Remark. The hypothesis of symmetry for p above is necessary, as is shown by [132].

CHAPTER THREE

Dynamics of Groups
of Diffeomorphisms

3.1 Denjoy's Theorem

In § 2.2.1 we have seen an example of a circle homeomorphism with irrational rotation number admitting an exceptional minimal set. In this
section we will see that such a homeomorphism cannot be a diffeomorphism of certain regularity. This is the content of a classical and very important theorem due to Denjoy. To state it properly, we will denote by
$\mathrm{Diff}_+^{1+\mathrm{bv}}(S^1)$ the group of C^1 circle diffeomorphisms whose derivatives
have bounded variation. We will say that such a diffeomorphism is of class
$C^{1+\mathrm{bv}}$, and we will denote by $V(f)$ the total variation of the logarithm of
its derivative, that is,

$$V(f) = \sup_{a_0 < \cdots < a_n = a_0} \sum_{i=0}^{n-1} |\log(f')(a_{i+1}) - \log(f')(a_i)| = \mathrm{var}(\log(f'); S^1).$$

For an interval I, we will use the notation $V(f; I) = \mathrm{var}(\log(f')|_I)$.

Theorem 3.1.1. *If f is a circle diffeomorphism of class $C^{1+\mathrm{bv}}$ with irrational rotation number, then f is topologically conjugate to the rotation of
angle $\rho(f)$.*

Every circle diffeomorphism of class C^2 or $C^{1+\mathrm{Lip}}$ belongs to
$\mathrm{Diff}_+^{1+\mathrm{bv}}(S^1)$. For instance, for every $f \in \mathrm{Diff}_+^2(S^1)$, one has

$$V(f) = \int_{S^1} \left| \frac{f''(s)}{f'(s)} \right| ds. \tag{3.1}$$

The same formula holds for diffeomorphisms of class $C^{1+\text{Lip}}$. (In this case the function $s \mapsto f''(s)/f'(s)$ is defined almost everywhere and essentially bounded.) However, let us point out that Denjoy's theorem no longer holds in class $C^{1+\tau}$ for any $\tau < 1$. This is a consequence of a construction due to Herman [115] that we will reproduce in § 4.1.4.

Theorem 3.1.2. *For every irrational angle θ, every $\tau < 1$, and every neighborhood of the rotation R_θ in $\text{Diff}_+^{1+\tau}(S^1)$, there exists an element in this neighborhood with rotation number θ that is not topologically conjugate to R_θ.*

Before passing to the proof of Theorem 3.1.1, we would like to give a "heuristic argument" that applies in a particular but very illustrative case. Suppose that f is a C^1 circle diffeomorphism with irrational rotation number and admitting a minimal invariant Cantor set Λ such that the derivative of f on Λ is identically equal to 1. Following [112], we will prove by contradiction that f cannot be of class $C^{1+\text{Lip}}$ (compare [195]; see also Exercise 3.1.4). To do this, let us fix a connected component I of the complementary set of Λ, and for each $n \in \mathbb{N}$, let us set $I_n = g^n(I)$. Notice that $|I_{n+1}| = g'(p)|I_n|$ for some $p \in I_n$, and since the derivative of g at the endpoints of I_n equals 1, we conclude that

$$\left| \frac{|I_{n+1}|}{|I_n|} - 1 \right| = |f'(p) - 1| \leq C|I_n|,$$

where C is the Lipschitz constant of the derivative of f. We then have

$$|I_{n+1}| \geq |I_n|(1 - C|I_n|). \tag{3.2}$$

Without loss of generality, we may assume that the length of I, as well as that of all of its (forward) iterates by f, is less than or equal to $1/2C$. The contradiction we seek then follows from the following elementary lemma by letting $d = 1$ (the case $d > 1$ is naturally related to § 4.1.41).

Lemma 3.1.3. *If $d \in \mathbb{N}$ and (ℓ_n) is a sequence of positive real numbers such that $\ell_n \leq 1/C^d(1 + 1/d)^d$ and $\ell_{n+1} \geq \ell_n(1 - C\ell_n^{1/d})$ for all $n \in \mathbb{N}$ and some positive constant C, then there exists $A > 0$ such that $\ell_n \geq A/n^d$ for all $n \in \mathbb{N}$. In particular, if $d = 1$, then $S = \sum \ell_n$ diverges.*

Proof. The function $s \mapsto s(1 - Cs^{1/d})$ is increasing on the interval $[0, \left(\frac{1}{C(1+1/d)}\right)^d]$. If we use this fact, the claim of the lemma easily follows by induction for the constant $A = \min\{\ell_1, d^d/2^{d^2}C^d\}$. We leave the details to the reader.   □

We will give two distinct proofs of Denjoy's theorem (naturally, both of them will strongly use the combinatorial properties of circle homeomorphisms with irrational rotation number). The first one, due to Denjoy himself, consists in controlling the distortion produced by the diffeomorphism on the affine structure of the circle. As we will see later, this is also the main idea behind the proof of many other results in one-dimensional dynamics of group actions, such as, for example, those of Sacksteder and Duminy. In the second proof, we will concentrate on the distortion with respect to the projective structure of the circle. This technique has been useful in many other cases. For instance, it was used by Yoccoz in [257] to extend Denjoy's theorem to C^∞ circle *homeomorphisms* whose critical points are not infinitely flat (in particular, for real-analytic circle homeomorphisms), and by Hu and Sullivan in [118] to obtain very fine results concerning the optimal regularity of Denjoy's theorem for diffeomorphisms. We will use a variation of this idea to give a general rigidity theorem in § 5.2.

First proof of Denjoy's theorem. Assume for a contradiction that (the group generated by) a C^{1+bv} circle diffeomorphism f admits an exceptional minimal set Λ. Let I be a connected component of $S^1 \setminus \Lambda$, let x_0 be an interior point of I, and let φ be the semiconjugacy of f to R_θ, where $\theta = \rho(f)$. Without loss of generality, we may suppose that $\varphi(x_0) = 0$. For $n \in \mathbb{N}$, let us consider the intervals I_n and J_n of the construction given at the end of § 2.2.1, and let us define $I_n(f) = \varphi^{-1}(I_n)$ and $J_n(f) = \varphi^{-1}(J_n)$. Notice that I is contained in each $I_n(f)$. From the combinatorial properties of I_n and J_n, one concludes that

(i) the intervals in $\{f^j(J_n(f)) : j \in \{0, 1, \ldots, q_{n+1} - 1\}\}$ cover the circle, and every point of S^1 is contained in at most two of them; and

(ii) the interval $f^{q_{n+1}}(I_n(f))$ is contained in $J_n(f)$ for every $n \in \mathbb{N}$, and the same holds for $f^{-q_{n+1}}(I_{-n}(f))$.

We claim that for each x and y in I and each $n \in \mathbb{N}$,

$$|(f^{q_{n+1}})'(x)(f^{-q_{n+1}})'(y)| \geq e^{-2V(f)}. \tag{3.3}$$

To show this, we let $\bar{y} = f^{-q_{n+1}}(y)$. Since I is contained in $I_{-n}(f)$, by property (ii) we have $\bar{y} \in J_n(f)$. From (1.1) and property (i) we obtain

$$\left| \log((f^{q_{n+1}})'(x)(f^{-q_{n+1}})'(y)) \right| = \left| \log\left(\frac{(f^{q_{n+1}})'(x)}{(f^{q_{n+1}})'(\bar{y})} \right) \right|$$

$$\leq \sum_{k=0}^{q_{n+1}-1} |\log(f')(f^k(x)) - \log(f')(f^k(\bar{y}))|$$

$$\leq \sum_{k=0}^{q_{n+1}-1} V(f; f^k(J_n(f))) \leq 2V(f),$$

from which one easily obtains inequality (3.3).

To conclude the proof, notice that for some x and y in I,

$$|f^{q_{n+1}}(I)| \cdot |f^{-q_{n+1}}(I)| = (f^{q_{n+1}})'(x)(f^{-q_{n+1}})'(y) \cdot |I|^2.$$

From inequality (3.3) one deduces that the right-hand expression is bounded from below by $\exp(-2V(f)) \cdot |I|^2$. However, this is absurd, since the intervals $f^k(I)$ are two-by-two disjoint for $k \in \mathbb{Z}$, and hence $|f^{q_{n+1}}(I)| \cdot |f^{-q_{n+1}}(I)|$ goes to zero as n goes to infinity. $\qquad\square$

In the second proof we will use the following notation: given an interval $I = [a, b]$ and a diffeomorphism f defined on I, let (compare (1.5))

$$M(f; I) = \frac{|f(b) - f(a)|}{|b - a|\sqrt{f'(a)f'(b)}}.$$

The reader may easily check the relation (compare (1.4))

$$M(f \circ g; I) = M(g; I) \cdot M(f; g(I)). \qquad (3.4)$$

Moreover, from the existence of a point $c \in I$ such that $|f(b) - f(a)| = f'(c)|b - a|$, one concludes that

$$|\log M(f; I)| \leq \frac{1}{2}[|\log(f'(c)) - \log(f'(a))| + |\log(f'(c)) - \log(f'(b))|]$$

$$\leq \frac{V(f; I)}{2}. \qquad (3.5)$$

Second proof of Denjoy's theorem. Suppose again that f admits an exceptional minimal set Λ, and let us use the notation introduced at the beginning of the first proof. For each $n \in \mathbb{N}$, let us fix two points $a \in f^{-(q_n+q_{n+1})}(I)$ and $b \in f^{-q_{n+1}}(I)$ such that

$$(f^{q_{n+1}})'(a) = \frac{|f^{-q_n}(I)|}{|f^{-(q_n+q_{n+1})}(I)|}, \qquad (f^{q_{n+1}})'(b) = \frac{|I|}{|f^{-q_{n+1}}(I)|}.$$

The open interval J with endpoints a and b is contained in $J_n(f)$ (recall the combinatorial properties at the end of § 2.2.1; in particular, see Figures 11 and 12). By (3.4), (3.5), and property (i) of the first proof,

$$|\log M(f^{q_{n+1}}; J)| \leq \sum_{k=0}^{q_{n+1}-1} |\log M(f, f^k(J))|$$

$$\leq \sum_{k=0}^{q_{n+1}-1} \frac{V(f; f^k(J))}{2} \leq V(f).$$

We then obtain

$$\left(\frac{|f^{q_{n+1}}(J)|}{|J|}\right)^2 \frac{1}{(f^{q_{n+1}})'(a)(f^{q_{n+1}})'(b)} \geq \exp(-2V(f)),$$

that is,

$$\left(\frac{|f^{q_{n+1}}(J)|}{|J|}\right)^2 \frac{|f^{-(q_n+q_{n+1})}(I)|}{|f^{-q_n}(I)|} \frac{|f^{-q_{n+1}}(I)|}{|I|} \geq \exp(-2V(f)).$$

Since $|f^{q_{n+1}}(J)| \leq 1$ for all $n \in \mathbb{N}$ and $|f^{-(q_n+q_{n+1})}(I)|$ tends to zero as n goes to infinity, this inequality implies that for every large-enough n,

$$|f^{-q_{n+1}}(I)| \geq |f^{-q_n}(I)|.$$

Nevertheless, this contradicts the fact that $|f^{-q_n}(I)|$ converges to zero as n goes to infinity.   □

Exercise 3.1.4. The proof proposed below for Denjoy's theorem in class $C^{1+\text{Lip}}$ is quite interesting because it does not use any control-of-distortion-type argument.

(i) Suppose that f is a $C^{1+\text{Lip}}$ counterexample to Denjoy's theorem, and denote by I one of the connected components of the complement of the exceptional minimal set. For each $j \in \mathbb{Z}$, choose a point $x_j \in f^j(I)$ such that $|f^{j+1}(I)| = f'(x_j)|f^j(I)|$. Denoting by C the Lipschitz constant of f' and fixing $n \in \mathbb{N}$, from the inequality $|f'(x_j) - f'(x_{j-q_n})| \leq C|x_j - x_{j-q_n}|$, conclude that for some constant $\bar{C} > 0$ that is independent of n, and for every $j \in \{0, \ldots, q_n - 1\}$,

$$\frac{|f^{j+1}(I)| \cdot |f^{-q_n+j}(I)|}{|f^j(I)| \cdot |f^{-q_n+j+1}(I)|} \geq 1 - \bar{C}|x_j - x_{j-q_n}|. \tag{3.6}$$

(ii) Using the combinatorial properties in § 2.2.1, from the above inequality, conclude the existence of a constant $D > 0$ such that for every $n \in \mathbb{N}$,

$$\frac{|f^{q_n}(I)| \cdot |f^{-q_n}(I)|}{|I|^2} \geq De^{-D}. \tag{3.7}$$

Hint. The left-side expression coincides with

$$\prod_{j=0}^{q_n-1} \frac{|f^{j+1}(I)| \cdot |f^{-q_n+j}(I)|}{|f^j(I)| \cdot |f^{-q_n+j+1}(I)|}.$$

Use inequality (3.6) to estimate each factor of this product for large-enough j, and then use the elementary inequality $1 - x \geq e^{-2x}$ (which holds for very small $x > 0$), as well as the fact that the intervals with end-points x_j and x_{j-q_n}, with $j \in \{0, \dots, q_n - 1\}$, are two-by-two disjoint.

(iii) Show that inequality (3.7) cannot hold for all n.

To construct smooth circle homeomorphisms admitting an exceptional minimal set, one may proceed in two distinct ways. The first consists in allowing critical points. In this direction, in [109] one may find examples of C^∞ circle homeomorphisms with a single critical point and a minimal invariant Cantor set (by Yoccoz's theorem already mentioned [257], such a homeomorphism cannot be real-analytic). The second consists in considering diffeomorphisms of differentiability class less than C^2. We will carefully study examples of this type in § 4.1.4.

Concerning the nature of the invariant Cantor set that may appear in differentiability class less than C^2, let us first notice that any Cantor set contained in S^1 may be exhibited as the exceptional minimal set of a circle *homeomorphism*. However, if those the homeomorphism has some regularity, then such a set must satisfy certain properties. For example, in [165] it is proved that the triadic Cantor set cannot appear as the exceptional minimal set of any C^1 circle diffeomorphism (see also [194]). On the other hand, in [196] it is shown that every "affine" Cantor set may appear as the exceptional minimal set of a Lipschitz circle homeomorphism.

From the viewpoint of group actions, Denjoy's theorem may be re-formulated by saying that there is no action of $(\mathbb{Z}, +)$ by $C^{1+\mathrm{bv}}$ circle diffeomorphisms that admits an exceptional minimal set. As we will next see, this property is shared by a large family of finitely generated groups.

Proposition 3.1.5. *Let* Γ *be a finitely generated group of* $C^{1+\text{Lip}}$ *circle diffeomorphisms. If* Γ *admits an exceptional minimal set, then it contains a free subgroup on two generators.*

Proof. We will show that Γ acts expansively on the topological circle S^1_Λ associated to the exceptional minimal set Λ. By the proof of Theorem 2.3.2, this implies that Γ contains a free group.

Suppose that the (minimal) action of Γ on S^1_Λ is not expansive. Then it is equicontinuous, and there is an invariant probability measure on S^1_Λ with total support and no atoms that allows conjugating Γ to a group of rotations of S^1_Λ. Let $g_1, \ldots, g_n$ be the generators of Γ. If each g_i has finite order, then every orbit in S^1_Λ is finite, which contradicts the minimality of Λ. On the other hand, if some generator g_i has infinite order, then the rotation number of g_i is irrational. Since g_i has nondense orbits (for instance, those contained in Λ), this contradicts Denjoy's theorem. $\quad\square$

The preceding proposition applies to finitely generated, amenable groups. (Indeed, such a group cannot contain $\mathbb{F}_2$; see Appendix B). However, this may be shown in a more direct way (i.e., without using Margulis's theorem).

Exercise 3.1.6. Let Γ be a finitely generated amenable group of $C^{1+\text{bv}}$ circle diffeomorphisms. Suppose that Γ admits an exceptional minimal set Λ, and consider an invariant probability measure μ for the action of Γ on the topological circle S^1_Λ associated to Λ. Prove that μ has total support and no atoms. Conclude that Γ is semiconjugate to a group of rotations, and obtain a contradiction using Denjoy's theorem.

Proposition 3.1.5 still holds for non–finitely generated groups provided that there exists a constant C such that $V(g) \leq C$ for every generator g. Without this hypothesis, it fails to hold for general non–finitely generated subgroups of $\text{Diff}^{1+\text{bv}}_+(S^1)$, as is shown by the following example, essentially due to Hirsch [117] (see also [127]).

Example 3.1.7. Let $H : \mathbb{R} \to \mathbb{R}$ be a homeomorphism satisfying the properties (i) and (ii) of § 1.5.2 and such that $\text{Fix}(H) = \{a, b\}$, with $a = 0$ and small $b > 0$. This homeomorphism H naturally induces a degree-2 circle map $\bar{H}$ (see Figure 10). Let us denote by I the interval $]a, b[$. Let us now define $g_1 : S^1 \to S^1$ by $g_1(x) = y$ if $\bar{H}(x) = \bar{H}(y)$ and $x \neq y$. In general, for each $n \in \mathbb{N}$, let us consider the degree-2^n circle map $\bar{H}^n$, and let us define $g_n : S^1 \to S^1$ by $g_n(x) = y$ if $\bar{H}^n(x) = \bar{H}^n(y)$ and $\bar{H}^n(x) \neq \bar{H}^n(y')$ for every $y' \in \,]x, y[$.

The group $\Gamma = \Gamma_H$ generated by the g_n's is Abelian and isomorphic to $\mathbb{Q}_2(S^1)$. Its action on the circle is semiconjugate to that of a group of rotations, and its exceptional minimal set is $\Lambda = S^1 \setminus \cup_{n \in \mathbb{N}} \cup_{i=0}^{2^n-1} g_n^i(C)$. Notice that if H is a real-analytic diffeomorphism of the real line, then the elements in Γ_H are real-analytic circle diffeomorphisms. The reader should notice the similarity between this construction and the last one of § 2.1.1. Indeed, the group Γ_H is a subgroup of (the realization $\Phi_H(G)$) of Thompson's group G (although $\Phi_H(G)$ is not a group of real-analytic diffeomorphisms of S^1).

Denjoy's theorem may also be seen as a consequence of the next proposition, which is the basis for dealing with the relevant problem of the regularity of the **linearization** of smooth circle diffeomorphisms (i.e., the conjugacy to the corresponding rotation).

Proposition 3.1.8. *Let $\theta \in \,]0, 1[$ be an irrational number, and let p/q be one of the rational approximations of θ (i.e., $|\theta - p/q| \leq 1/q^2$). For a circle homeomorphism f with rotation number θ, denote by μ the unique probability measure on S^1 that is invariant under f. If $\psi : S^1 \to \mathbb{R}$ is a (nonnecessarily continuous) function with finite total variation, then for every $x \in S^1$, one has*

$$\left| \sum_{i=0}^{q-1} \psi(f^i(x)) - q \int_{S^1} \psi \, d\mu \right| \leq \mathrm{var}(\psi; S^1). \tag{3.8}$$

In particular, if f is differentiable and its derivative has bounded variation, then for every $x \in S^1$ and every $n \in \mathbb{N}$, one has

$$\exp(-V(f)) \leq (f^{q_n})'(x) \leq \exp(V(f)). \tag{3.9}$$

To prove this proposition, we will use the following combinatorial lemma.

Lemma 3.1.9. *If $\theta \in \,]0, 1[$ is irrational and p and q are two relatively prime integers such that $q \neq 0$ and $|\theta - p/q| \leq 1/q^2$, then for each $i \in \{0, \ldots, q-1\}$, the interval $]i/q, (i+1)/q[$ contains a unique point of the set $\{j\theta \pmod 1 : j \in \{1, \ldots, q\}\}$.*

Proof. By the hypothesis, we have either $0 < \theta - p/q < 1/q^2$ or $-1/q^2 < \theta - p/q < 0$. Because both cases are analogous, let us consider only the first one. For every $j \in \{1, \ldots, q\}$, one has

$$0 < j\theta - \frac{jp}{q} < \frac{j}{q^2} \leq \frac{1}{q},$$

and hence $j\theta \mod 1$ belongs to $]jp/q, (j12p+1)/q[$. The claim of the lemma follows from the fact that the intervals $[jp/q, (jp+1)/q[$ cover the circle without intersection. □

Proof of Proposition 3.1.8. It is evident that the claim to be proved is equivalent to the claim that the inequality

$$\left| \sum_{i=1}^{q} \psi(f^i(x)) - q \int_{S^1} \psi \, d\mu \right| \leq \mathrm{var}(\psi; S^1)$$

holds for all $x \in S^1$ and every function of bounded variation ψ. To show this inequality, we consider the map from the circle into itself defined by

$$\varphi(y) = \int_x^y d\mu \qquad \mod 1.$$

From the relation $\varphi \circ f = R_\theta \circ \varphi$ we deduce that if we let $x_j = f^j(x)$, then we have $\varphi(x_j) = j\theta \mod 1$. By the preceding lemma, if for each $j \in \{1, \ldots, q\}$ we choose an interval $I_j =]i_j/q, (i_j+1)/q[$ containing $\varphi(x_j)$, then the intervals that appear are two-by-two disjoint. Therefore, denoting by J_j the interval $\varphi^{-1}(\bar{I}_j)$ (whose μ-measure equals $1/q$), from the equality

$$\left| \sum_{i=1}^{q} \psi(f^i(x)) - q \int_{S^1} \psi \, d\mu \right| = \left| \sum_{j=1}^{q} \left(\psi(f^j(x)) - q \int_{J_j} \psi \, d\mu \right) \right|$$

one concludes that

$$\left| \sum_{i=1}^{q} \psi(f^i(x)) - q \int_{S^1} \psi \, d\mu \right| \leq q \sum_{j=1}^{q} \int_{J_j} |\psi(f^j(x)) - \psi| \, d\mu$$

$$\leq \sum_{j=1}^{q} \sup_{y \in J_j} |\psi(f^j(x)) - \psi(y)| \leq \mathrm{var}(\psi; S^1).$$

The second part of the proposition follows from the first part applied to the function $\psi(x) = \log(f'(x))$, thanks to the equality

$$\int_{S^1} \log(f') \, d\mu = 0. \qquad (3.10)$$

However, we point out that this equality is not at all evident. For the proof, notice that from (3.8) one deduces that for every $x \in S^1$,

$$\exp(-V(f)) \leq \frac{(f^{q_n})'(x)}{\exp(q_n \int_{S^1} \log(f')d\mu)} \leq \exp(V(f)).$$

If we integrate with respect to the Lebesgue measure, this allows us to conclude that

$$\exp(-V(f)) \leq \frac{1}{\exp(q_n \int_{S^1} \log(f')d\mu)} \leq \exp(V(f)),$$

which cannot hold for every $n \in \mathbb{N}$ unless (3.10) holds (compare Exercise 3.1.11). $\qquad\square$

Relation (3.8) (resp. (3.9)) is known as the **Denjoy-Koksma inequality** (resp. **Denjoy inequality**). Let us point out that similarly to Denjoy's theorem and equality (3.10), these inequalities still hold for piecewise affine circle homeomorphisms, and the arguments of the proofs are similar to those of the C^{1+bv} case (of course, in the piecewise affine case, one considers a lateral derivative instead of the usual one).

Remark 3.1.10. For every $C^{2+\tau}$ circle diffeomorphism f with irrational rotation number, a stronger conclusion than that of Proposition 3.1.8 holds: the sequence (f^{q_n}) converges to the identity in the C^1 topology, where p_n/q_n denotes the nth rational approximation of $\rho(f)$ (see [115, 136, 256]).

Exercise 3.1.11. Prove that equality (3.10) holds for every C^1 circle diffeomorphism by using the Birkhoff ergodic theorem and the unique ergodicity of f (see [249]).

Notice that Denjoy's theorem implies that if Γ is a subgroup of $\mathrm{Diff}_+^{1+bv}(S^1)$ admitting an exceptional minimal set, then each of its elements has periodic points. Indeed, if $g \in \Gamma$ has no periodic point, then its rotation number is irrational, which implies the density of the orbits under g (and hence the density of the orbits under Γ). An interesting consequence of this fact is the rationality of the rotation number of the elements in Thompson's group G. Indeed, in § 2.1.1 we constructed an action of G by C^∞ circle diffeomorphisms that is semiconjugate to the standard action and admits an exceptional minimal set. The fact that $\rho(g)$ belongs to $\mathbb{Q}$ for each $g \in G$ then follows from the preceding remark together with the invariance of the rotation number under semiconjugacy. Following [152], we invite the reader to develop an alternative and "more direct" proof only using Denjoy's inequality (alternative proofs may be found in [40] and [138]).

Exercise 3.1.12. Given $g \in G$, let us define $M_n \in \mathbb{N}$ and $N_n \in \mathbb{N} \cup \{0\}$ such that for each $n \in \mathbb{N}$, one has $g^n(0) = M_n/2^{N_n}$, where M_n is either odd or zero. Suppose that 0 is not a periodic point of g, and thus $M_n \neq 0$ for every $n \in \mathbb{N}$.

(i) Prove that N_n tends to infinity together with n.

Hint. Notice that for every $N \in \mathbb{N}$, the set of dyadic rational numbers with denominator less than or equal to 2^N is finite.

(ii) Using the convergence of N_n to infinity, conclude that $\lim_{n \to \infty} (g^n)'(0) = 0$ (where we consider the right derivative of the map).

Hint. Write $g(x) = 2^{\lambda(x)} + M(x)/2^{N(x)}$ for some integer-valued uniformly bounded functions λ, M, and N, and show that for large-enough $n \in \mathbb{N}$, one has

$$M_{n+1} = M_n + 2^{N_n - \lambda(g^n(0)) - N(g^n(0))} M(g^n(0)), \qquad N_{n+1} = N_n - \lambda(g^n(0)).$$

(iii) Show that $\rho(g)$ is rational by applying Denjoy's inequality.

3.2 Sacksteder's Theorem

Let us begin by recalling the notion of a *pseudogroup* of homeomorphisms.

Definition 3.2.1. A family $\Gamma = \{g : dom(g) \to ran(g)\}$ of homeomorphisms between subsets of a topological space M is a **pseudogroup** if the following properties are satisfied:

- The domain $dom(g)$ and the image $ran(g)$ of each $g \in \Gamma$ are open sets.
- If g and h belong to Γ and $ran(h) \subset dom(g)$, then gh belongs to Γ.
- If $g \in \Gamma$, then $g^{-1} \in \Gamma$.
- The identity on M is an element of Γ.
- If g belongs to Γ and $A \subset dom(g)$ is an open subset, then the restriction $g|_A$ of g to A is an element of Γ.

- If $g : dom(g) \to ran(g)$ is a homeomorphism between open subsets of M such that for every $x \in dom(g)$ there exists a neighborhood V_x such that $h|_{V_x}$ is an element of Γ, then g belongs to Γ.

The notions of orbit and invariant set for a pseudogroup are naturally defined. We will say that Γ is **finitely generated** (resp. **countably generated**) if there exists a finite (resp. countable) subset $\mathcal{G}$ of Γ such that every element in Γ may be written as (the restriction to its domain of) a product of elements in $\mathcal{G}$. A measure μ on the Borel sets of M is invariant under Γ if for every Borel set A and every $g \in \Gamma$, one has $\mu(A \cap dom(g)) = \mu(g(A) \cap ran(g))$. Finally, for $a \in M$, we denote by $\Gamma(a)$ the orbit of a, and we define the **order** of each $p \in \Gamma(a)$ by letting

$$\text{ord}(p) = \min\{n \in \mathbb{N} : \text{ there exist } g_{i_j} \in \mathcal{G} \text{ such that } g_{i_n} \cdots g_{i_1}(a) = p\}.$$

Pseudogroups naturally appear in foliation theory. Indeed, after one prescribes a set of open transversals, the holonomies along leaves form a pseudogroup under composition. But as we will see, even in the context of group actions, it is sometimes important to keep in mind the pseudogroup approach, since many results are proved by restricting the original action to certain open sets where the group structure is loosened.

3.2.1 The classical version in class $\mathbf{C^{1+Lip}}$

In this section we will prove the most classical version of a theorem due to Sacksteder [220] about the existence of elements with **hyperbolic fixed points** in some pseudo-groups of diffeomorphisms of one-dimensional compact manifolds.

Theorem 3.2.2. *Let Γ be a pseudogroup of C^{1+Lip} diffeomorphisms of a one-dimensional compact manifold generated by a finite family of elements $\mathcal{G}$. Suppose that for each generator $g \in \mathcal{G}$ there exists $\bar{g} \in \Gamma$ whose domain contains the closure of $dom(g)$ and that coincides with g when restricted to $dom(g)$.[1] Suppose, moreover, that there exists an invariant closed set Λ such that for some connected component $I =]a, b[$ of its complement,*

---

1. This is the **compact generation property** for the pseudogroup, whose relevance was cleverly noticed by Haefliger [108] (see Remark 3.2.13 on this). It is always satisfied by the holonomy pseudogroup of a codimension-one foliation on a compact manifold.

either a is an accumulation point of the orbit $\Gamma(a)$, *or b is an accumulation point of* $\Gamma(b)$. *Then there exist* $p \in \Lambda$ *and* $h \in \Gamma$ *such that* $h(p) = p$ *and* $h'(p) < 1$.

A set Λ satisfying these hypotheses (with the possible exception of the one concerning the regularity) is called a **local exceptional set**. To show the theorem, the main technical tool will be the next lemma, whose $C^{1+\mathrm{Lip}}$ version is essentially due to Schwartz [223]. We give a general version including the $C^{1+\tau}$ case for future reference; a slightly refined $C^{1+\mathrm{Lip}}$ version will be discussed in § 3.4.2 (see also Exercise 3.5.11).

Lemma 3.2.3. *Let* Γ *be a pseudogroup of diffeomorphisms of class* $C^{1+\tau}$ *(resp.* $C^{1+\mathrm{Lip}}$*) of a one-dimensional, compact manifold. Suppose that there exist a finite subset* $\mathcal{G}$ *of* Γ, *a constant* $S \in [1, \infty[$, *and an open interval* I *such that to each* $g \in \mathcal{G}$ *one may associate a compact interval* L_g *contained in an open set where g is defined in such a way that for each* $m \in \mathbb{N}$, *there exists* $g_{i_m} \in \mathcal{G}$ *such that for all* $n \in \mathbb{N}$, *the element* $h_n = g_{i_n} \cdots g_{i_1} \in \Gamma$ *satisfies the following properties:*

- *The interval* $g_{i_{k-1}} \cdots g_{i_1}(I)$ *is contained in* $L_{g_{i_k}}$ *(where* $g_{i_{k-1}} \cdots g_{i_1} = \mathrm{Id}$ *for* $k = 1$*).*
- *One has the inequality* $\sum_{k=0}^{n-1} |g_{i_k} \cdots g_{i_1}(I)|^{\tau} \leq S$ *(resp.* $\sum_{k=0}^{n-1} |g_{i_k} \cdots g_{i_1}(I)| \leq S$*).*

Then there exists a positive constant $\ell = \ell(\tau, S, |I|; \mathcal{G})$ *such that if for some* $n \in \mathbb{N}$, *the interval* $h_n(I)$ *is contained in an* ℓ-*neighborhood of* I *and does not intersect the interior of* I, *then the map* h_n *has a hyperbolic fixed point.*

Proof. To simplify, we will think of the case $\tau = 1$ as the Lipschitz case in order to treat it simultaneously with the Hölder case $\tau \in]0, 1[$. Let $\varepsilon > 0$ be a constant such that each $g \in \mathcal{G}$ is defined in a 2ε-neighborhood of L_g. We fix $C > 0$ in such a way that for all $g \in \mathcal{G}$ and all x and y in the ε-neighborhood of L_g, one has

$$|\log(g'(x)) - \log(g'(y))| \leq C|x - y|^{\tau}.$$

We will show that the claim of the lemma is satisfied for

$$\ell = \min \left\{ \frac{|I|}{2 \exp(2^{\tau} C S)}, \frac{|I| \varepsilon}{2 \exp(2^{\tau} C S)}, \frac{\varepsilon}{2} \right\}.$$

We denote by J the 2ℓ-neighborhood of I, and we let I' (resp. I'') be the connected component of $J \setminus I$ to the right (resp. to the left) of I. We will show by induction that the following properties are simultaneously satisfied:

(i)$_k$ The interval $g_{i_k} \cdots g_{i_1}(I')$ is contained in the ε-neighborhood of L_g.

(ii)$_k$ $|g_{i_k} \cdots g_{i_1}(I')| \leq |g_{i_k} \cdots g_{i_1}(I)|$.

(iii)$_k$ $\sup_{\{x,y\} \subset I \cup I'} \frac{(g_{i_k} \cdots g_{i_1})'(x)}{(g_{i_k} \cdots g_{i_1})'(y)} \leq \exp(2^\tau CS)$.

Condition (iii)$_0$ is obviously satisfied, while (i)$_0$ and (ii)$_0$ follow from the hypothesis $|I'| = 2\ell$ and the inequalities $\ell \leq \varepsilon/2$ and $\ell \leq |I|/2$. Suppose that (i)$_j$, (ii)$_j$, and (iii)$_j$ are true for every $j \in \{0, \ldots, k-1\}$. In this case, for every x and y in $I \cup I'$, we have

$$
\left| \log \left(\frac{(g_{i_k} \cdots g_{i_1})'(x)}{(g_{i_k} \cdots g_{i_1})'(y)} \right) \right|
$$

$$
\leq \sum_{j=0}^{k-1} |\log(g'_{i_{j+1}}(g_{i_j} \cdots g_{i_1}(x))) - \log(g'_{i_{j+1}}(g_{i_j} \cdots g_{i_1}(y)))|
$$

$$
\leq C \sum_{j=0}^{k-1} |g_{i_j} \cdots g_{i_1}(x) - g_{i_{j,n}} \cdots g_{i_1}(y)|^\tau
$$

$$
\leq C \sum_{j=0}^{k-1} (|g_{i_j} \cdots g_{i_1}(I)| + |g_{i_j} \cdots g_{i_1}(I')|)^\tau
$$

$$
\leq C 2^\tau S.
$$

This inequality shows (iii)$_k$. Concerning (i)$_k$ and (ii)$_k$, notice that there exist $x \in I$ and $y \in I'$ such that

$$
|g_{i_k} \cdots g_{i_1}(I)| = |I| \cdot (g_{i_k} \cdots g_{i_1})'(x),
$$
$$
|g_{i_k} \cdots g_{i_1}(I')| = |I'| \cdot (g_{i_k} \cdots g_{i_1})'(y).
$$

Hence by (iii)$_k$,

$$
\frac{|g_{i_k} \cdots g_{i_1}(I')|}{|g_{i_k} \cdots g_{i_1}(I)|} = \frac{(g_{i_k} \cdots g_{i_1})'(x)}{(g_{i_k} \cdots g_{i_1})'(y)} \cdot \frac{|I'|}{|I|} \leq \exp(2^\tau CS) \frac{|I'|}{|I|},
$$

which shows (i)$_k$ and (ii)$_k$ by the definition of ℓ. Of course, similar properties also hold for I''.

Now suppose that $h_n(I)$ is contained in the ℓ-neighborhood of the interval I and does not intersect the interior of I. Property (ii)$_n$ then gives $h_n(J) \subset J$. Moreover, if $h_n(I)$ is to the right (resp. to the left) of I, then $h_n(I \cup I') \subset I'$ (resp. $h_n(I'' \cup I) \subset I''$). Because both cases are analogous, we will deal only with the first one. Clearly, h_n has a fixed point x in I', and we need to verify that this fixed point is hyperbolic (and contracting). To do this, choose $y \in I$ such that $h'_n(y) = |h_n(I)|/|I| \leq \ell/|I|$. By (iii)$_n$,

$$h'_n(x) \leq h'_n(y) \exp(2^\tau C S) \leq \frac{\ell \exp(2^\tau C S)}{|I|} \leq \frac{1}{2},$$

and this concludes the proof.   □

An inequality of type $\sum_{k \geq 0} |g_{i_k} \cdots g_{i_1}(I)| \leq S$ obviously holds when the intervals $g_{i_k} \cdots g_{i_1}(I)$ are two-by-two disjoint. In this case, the issue of the preceding lemma lies in the possibility of controlling the distortion of the compositions in the 2ℓ-neighborhood of I, despite the fact that the images of this neighborhood are no longer disjoint.

Proof of Theorem 3.2.2.   Suppose that b is an accumulation point of $\Gamma(b)$ (the case where a is an accumulation point of $\Gamma(a)$ is analogous). Let $(h_n) = (g_{i_n} \cdots g_{i_1})$ be a sequence in Γ such that $h_n(b)$ converges to b, such that $h_n(b) \neq b$ for every $n \in \mathbb{N}$, and such that the order of the point $h_n(b) \in \Gamma(b)$ is realized by h_n (we leave to the reader the task of showing the existence of such a sequence). Clearly, the hypotheses of the preceding lemma are satisfied for the connected component I of the complement of Λ containing b and every constant $S \geq 1$ greater than the total length of the underlying one-dimensional manifold. We then conclude that for large-enough n, the map $h_n \in \Gamma$ contracts (hyperbolically) into itself the interval I' of length ℓ situated immediately to the right of I. Therefore, h_n has a unique fixed point in I (namely, the point $p = \cap_{k \in \mathbb{N}} (h_n)^k(I')$), which is hyperbolic (see Figure 15). Remark finally that since Λ is a compact invariant set and I' intersects Λ, the point p must belong to it.   □

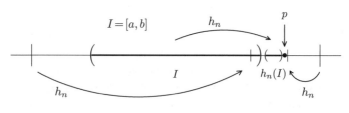

From what precedes, one immediately deduces the following:

Corollary 3.2.4. *If Γ is a finitely generated subgroup of $\mathrm{Diff}_+^{1+\mathrm{Lip}}(\mathrm{S}^1)$ having an exceptional minimal set Λ, then there exist $p \in \Lambda$ and $h \in \Gamma$ such that $h(p) = p$ and $h'(p) < 1$.*

Notice that Denjoy's theorem for $C^{1+\mathrm{Lip}}$ diffeomorphisms may be deduced from this corollary. Actually, Sacksteder's theorem is customarily presented as a generalization of Denjoy's theorem for groups of diffeomorphisms and codimension-one foliations. Nevertheless, this approach is not appropriate from a dynamical point of view. Indeed, Denjoy's theorem strongly uses the very particular combinatorics of circle homeomorphisms with irrational rotation number, which is quite distinct (even locally) from that of pseudogroups acting without invariant probability measure.[2] In the next section we will develop this view to obtain finer results than those of this section. Here we content ourselves with showing how to extend Sacksteder's theorem to non-Abelian groups acting minimally on the circle. We begin with a clever remark due to Ghys and taken from [72].

Proposition 3.2.5. *Let Γ be a pseudogroup of $C^{1+\mathrm{Lip}}$ diffeomorphisms of the circle or of a bounded interval, all of whose orbits are dense. Suppose that there exists an element $f \in \Gamma$ fixing a single point on a nondegenerate (nonnecessarily open) interval containing this point. Then Γ contains an element with a hyperbolic fixed point.*

Proof. Let a be the fixed point given by the hypothesis, and let $a' > a$ be such that the restriction of f to an interval $[a, a']$ has no other fixed point than a (if such a point a' does not exist, then one may consider an interval of type $[a', a]$ with a similar property). Replacing f by f^{-1} if necessary, we may assume that $f(y) < y$ for all $y \in \,]a, a']$. Since the orbits under Γ are dense, there must exist $h \in \Gamma$ such that $h(a) \in \,]a, a'[$. Let $n \in \mathbb{N}$ be large enough that $[a, f^n(a')] \subset dom(h)$ and $h([a, f^n(a')]) \subset \,]a, a']$. $\quad\square$

Let us consider the intervals $A_0 = [a, f^n(a')]$ and $B_0 = h([a, f^n(a')])$, and let us define by induction the sets $A_{j+1} = f^n(A_j \cup B_j)$ and $B_{j+1} = h(A_{j+1})$. For very small $\epsilon > 0$, consider the pseudogroup generated by the restrictions of f^n and h to the intervals $]a - \epsilon, hf^n(a') + \epsilon[$ and

---

2. The reader should notice that an analogous dichotomy was exploited in dealing with orderable groups in § 2.2.6.

$]a - \varepsilon, f^n(a') + \varepsilon[$, respectively. With respect to this pseudogroup, the Cantor set $\Lambda = \cap_{j\in\mathbb{N}}(A_j \cup B_j)$ satisfies the hypotheses of Sacksteder's theorem. The result then follows. □

By Hölder's theorem, a group of circle homeomorphisms that is not semiconjugate to a group of rotations contains an element f satisfying the hypothesis of the preceding proposition. Since every semiconjugacy from a group whose orbits are dense to a group of rotations is necessarily a conjugacy, as a corollary we obtain the following:

Corollary 3.2.6. *Let Γ be a subgroup of $\mathrm{Diff}_+^{1+\mathrm{Lip}}(\mathrm{S}^1)$. Suppose that the orbits of Γ are dense and that Γ is not conjugate to a group of rotations (equivalently, suppose that Γ acts minimally and without invariant probability measure). Then Γ contains an element with a hyperbolic fixed point.*

In the next section we will see that this corollary still holds in class C^1.

Exercise 3.2.7. The fact that some nontrivial holonomy is hyperbolic is relevant when one is studying stability and/or rigidity properties (see, for instance, § 3.6.1). However, knowing that some holonomy is nontrivial (but not necessarily hyperbolic) may be also useful. The next proposition corresponds to a clever remark (due to Hector) in this direction. Before passing to its proof, the reader should check its validity for the modular group and for Thompson's group G.

Proposition 3.2.8. *Let Γ be a finitely generated group of $\mathrm{C}^{1+\mathrm{Lip}}$ circle diffeomorphisms admitting an exceptional minimal set Λ. If p is the endpoint of one of the connected components I of $\mathrm{S}^1 \setminus \Lambda$, then there exists $g \in \Gamma$ fixing p and whose restriction to $I \cap V$ is nontrivial for every neighborhood V of p.*

Hint. Suppose that this is not the case and obtain a contradiction to the fact that for every neighborhood of p, there exist elements in Γ having a hyperbolically *repelling* fixed point inside.

3.2.2 The C^1 version for pseudogroups

The main object of this section and the next one consists in formulating and proving several generalizations of Sacksteder's theorem in class C^1. Some closely related versions were obtained by Hurder in [120, 121, 123, 124] via dynamical methods coming from the so-called foliated Pesin's theory. The optimal and definitive versions (obtained by the author in collaboration with Deroin and Kleptsyn in [65]) make strong use of probabilistic arguments.

Theorem 3.2.9. *If Γ is a pseudogroup of C^1 diffeomorphisms of a compact one-dimensional manifold without invariant probability measure, then Γ has elements with hyperbolic fixed points.*

We will show this theorem inspired by the proof of Proposition 3.2.5. Let us suppose that Γ contains two elements f and h satisfying the following conditions (compare Definition 2.2.43):

(i) The domain of definition of f contains an interval $[a, a'[$ such that $f(a) = a$ and f topologically contracts toward a.
(ii) The element h is defined on a neighborhood of a, and $h(a) \in]a, a'[$.

Put $c = h(a)$ and fix $d' \in]c, a'[$. Replacing f by f^n for large-enough $n \in \mathbb{N}$ if necessary, we may suppose that $f(d') < c$, that $f(d')$ belongs to the domain of definition of h, and that $hf(d') \in]c, d'[$. The last condition implies in particular that hf has fixed points in $]c, d'[$. Let d be the first fixed point of hf to the right of c, and let $b = f(d)$. The interval $I =]b, c[$ corresponds to the "first gap" (i.e., the "central" connected component of the complement) of a Cantor set Λ that is invariant under f and $g = hf$ (see Figure 16).

Proposition 3.2.10. *With the preceding notation, the pseudogroup generated by f and g contains elements with hyperbolic fixed points in Λ.*

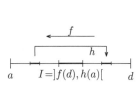

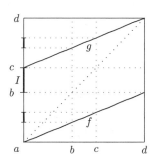

FIGURE 16

The proof of this proposition is particularly simple in class $C^{1+\tau}$ (where $\tau > 0$). Moreover, it allows illustrating the usefulness of probabilistic methods in the theory. The (slightly more technical) C^1 case will be discussed later. For an alternative "purely deterministic" proof of a closely related result, we refer the reader to [134].

Proof of Proposition 3.2.10 in class $C^{1+\tau}$. On $\Omega = \{f, g\}^{\mathbb{N}}$ let us consider the Bernoulli measure $\mathbb{P}$ giving mass $1/2$ to each "random variable" f and g. For $\omega = (g_1, g_2, \ldots)$ in Ω and $n \in \mathbb{N}$, let $h_n(\omega) = g_n \cdots g_1$, and let $h_0(\omega) = id$. Since the interval I is wandering in the sense that the intervals in the family $\{g_n \cdots g_1(I) : n \geq 0, (g_1, \ldots, g_n) \in \{f, g\}^n\}$ are two-by-two disjoint,

$$\sum_{n \geq 0} \sum_{(g_1, \ldots, g_n) \in \{f, g\}^n} |g_n \cdots g_1(I)| \leq d - a < \infty.$$

Moreover, Fubini's theorem gives

$$\sum_{n \geq 0} \sum_{(g_1, \ldots, g_n) \in \{f, g\}^n} |g_n \cdots g_1(I)| = \sum_{n \geq 0} 2^n \left(\int_{\Omega} |h_n(\omega)(I)| \, d\mathbb{P}(\omega) \right)$$

$$= \int_{\Omega} \left(\sum_{n \geq 0} 2^n |h_n(\omega)(I)| \right) d\mathbb{P}(\omega).$$

This allows us to conclude that for $\mathbb{P}$-almost every random sequence $\omega \in \Omega$, the series $\sum 2^n |h_n(\omega)(I)|$ converges. For each $B > 0$, let us consider the set

$$\Omega(B) = \{\omega \in \Omega : |h_n(\omega)(I)| \leq B/2^n \text{ for all } n \geq 0\}.$$

The probability $\mathbb{P}[\Omega(B)]$ converges to 1 as B goes to infinity. In particular, we may fix B such that $\mathbb{P}[\Omega(B)] > 0$. Notice that if ω belongs to $\Omega(B)$, then

$$\sum_{n \geq 0} |h_n(\omega)(I)|^{\tau} \leq B^{\tau} \sum_{n \geq 0} \frac{1}{2^{n\tau}} = S < \infty. \tag{3.11}$$

Let us consider the interval $J' = [b - \ell, c + \ell]$ containing the wandering interval I, where $\ell = \ell(\tau, S, |I|; \{f, g\})$ is the constant that appears in Lemma 3.2.3. If $N \in \mathbb{N}$ is large enough, then both $f^N g$ and $g^N f$ send the whole interval $[0, 1]$ into $J' \setminus I$. A direct application of the Borel-Cantelli lemma then gives

$$\mathbb{P}[h_n(\omega)(I) \subset J' \setminus I \text{ infinitely many times}] = 1.$$

If $\omega \in \Omega(B)$ and $n \in \mathbb{N}$ satisfy $h_n(\omega)(I) \subset J' \setminus I$, then Lemma 3.2.3 shows that $h_n(\omega)$ has a hyperbolic fixed point p. Finally, since Λ is invariant under the pseudogroup and the fixed point of $h_n(\omega)$ that we found contracts part of this set, p must belong to Λ. □

The proof of the general version (in class C^1) of Proposition 3.2.10 needs some technical improvements for finding hyperbolic fixed points in the absence of control of distortion. The main idea lies in the fact that when one knows *a priori* that the dynamics is differentiably contracting somewhere, the continuity of the derivatives forces this contraction to persists (perhaps at a smaller rate) on a larger domain.

Keeping the notation of the preceding proof, fix once and for all a constant $\varepsilon \in \,]0, 1/2[$. We know that for $\mathbb{P}$-almost every $\omega \in \Omega$, there exists $B = B(\omega) \geq 1$ such that

$$|h_n(\omega)(I)| \leq \frac{B}{2^n} \quad \text{for all} \quad n \geq 0. \tag{3.12}$$

Lemma 3.2.11. *There exists a constant $\bar{C}$ depending only on f and g such that if $\omega = (g_1, g_2, \ldots) \in \Omega$ satisfies (3.12), then for all $x \in I$ and every integer $n \geq 0$, one has*

$$h_n(\omega)'(x) \leq \frac{B\bar{C}}{(2-\varepsilon)^n}. \tag{3.13}$$

Proof. Let us fix $\varepsilon_0 > 0$ small enough that for every pair of points y and z in $[a, d]$ at distance less than or equal to ε_0, one has

$$\frac{f'(y)}{f'(z)} \leq \frac{2}{2-\varepsilon}, \qquad \frac{g'(y)}{g'(z)} \leq \frac{2}{2-\varepsilon}. \tag{3.14}$$

It is easy to check the existence of $N \in \mathbb{N}$ such that for every $\omega \in \Omega$ and every $i \geq 0$, the length of the interval $h_{N+i}(\omega)(I)$ is less than or equal to ε_0. We claim that (3.13) holds for $\bar{C} = \max\{A, \bar{A}\}$, where

$$A = \sup_{x \in I, n \leq N, \omega \in \Omega} \frac{h_n(\omega)'(x)(2-\varepsilon)^n}{B},$$

$$\bar{A} = \sup_{x, y \in I, \omega \in \Omega} \frac{h_N(\omega)'(x)}{h_N(\omega)'(y)|I|} \left(\frac{2-\varepsilon}{2}\right)^N.$$

Indeed, if $n \leq N$, then (3.13) holds because of the condition $\bar{C} \geq A$. For $n > N$, let us fix $y = y(n) \in I$ such that $|h_n(\omega)(I)| = h_n(\omega)'(y)|I|$. For

$x \in I$, the distance between $h_{N+i}(\omega)(x)$ and $h_{N+i}(\omega)(y)$ is less than or equal to ε_0 for all $i \geq 0$. Hence by (3.14),

$$
\begin{aligned}
\frac{h_n(\omega)'(x)}{h_n(\omega)'(y)} &= \frac{h_N(\omega)'(x)}{h_N(\omega)'(y)} \frac{g'_{N+1}(h_N(\omega)(x))}{g'_{N+1}(h_N(\omega)(y))} \cdots \frac{g'_n(h_{n-1}(\omega)(x))}{g'_n(h_{n-1}(\omega)(y))} \\
&\leq \frac{h_N(\omega)'(x)}{h_N(\omega)'(y)} \left(\frac{2}{2-\varepsilon} \right)^{n-N},
\end{aligned}
$$

and therefore

$$
\begin{aligned}
h_n(\omega)'(x) &\leq \frac{h_N(\omega)'(x)}{h_N(\omega)'(y)} \frac{|h_n(\omega)(I)|}{|I|} \left(\frac{2}{2-\varepsilon} \right)^{n-N} \\
&\leq \frac{h_N(\omega)'(x)}{h_N(\omega)'(y)} \frac{B}{|I|2^n} \left(\frac{2}{2-\varepsilon} \right)^{n-N} \leq \frac{B\bar{C}}{(2-\varepsilon)^n},
\end{aligned}
$$

where the last inequality follows from the condition $\bar{C} \geq \bar{A}$. $\square$

To verify the persistence of the contraction beyond the interval I, we will use a kind of "dual" argument. Fix a small constant $\varepsilon_1 > 0$ such that for every y and z in $[a, d]$ at distance less than or equal to ε_1, one has

$$
\frac{f'(y)}{f'(z)} \leq \frac{2-\varepsilon}{2-2\varepsilon}, \qquad \frac{g'(y)}{g'(z)} \leq \frac{2-\varepsilon}{2-2\varepsilon}. \tag{3.15}
$$

Lemma 3.2.12. *Let $C \geq 1$, $\omega = (g_1, g_2, \ldots) \in \Omega$, and $x \in [a, d]$ be such that*

$$
h_n(\omega)'(x) \leq \frac{C}{(2-\varepsilon)^n} \quad \text{for all } n \geq 0. \tag{3.16}
$$

If $y \in [a, d]$ is such that $dist(x, y) \leq \varepsilon_1/C$ and $n \geq 0$, then

$$
h_n(\omega)'(y) \leq \frac{C}{(2-2\varepsilon)^n}. \tag{3.17}
$$

Proof. We will verify inequality (3.17) by induction. For $n = 0$, it holds because of the condition $C \geq 1$. Let us assume that it holds for every $j \in \{0, \ldots, n\}$, and let $y_j = h_j(\omega)(y)$ and $x_j = h_j(\omega)(y)$. Suppose that $y \leq x$ (the case $y \geq x$ is analogous). Each point y_j belongs to the interval $h_j(\omega)([x - \varepsilon_1/C, x])$. By the induction hypothesis,

$$
|h_j(\omega)([x - \varepsilon_1/C, x])| \leq \frac{C}{(2-2\varepsilon)^j}|[x - \varepsilon_1/C, x]| \leq C\frac{\varepsilon_1}{C} = \varepsilon_1.
$$

By the definition of ε_1, for every $j \leq n$, one has $g'_{j+1}(y_j) \leq g'_{j+1}(x_j)\left(\frac{2-\varepsilon}{2-2\varepsilon}\right)$. Therefore, because of (3.16),

$$h_{n+1}(\omega)'(y) = g'_1(y_0) \cdots g'_{n+1}(y_n)$$

$$\leq g'_1(x_0) \cdots g'_{n+1}(x_n)\left(\frac{2-\varepsilon}{2-2\varepsilon}\right)^{n+1}$$

$$\leq \frac{C}{(2-\varepsilon)^{n+1}}\left(\frac{2-\varepsilon}{2-2\varepsilon}\right)^{n+1} \leq \frac{C}{(2-2\varepsilon)^{n+1}},$$

which concludes the inductive proof of (3.17). $\qquad\square$

We can now complete the proof of Proposition 3.2.10. To do this, let us notice that by Lemma 3.2.11, if C is large enough, then the probability of the set

$$\Omega(C, \varepsilon) = \left\{\omega \in \Omega : h_n(\omega)'(x) \leq \frac{C}{(2-\varepsilon)^n} \text{ for every } n \geq 0 \text{ and every } x \in I\right\}$$

is positive. Fix such a $C \geq 1$, let $\ell = \min\{\varepsilon_1/2C, |I|/2\}$, and let us denote by J the 2ℓ-neighborhood of I. Lemma 3.2.12 implies that for every $\omega \in \Omega(C, \varepsilon)$, every $n \geq 0$, and every $y \in J$,

$$h_n(\omega)'(y) \leq \frac{C}{(2-2\varepsilon)^n}. \qquad (3.18)$$

If J' denotes the ℓ-neighborhood of the interval I, then the probability of the set

$$\{\omega : h_n(\omega)(I) \subset J' \setminus I \text{ infinitely many times}\}$$

equals 1. Since $\varepsilon < 1/2$, we deduce the existence of $\omega \in \Omega(C, \varepsilon)$ and $m \in \mathbb{N}$ such that

$$h_m(\omega)(I) \subset J' \setminus I, \qquad (2-2\varepsilon)^m > C.$$

By (3.18) and the inequality $\ell \leq |I|/2$, this implies that $h_m(\omega)$ sends J into one of the two connected components of $J \setminus I$ in such a way that $h_m(\omega)$ has a fixed point in this component, which is necessarily hyperbolic and belongs to Λ. The proof of Proposition 3.2.10 is then concluded.

From the preceding discussion we see that in order to prove Theorem 3.2.9, it suffices to show that every pseudogroup of C^1 diffeomorphisms of a one-dimensional compact manifold without invariant probability measure must contain elements f and h that satisfy properties (i) and (ii) at the beginning of this section (compare Proposition 2.2.45). Although this is actually true, the proof uses quite involved techniques related to the construction of the Haar measure, and we do not want to enter into the details here to avoid overloading the presentation; the interested reader may find the proof in [65].

Remark 3.2.13. An interesting aspect of the preceding proof is the fact that it allows suppressing the hypothesis of compact generation for the pseudogroup (see note 1 in § 3.2.1).

In the case of *groups* acting on the circle without invariant probability measure, getting a pair of elements f and h as before is relatively simple: for minimal actions, this may be done by using Hölder's theorem, whereas for actions with an exceptional minimal set, a similar argument applies to the action induced on the topological circle obtained after collapsing the connected components of its complement. However, in the next section we will see that for groups of C^1 circle diffeomorphisms without invariant probability measure, a much stronger conclusion than the one of Sacksteder's theorem holds: these groups have elements that have finitely many fixed points, all of them hyperbolic.

Exercise 3.2.14. Anticipating the main result of the next section, show that every group Γ of $C^{1+\text{Lip}}$ circle diffeomorphisms without invariant probability measure contains elements with at least $2d(\Gamma)$ hyperbolic fixed points, half of them contracting and half of them repelling.

Hint. Consider the minimal and strongly expansive case (the general case easily reduces to this one). Prove that Γ contains two elements f and g such that for some interval $I = [a, b]$, one has $g(I) \cup g^{-1}(I) \subset S^1 \setminus I$, as well as $\lim_{n \to \infty} f^n(x) = b$ and $\lim_{n \to \infty} f^{-n}(x) = a$ for all $x \in S^1 \setminus I$. Then, for very small $\varepsilon > 0$, apply Schwartz's technique to the sequences of compositions $(f^{-n}g)$ and $(g^{-1}f^n)$ over the intervals $[a, a + \varepsilon]$ and $g^{-1}([b - \varepsilon, b])$.

Remark that the hypothesis of the nonexistence of an invariant probability measure assumed throughout this section is necessary for the validity of Theorem 3.2.9. An easy (but uninteresting) example is given by (the group generated by) a diffeomorphism having fixed points, all of them parabolic. More interesting examples correspond to C^1 (and even $C^{1+\tau}$) circle diffeomorphisms with irrational rotation number and nondense orbits (see § 4.1.4 for the construction of such diffeomorphisms). These last examples show that the statement of Theorem 3.2.2 is not valid for groups of $C^{1+\tau}$ circle diffeomorphisms. The C^{1+bv} case is very special. The following proposition is still true for pseudogroups of C^{1+bv} diffeomorphisms having a minimal invariant Cantor set (compare [71]).

Proposition 3.2.15. *If* Γ *is a finitely generated group of* C^{1+bv} *circle diffeomorphisms admitting an exceptional minimal set, then* Γ *has elements with hyperbolic fixed points.*

Proof. It suffices to show that Γ cannot preserve a probability measure, because this allows applying Theorem 3.2.9. However, if Γ preserves a probability measure, then it is semiconjugate to a group of rotations. Since Γ is finitely generated, at least one of its generators must have irrational rotation number. Nevertheless, the existence of such an element contradicts Denjoy's theorem. □

As we will see in the next section, under the hypothesis of Proposition 3.2.15 the group Γ has elements that have finitely many fixed points, all of them hyperbolic.

Remark 3.2.16. We do not know whether Proposition 3.2.15 extends to groups of piecewise affine homeomorphisms. Actually, it seems to be unknown whether some version of Sacksteder's theorem holds in this context. This problem is certainly related to that of knowing what are the subgroups of $\mathrm{PAff}_+(S^1)$ that are conjugate to groups of C^1 circle diffeomorphisms (compare § 1.5.2). A seemingly related problem consists in finding a purely combinatorial proof of Denjoy's theorem in the piecewise affine case.

3.2.3 A sharp C^1 version via Lyapunov exponents

For groups of C^1 circle diffeomorphisms, it is possible to give a global and optimal version of Sacksteder's theorem. Recall that for every group Γ of circle homeomorphisms without invariant probability measure, there is an associated *degree* $d(\Gamma) \in \mathbb{N}$ (see § 2.3.1).

Theorem 3.2.17. *If* Γ *is a finitely generated subgroup of* $\mathrm{Diff}_+^1(S^1)$ *without invariant probability measure, then* Γ *contains elements having finitely many fixed points, all of them hyperbolic. More precisely,* Γ *contains elements* *having $2d(\Gamma)$ fixed points, half of them hyperbolically contracting and half of them hyperbolically repelling.*

Probabilistic methods will again be fundamental for the proof of this theorem. Let Γ be a finitely generated subgroup of $\mathrm{Diff}^1_+(S^1)$ endowed with a nondegenerate finitely supported probability measure p. Let μ be a probability on S^1 that is stationary with respect to p, and let T be the skew-product map from $\Omega \times S^1$ into itself given by

$$T(\omega, x) = (\sigma(\omega), h_1(\omega)(x)).$$

This map T preserves $\mathbb{P} \times \mu$, and hence the Birkhoff ergodic theorem [249] applied to the function $(\omega, x) \mapsto \log(h_1(\omega)'(x))$ shows that for μ-almost every point $x \in S^1$ and $\mathbb{P}$-almost every random sequence $\omega \in \Omega$, the following limit exists:

$$\lambda_{(\omega,x)}(\mu) = \lim_{n \to \infty} \frac{\log(h_n(\omega)'(x))}{n}.$$

This value will be called the ***Lyapunov exponent*** of (ω, x).

To continue with our discussion, the following Kakutani random ergodic theorem will be essential (the reader may find more details on this in [82, 137]).

Theorem 3.2.18. *If the stationary measure μ cannot be written as a nontrivial convex combination of two stationary probabilities, then the map T is ergodic with respect to $\mathbb{P} \times \mu$.*

Proof. Let $\psi : \Omega \times S^1 \to \mathbb{R}$ be a T-invariant integrable function. We need to show that ψ is almost everywhere constant. For each $n \geq 0$, let ψ_n be the conditional expectation of ψ given the first n entries of $\omega \in \Omega$. We have $\psi_0(\omega, x) = \mathbb{E}(\psi_1)(\omega, x)$. Since ψ is invariant under T, for each $n \in \mathbb{N}$, we have

$$\psi(\omega, x) = \psi(\sigma^n(\omega), h_n(\omega)(x)),$$

which allows us to deduce easily that

$$\psi_n(\omega, x) = \psi_{n-1}(\sigma(\omega), h_1(\omega)(x)) = \ldots = \psi_0(\sigma^n(\omega), h_n(\omega)(x)). \quad (3.19)$$

We then conclude that

$$\psi_0(\omega, x) = \mathbb{E}(\psi_0)(\sigma(\omega), h_1(\omega)(x)).$$

Now, since $\psi_0(\omega, x)$ does not depend on ω, we may write $\psi_0(\omega, x) = \bar{\psi}(x)$, and because of the preceding equality, $\bar{\psi}$ is invariant under the diffusion operator. We claim that this implies that the function $\bar{\psi}$ is μ-almost everywhere constant. Before showing this, let us remark that by (3.19), this implies that all the functions ψ_n are constant, and hence the limit function ψ is also constant, thus concluding the proof of the ergodicity of T.

To prove that $\bar{\psi}$ is constant, let us suppose for a contradiction that for some $c \in \mathbb{R}$, the sets $S_+ = \{\bar{\psi} \geq c\}$ and $S_- = \{\bar{\psi} < c\}$ have positive μ-measure. The invariance of $\bar{\psi}$ and μ under diffusion allows us to show easily that these sets are invariant under the action. This implies that $\mu = \mu \mathcal{X}_{S_+} + \mu \mathcal{X}_{S_-}$ is a nontrivial convex decomposition of μ into stationary probabilities, thus contradicting our hypothesis. $\qquad \square$

Remark 3.2.19. A stationary measure that cannot be written as a nontrivial convex combination of two stationary probabilities is said to be *ergodic*. Obviously, if the stationary measure is unique, then it is necessarily ergodic. In general, every stationary measure may be written as a "weighted mean" of ergodic ones (the *ergodic decomposition*; see, for instance, [82, 137]).

Whenever T is ergodic, the Lyapunov exponent is $\mathbb{P} \times \mu$-almost everywhere equal to

$$\lambda(\mu) = \int_\Omega \int_{S^1} \log(h_1(\omega)'(x)) \, d\mu(x) \, d\mathbb{P}(\omega)$$
$$= \int_\Gamma \int_{S^1} \log(g'(x)) \, d\mu(x) \, dp(g).$$

Moreover, if p is symmetric and the stationary measure μ is invariant under the action of Γ, then the Lyapunov exponent $\lambda_{(\omega,x)}(\mu)$ is equal to zero for almost every (ω, x). Indeed, if μ is invariant, then the map from $\Omega \times S^1$ into itself defined by $(\omega, x) \mapsto (\omega, h_1(\omega)^{-1}(x))$ preserves $\mathbb{P} \times \mu$. Because of the symmetry of p, this implies that $\lambda(\mu)$ coincides with

$$\int_\Omega \int_{S^1} \log(h_1(\omega)'(x)) \, d\mu(x) \, d\mathbb{P}(\omega)$$
$$= \int_\Omega \int_{S^1} \log(h_1(\omega)'(h_1(\omega)^{-1}(x))) \, d\mu(x) \, d\mathbb{P}(\omega)$$
$$= -\int_\Omega \int_{S^1} \log((h_1(\omega)^{-1})'(x)) \, d\mu(x) \, d\mathbb{P}(\omega)$$
$$= -\int_\Gamma \int_{S^1} \log((g^{-1})'(x)) \, d\mu(x) \, dp(g),$$

that is, $\lambda(\mu) = -\lambda(\mu)$. To summarize, if μ is invariant and ergodic, then its Lyapunov exponent equals zero. The general case may be deduced from this by an ergodic-decomposition-type argument.

The next proposition (which admits a more general version for codimension-one foliations; see [63]) may be seen as a kind of converse to the preceding remark.

Proposition 3.2.20. *If p is nondegenerate and symmetric and has finite support, and if Γ does not preserve any probability measure on the circle, then the Lyapunov exponent of the unique stationary measure is negative.*[3]

Before giving the proof of this proposition, we will explain how to obtain Theorem 3.2.17 from it. Suppose, for instance, that the action of Γ is minimal and strongly expansive (i.e., $d(\Gamma) = 1$; see § 2.3.1). In this case we know that for every ω in a subset Ω^* of total probability of Ω, the contraction coefficient contr$(h_n(\omega))$ converges to 0 as n goes to infinity. This means that for every $\omega \in \Omega^*$ there exist two closed intervals $I_n(\omega)$ and $J_n(\omega)$ whose lengths tend to 0 and such that $h_n(\omega)(\overline{S^1 \setminus I_n(\omega)}) = J_n(\omega)$. Moreover, $I_n(\omega)$ converges to a point $\varsigma_+(\omega)$, and the map $\varsigma_+ : \Omega^* \to S^1$ is measurable. If $I_n(\omega)$ and $J_n(\omega)$ are disjoint, then the fixed points of $h_n(\omega)$ are in the interior of these two intervals. To show the uniqueness and the hyperbolicity of at least the fixed point in $J_n(\omega)$ (i.e., of the contracting fixed point), one uses the fact that the Lyapunov exponent is negative. However, two technical difficulties immediately appear: the intervals $I_n(\omega)$ and $J_n(\omega)$ are not necessarily disjoint, and the "deviation" from the local contraction rate depends on the initial point (ω, x). To overcome the first problem, it suffices to notice that the "times" $n \in \mathbb{N}$ for which $I_n(\omega) \cap J_n(\omega) \neq \emptyset$ are rare (the density of this set of integers is generically equal to 0). The second difficulty is overcome by using the fact that the map T is ergodic, which implies that almost every initial point (ω, x) will enter into a set where the deviation from the local contraction rate is well controlled. Let us finally remark that to show the uniqueness and the hyperbolicity of the repelling fixed point, one may apply an analogous argument to the compositions of the inverse maps in the opposite order (compare Exercise 3.2.14). To do this, it suffices to consider the finite time distributions. Indeed, since the probability p is supposed to be symmetric, the distributions of both Markov processes coincide for every finite time.

---

3. The uniqueness of the stationary measure comes from Theorem 2.3.20.

If the action is minimal and $d(\Gamma) > 1$, the preceding arguments may be applied "after passing to a finite quotient". Finally, if there is an exceptional minimal set, we may argue in an analogous way "over this set". Making these arguments formal is just a technical issue, and hence we will not develop the details here; the interested reader is referred to [65]. However, we point out that from the arguments contained in [65] one may deduce a more concise statement: if for $d \in \mathbb{N}$ we denote by $D_d(S^1)$ the set of C^1 circle diffeomorphsims having exactly $2d$ periodic points, all of them hyperbolic, then

$$\mathbb{P}\left[\lim_{n \to \infty} \frac{1}{N} card\{n \in \{1, \ldots, N\} : h_n(\omega) \in D_{d(\Gamma)}(S^1)\} = 1\right] = 1.$$

Exercise 3.2.21. Using some of the preceding arguments, prove that every group Γ of circle homeomorphisms without invariant probability measure contains elements with at least $2d(\Gamma)$ fixed points. Give an example of such a group in which, for every element having fixed points, the number of these points is greater than $2d(\Gamma)$.

Remark. We do not know whether the group of piecewise affine circle homeomorphisms contains subgroups like the example asked for above (compare Remark 3.2.16).

We now pass to the proof of Proposition 3.2.20. Let

$$\psi(x) = \int_{\Gamma} \log(g'(x)) \, dp(g),$$

and let us suppose for a contradiction that $\lambda(\mu) \geq 0$, that is,

$$\int_{S^1} \psi(x) \, d\mu(x) \geq 0. \tag{3.20}$$

In this case we will show that Γ preserves a probability measure. To do this, we will strongly use a lemma that is inspired by Sullivan's theory of foliated cycles [236] (see also [90]). Recall that the **Laplacian** $\Delta\zeta$ of a continuous real-valued function ζ is defined by $\Delta\zeta = D\zeta - \zeta$, where D denotes the diffusion operator (2.7). A function ψ is **harmonic** if its Laplacian is identically equal to 0 (c.f. Exercise 2.3.18).

Lemma 3.2.22. *Under hypothesis (3.20), there exists a sequence of continuous functions ζ_n defined on the circle such that for every integer $n \in \mathbb{N}$ and* *every point $x \in S^1$,*

$$\psi(x) + \Delta\zeta_n(x) \geq -\frac{1}{n}. \tag{3.21}$$

Proof. Let us denote by $C(S^1)$ the space of continuous functions on the circle. Let E be the subspace of functions arising as the Laplacian of some function in $C(S^1)$, and let C_+ be the convex cone in $C(S^1)$ formed by the nonnegative functions. We need to show that if ψ satisfies (3.20), then its image under the projection $\pi : C(S^1) \rightarrow C(S^1)/\bar{E}$ is contained in $\pi(C_+)$. Suppose that this is not the case. Then the Hahn-Banach separation theorem provides us with a continuous linear functional $\bar{I} : C(S^1)/\bar{E} \rightarrow \mathbb{R}$ such that $\bar{I}(\pi(\psi)) < 0 \leq \bar{I}(\pi(\eta))$ for every $\eta \in C_+$. Obviously, $\bar{I}$ induces a continuous linear functional $I : C(S^1) \rightarrow \mathbb{R}$ that is identically zero on E and such that $I(\psi) < 0 \leq I(\eta)$ for every $\eta \in C_+$. We claim that there exists a constant $c \in \mathbb{R}$ such that $I = c\mu$ (we identify the probability measures with the linear functional induced on the space of continuous functions by integration). To show this, let us begin by noticing that since I is zero on E, for every $\zeta \in C(S^1)$, one has

$$\langle DI, \zeta \rangle = \langle I, D\zeta \rangle = \langle I, \Delta\zeta + \zeta \rangle = \langle I, \zeta \rangle,$$

that is, I is invariant under the (dual operator of the) diffusion. Suppose that the Hahn decomposition of I may be expressed in the form $I = \alpha v_1 - \beta v_2$, where v_1 and v_2 are probability measures with disjoint support, $\alpha > 0$, and $\beta > 0$. In this case, the equality $DI = I$ and the uniqueness of the Hahn decomposition of DI show that v_1 and v_2 are also invariant under the diffusion. Consequently, $v_1 = v_2 = \mu$, which contradicts the fact that the supports of v_1 and v_2 are disjoint. The functional I may then be expressed in the form $I = cv$ for some probability measure v on the circle, and the equality $I = DI$ implies that $v = \mu$.

Now notice that since

$$0 > I(\psi) = c\mu(\psi) = c\int_{S^1} \psi(x)d\mu(x),$$

hypothesis (3.20) implies that $\mu(\psi) > 0$ and $c < 0$. Nevertheless, since the constant function 1 belongs to C_+, we must have $c = I(1) \geq 0$. This contradiction concludes the proof. $\qquad\square$

Coming back to the proof of Proposition 3.2.20, we begin by noticing that, adding a constant c_n to each ζ_n if necessary, we may suppose that the mean of $\exp(\zeta_n)$ equals 1 for every $n \in \mathbb{N}$. Let us consider the probability measures v_n on the circle defined by

$$\frac{dv_n(s)}{d\,Leb} = \exp(\zeta_n(s)),$$

and let us fix a subsequence v_{n_i} converging to a probability measure v on S^1. We will show that this measure v is invariant under Γ.

Let us first check that v is stationary. To do this, notice that if we denote by $Jac_n(g)$ the Radon-Nikodym derivative of $g \in \Gamma$ with respect to v_n, then relation (3.21) implies that for every $x \in S^1$, the value of

$$\int_\Gamma \log(Jac_n(g)(x))\,dp(g)$$

equals

$$\int_\Gamma \log(g'(x))\,dp(g) + \int_\Gamma [\zeta_n(g(x)) - \zeta_n(x)]\,dp(g) = \psi(x) + \Delta\zeta_n(x) \geq -\frac{1}{n}.$$

Now notice that since the dual of the diffusion operator acts continuously on the space of probability measures on the circle, the sequence of measures Dv_{n_i} converges in the weak* topology to the measure Dv. Moreover, the diffusion of v_n is a measure that is absolutely continuous with respect to v_n, and whose density may be written in the form

$$\frac{d\,Dv_n(x)}{d\,v_n(x)} = \int_\Gamma Jac_n(g^{-1})(x)\,dp(g) = \int_\Gamma Jac_n(g)(x)\,dp(g).$$

By the concavity of the logarithm, we have

$$\frac{d\,Dv_{n_i}(x)}{d\,v_{n_i}(x)} \geq \exp\left(\int_\Gamma \log(Jac_{n_i}(g)(x))\,dp(g)\right) \geq \exp(-1/n_i),$$

that is, $Dv_{n_i} \geq \exp(-1/n_i)\,v_{n_i}$. Passing to the limit, we obtain $Dv \geq v$, and since both v and Dv have total mass 1, they must be equal. Hence v is stationary.

We may now complete the proof of the invariance of v. To do this, fix an interval J such that $v(J) > 0$, and consider the function $\psi_{n,J} : \Gamma \to]0, 1]$ defined by $\psi_{n,J}(g) = v_n(g(J))$. From

$$\Delta \log(\psi_{n,J})(id) = \int_\Gamma \log\left(\frac{v_n(g(J))}{v_n(J)}\right)\,dp(g)$$

$$= \int_\Gamma \log\left(\int_J Jac_n(g)(x)\,\frac{dv_n(x)}{v_n(J)}\right)\,dp(g)$$

one deduces that

$$\Delta \log(\psi_{n,J})(id) \geq \int_{\Gamma} \left(\int_{J} \log(Jac_n(g)(x)) \frac{dv_n(x)}{v_n(J)} \right) dp(g)$$

$$= \int_{J} \left(\int_{\Gamma} \log(Jac_n(g)(x)) \, dp(g) \right) \frac{dv_n(x)}{v_n(J)} \geq -\frac{1}{n}.$$

Notice that this is true for every interval J satisfying $v(J) > 0$. Because of the relation $\psi_{n,J}(gf) = \psi_{n,f(J)}(g)$, this implies that the Laplacian of $\log(\psi_{n,J})$ is bounded from below by $-1/n$ at every element of Γ. Therefore, if ψ_J is the limit of the sequence of the functions $\psi_{n_i,J}$, that is, $\psi_J(g) = v(g(J))$, then the function $\log(\psi_J)$ is superharmonic (i.e., its Laplacian is nonnegative). On the other hand, since p is symmetric, ψ_J is harmonic. As a consequence, for every element $f \in \Gamma$, the inequalities below are forced to be equalities:

$$\log(\psi_J)(f) \leq \int_{\Gamma} \log(\psi_J)(gf) \, dp(g)$$

$$\leq \log \left(\int_{\Gamma} \psi_J(gf) \, d\mu(g) \right) = \log(\psi_J)(f).$$

This implies that the function ψ_J is constant. Now since the interval satisfying $v(J) > 0$ was arbitrary, we deduce that the measure v is invariant under all the elements in Γ. Proposition 3.2.20 is then proved.

3.3 Duminy's First Theorem: On the Existence of Exceptional Minimal Sets

3.3.1 The statement of the result

In general, the subgroups of nondiscrete groups that are generated by elements near the identity behave nicely. A classical result well illustrating this phenomenon is the *Zazenhäus lemma*: in every Lie group there exists a neighborhood of the identity such that every discrete group generated by elements inside is nilpotent.

Another well-known result of this kind is *Jørgensen's inequality* [12, 135]: if f and g are elements in PSL$(2, \mathbb{R})$ generating a Fuchsian group of second kind, then

$$|\text{tr}^2(f) - 4| + |\text{tr}([f, g]) - 2| > 1,$$

where tr denotes the trace of the (equivalence class of the) corresponding matrix. In the same direction, Marden [156] showed the existence of a universal constant $\varepsilon_0 > 0$ such that for every nonelementary Fuchsian group Γ, every system of generators $\mathcal{G}$ of Γ, and every point P in the Poincaré disk, there exists $g \in \mathcal{G}$ satisfying $dist(P, g(P)) \geq \varepsilon_0$. This result was extended by Margulis to discrete isometry groups of hyperbolic spaces of arbitrary dimension (*Margulis's inequality* still holds for more general spaces of negative curvature; see ⌊9⌋).

For groups of circle diffeomorphisms, there is a large variety of results in the same spirit, especially in the real-analytic case [20, 67, 92, 212, 213]. One of the main motivations is a beautiful theorem obtained by Duminy at the end of the 1970s. Unfortunately, Duminy never published the proof of his result.

Theorem 3.3.1. *There exists a universal constant $V_0 > 0$ satisfying the following property: if Γ is a subgroup of $\mathrm{Diff}_+^{1+\mathrm{bv}}(\mathrm{S}^1)$ generated by a (non-necessarily finite) family $\mathcal{G}$ of diffeomorphisms such that at least one of them has finitely many periodic points and $V(g) < V_0$ for every $g \in \mathcal{G}$, then Γ does not admit an exceptional minimal set.*

Notice that the condition $V(g) < V_0$ means that the elements of $\mathcal{G}$ are close to rotations: the equality $V(g) = 0$ is satisfied if and only if g is a rotation. Concerning the hypothesis of the existence of a generator with isolated periodic points, let us point out that it is "generically" satisfied [166]. It is very plausible that the theorem is still true without this assumption. This is known, for instance, in the real-analytic case; see Exercise 3.3.7.

Recall that if g is a $\mathrm{C}^{1+\mathrm{bv}}$ diffeomorphism and I is an interval contained in the domain of definition of g, then $V(g; I)$ denotes the total variation of the logarithm of the derivative of the restriction of g to I. Notice that $V(g^{-1}; I) = V(g; g^{-1}(I))$. This relation implies in particular that $V(g) = V(g^{-1})$ for a circle diffeomorphism g. Because of this, there is no loss of generality for the proof of Theorem 3.3.1 if we assume that $\mathcal{G}$ is symmetric. Moreover, for every circle diffeomorphism g, there must exist a point $p \in \mathrm{S}^1$ such that $g'(p) = 1$, which allows us to conclude that

$$\inf_{x \in \mathrm{S}^1} f'(x) \geq e^{-V(f)}, \qquad \sup_{x \in \mathrm{S}^1} f'(x) \leq e^{V(f)}. \tag{3.22}$$

From (3.1) and (3.22) one deduces that if $g \in \mathrm{Diff}_+^2(\mathrm{S}^1)$ satisfies $|g''(x)| < \delta$ for every $x \in \mathrm{S}^1$, then $V(g) < 2\pi\delta/(1 - 2\pi\delta)$. This implies that for some universal constant δ_0, the conclusion of Theorem 3.3.1 applies to subgroups of $\mathrm{Diff}_+^2(\mathrm{S}^1)$ generated by elements g satisfying $|g''(x)| < \delta_0$ for every $x \in S^1$ (at least when one of the generators has finitely many periodic points). In the same way, the reader should have no problem adapting the following arguments to groups of piecewise affine circle homeomorphisms generated by elements near rotations.

Before we begin the proof of Theorem 3.3.1, we record two important properties. On the one hand, for every pair of intervals I_1 and I_2 inside an interval I contained in the domain of definition of g, one has

$$\frac{|g(I_2)|}{|I_2|}e^{-V(g;I)} \leq \frac{|g(I_1)|}{|I_1|} \leq \frac{|g(I_2)|}{|I_2|}e^{V(g;I)}. \tag{3.23}$$

On the other hand, for every pair of C^{1+bv} diffeomorphisms f and g, the cocycle relation (1.1) implies the inequality

$$V(f \circ g; I) \leq V(g; I) + V(f; g(I)). \tag{3.24}$$

3.3.2 An expanding first-return map

The proof of Theorem 3.3.1 is based on Duminy's work on semiexceptional leaves of codimension-one foliations, which will be studied later. The following lemma appears in one of his unpublished manuscripts (see, however, [185]).

Lemma 3.3.2. *Let $a, b, b',$ and c be points in the line such that $a < c < b$ and $a < b' < b$. Let $f : [a, c] \to [a, b]$ and $g : [c, b] \to [a, b']$ be C^{1+bv} diffeomorphisms such that $f(x) > x$ for every $x \neq a$, and $g(x) < x$ for every x (see Figure 17). Suppose that for some m and n in $\mathbb{N}$ and $[u, v] \subset [c, b]$, the map $\bar{H} = g^{-n} \circ f^{-m}$ is defined on the whole interval $[u, v]$. Then one has the inequality*

$$\frac{\bar{H}(v) - \bar{H}(u)}{v - u} \leq \frac{\bar{H}(v) - f^{-1}(\bar{H}(v))}{v - f^{-1}(v)}\left(1 - \frac{1}{\sup\limits_{x \in [a,c]} f'(x)}\right)e^{V(f;[a,c]) + V(g;[c,b])}. \tag{3.25}$$

Proof. To simplify, let us denote $\bar{f} = f^{-1}$ and $\bar{g} = g^{-1}$ (see Figure 18). Recall that $V(\bar{f}; [a, b]) = V(f; [a, c])$ and $V(\bar{g}; [a, b']) = V(g; [c, b])$. We then have to show the inequality

$$\frac{\bar{H}(v) - \bar{H}(u)}{v - u} \leq \frac{\bar{H}(v) - \bar{f}(\bar{H}(v))}{v - \bar{f}(v)}\left(1 - \inf\limits_{x \in [a,b]} \bar{f}'(x)\right)e^{V(\bar{f};[a,b]) + V(\bar{g};[a,b'])}. \tag{3.26}$$

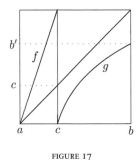

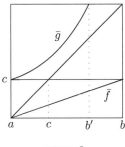

FIGURE 17 FIGURE 18

To do this, first remark that

$$V(\bar{f}^m; [\bar{f}(v), v]) \le \sum_{k=0}^{m-1} V(\bar{f}; [\bar{f}^{k-1}(v), \bar{f}^k(v)]) \le V(\bar{f}; [a, b]).$$

Since $\bar{f}(v) \le c \le u < v$, we have

$$\frac{\bar{f}^m(v) - \bar{f}^m(u)}{v - u} \le \frac{\bar{f}^m(v) - \bar{f}^{m+1}(v)}{v - \bar{f}(v)} e^{V(\bar{f}^m; [\bar{f}(v), v])}$$

$$\le \frac{\bar{f}^m(v) - \bar{f}^{m+1}(v)}{v - \bar{f}(v)} e^{V(\bar{f}; [a, b])}.$$

Moreover,

$$\bar{f}^{m+1}(v) - a = \bar{f}(\bar{f}^m(v)) - \bar{f}(a) \ge (\bar{f}^m(v) - a) \inf_{x \in [a, b]} \bar{f}'(x),$$

from which one obtains

$$\bar{f}^m(v) - \bar{f}^{m+1}(v) \le (\bar{f}^m(v) - a)\left(1 - \inf_{x \in [a, b]} \bar{f}'(x)\right).$$

We then deduce that

$$\frac{\bar{f}^m(v) - \bar{f}^m(u)}{v - u} \le \frac{\bar{f}^m(v) - a}{v - \bar{f}(v)}\left(1 - \inf_{x \in [a, b]} \bar{f}'(x)\right)e^{V(\bar{f}; [a, b])}. \qquad (3.27)$$

Since $a \le \bar{f}^m(u) < \bar{f}^m(v) \le c = \bar{g}(a)$ and $\bar{g}^n$ is well defined on the interval $[\bar{f}^m(u), \bar{f}^m(v)]$, analogous arguments show that

$$\frac{\bar{g}^n(\bar{f}^m(v)) - \bar{g}^n(\bar{f}^m(u))}{\bar{f}^m(v) - \bar{f}^m(u)} \le \frac{\bar{g}^n(\bar{f}^m(v)) - \bar{g}^n(a)}{\bar{f}^m(v) - a} e^{V(\bar{g}; [a, b'])}. \qquad (3.28)$$

From (3.27) and (3.28) one deduces that $(\bar{H}(v) - \bar{H}(u))/(v - u)$ is less than or equal to

$$\frac{\bar{H}(v) - \bar{f}(\bar{H}(v))}{v - \bar{f}(v)} \frac{\bar{H}(v) - \bar{g}(a)}{\bar{H}(v) - \bar{f}(\bar{H}(v))} \left(1 - \inf_{x \in [a,b]} \bar{f}'(x)\right) e^{V(\bar{f};[a,b]) + V(\bar{g};[a,b'])}.$$

Inequality (3.26) follows from this by using the fact that $\bar{f}(\bar{H}(v)) \leq c = \bar{g}(a)$.   ☐

Given two maps f and g as in Lemma 3.3.2, for each $x \in \,]c, b]$ there exists a positive integer $n = n(x)$ such that $g^{n-1}(x) \in \,]c, b]$ and $g^n(x) \in \,]a, c]$. Analogously, for $y \in \,]a, c]$, there exists $m = m(y)$ such that $f^{m-1}(y) \in \,]a, c]$ and $f^m(y) \in \,]c, b]$. We define the **first-return map** $H :]c, b] \to \,]c, b]$ by

$$H(x) = f^{m(g^{n(x)}(x))} \circ g^{n(x)}(x).$$

Notice that the set of discontinuity points of this map is $\{g^{-1}(f^{-j}(c)) : j \in \mathbb{N}\}$, and for every $\varepsilon > 0$, the intersection of this set with $[c + \varepsilon, b]$ is finite. We let

$$C(f) = \frac{\sup_{x \in [c,b]}(x - \bar{f}(x))}{\inf_{x \in [c,b]}(x - \bar{f}(x))}. \tag{3.29}$$

Proposition 3.3.3. *Under the hypothesis of Lemma 3.3.2, suppose, moreover, that*

$$\left(1 - \frac{1}{\sup\limits_{x \in [a,c]} f'(x)}\right) e^{V(f;[a,c]) + V(g;[c,b])} < 1. \tag{3.30}$$

Then for every $\kappa > 1$, there exists $N \in \mathbb{N}$ such that $(H^N)'(x) > \kappa$ for every x at which H^N is differentiable.

Proof. Let $N \in \mathbb{N}$ be such that

$$\left(1 - \frac{1}{\sup\limits_{x \in [a,c]} f'(x)}\right)^N e^{N(V(f;[a,c]) + V(g;[c,b]))} < \frac{1}{\kappa \cdot C(f)}. \tag{3.31}$$

We claim that each branch of the map H^N is κ-expanding. To show this, let x be a differentiability point of H^N. Let us fix an interval $[x - \varepsilon, x]$ contained in this component. Set $[u_0, v_0] = [x - \varepsilon, x]$, and for $j \in \{1, \ldots, N - 1\}$, let $[u_j, v_j] = H^j([u_0, v_0])$. Inequality (3.25) applied to $\bar{H} = H^{-1}$ gives

$$\frac{v_j - u_j}{v_{j+1} - u_{j+1}} \leq \frac{v_j - f^{-1}(v_j)}{v_{j+1} - f^{-1}(v_{j+1})} \left(1 - \frac{1}{\sup_{x \in [a,c]} f'(x)}\right) e^{V(f;[a,c]) + V(g;[c,b])}.$$

Taking the product from $j = 0$ to $j = N - 1$, we obtain

$$\frac{v_0 - u_0}{v_N - u_N} \leq \frac{v_0 - f^{-1}(v_0)}{v_N - f^{-1}(v_N)} \left(1 - \frac{1}{\sup_{x \in [a,c]} f'(x)}\right)^N e^{N(V(f;[a,c]) + V(g;[c,b]))}.$$

From this and (3.31), one deduces that

$$\frac{(H^N(x) - H^N(x - \varepsilon))}{\varepsilon} > \kappa.$$

Since this inequality holds for every small-enough $\varepsilon > 0$, this allows us to conclude that $(H^N)'(x) > \kappa$.   □

Proposition 3.3.4. *If f and g satisfy the hypothesis of Proposition 3.3.3, then all the orbits of the pseudogroup generated by these maps are dense in $]a, b[$.*

Proof.   Suppose that there exists an orbit that is nondense in $]a, b[$. Then it is easy to see that the points a, c, and b belong to the closure of this orbit. Among the connected components of the complementary set of this closure contained in $]c, b[$, let us choose one, say $]u, v[$, having maximal length. Then Proposition 3.3.3 shows that for large-enough N, the interval $H^N(]u, v[)$ has greater length, which is a contradiction.   □

At first glance, the preceding estimates may seem too technical. To understand them better, one may consider the particular case where the maps f and g are affine, say, $f(x) = \lambda x$, with $\lambda > 1$, and $g(x) = \eta(x - 1/\lambda)$. In this case, for every $n \in \mathbb{N}$, the restriction of the first-return map H : $[1/\lambda, 1] \to [1/\lambda, 1]$ to the interval $g^{-1}([1/\lambda^{n+1}, 1/\lambda^n])$ is given by $H(x) = f^n \circ g(x)$. Since for every $x \in g^{-1}([1/\lambda^{n+1}, 1/\lambda^n])$, one has

$$\frac{1}{\lambda^n} \geq \eta\left(1 - \frac{1}{\lambda}\right) \geq \frac{1}{\lambda^{n+1}},$$

this gives

$$H'(x) = \eta\lambda^n \geq \frac{1}{\lambda - 1} = \frac{1}{\lambda(1 - 1/\lambda)},$$

and since $C(f) = \lambda$, this inequality may be written in the form

$$H'(x) \geq \frac{1}{C(f)\left(1 - \frac{1}{\sup\limits_{x\in[0,1/\lambda]} f'(x)}\right)}.$$

Now because the values of $V(f)$ and $V(g)$ are equal to zero, the similarity between the last inequality and those appearing throughout this section becomes evident.

Exercise 3.3.5. State and show a proposition making formal the following claim: if the derivative of f is near 1 and the total variation $V(f; [a, c])$ is small, then the constant $C(f)$ defined by (3.29) is near 1.

Exercise 3.3.6. By modifying the first example in § 2.1.1 slightly, prove that Proposition 3.3.4 is optimal when one imposes a priori the condition $V(g) = 0$; show that in this case the critical parameter corresponds to (an integer multiple of) log(2). Moreover, prove that the proposition is optimal under the condition $V(f) = V(g)$; show that in this case the critical parameter is (an integer multiple of) the logarithm of the golden number (see [185] for more details on this).

3.3.3 Proof of the theorem

In order to illustrate the idea of the proof of Theorem 3.3.1, suppose that a group acts on the circle preserving an exceptional minimal set Λ, and two generators f and g are as in Figure 17 over an interval $[a, b]$ in S^1 whose interior intersects Λ. In this case, by an argument similar to that of Corollary 3.3.4, we conclude that the opposite inequality to (3.30) must be satisfied, that is,

$$\left(1 - \frac{1}{\sup\limits_{x\in[a,c]} f'(x)}\right) e^{V(f;[a,c])+V(g;[c,b])} \geq 1. \qquad (3.32)$$

By (3.22), this inequality implies that

$$(1 - e^{-V(f)}) e^{V(f;[a,c])+V(g;[c,b])} \geq 1.$$

If $V(f)$ and $V(g)$ are strictly smaller than some positive constant V_0, then this gives $(1 - e^{-V_0}) e^{2V_0} > 1$, that is,

$$e^{2V_0} - e^{V_0} - 1 > 0,$$

and hence $V_0 > \log((\sqrt{5}+1)/2)$. (The occurrence of the golden number here is not mysterious; actually, it seems to be related to the optimal constant for the theorem; see Exercise 3.3.6.)

Unfortunately, the proof in the general case involves a major technical problem: it is not always the case that there are generators to which one may directly apply the preceding argument. For instance, the modular group acts on S^1 admitting an exceptional minimal set, although its generators have finite order... This is the main reason that we assume the hypothesis that one of the generators has isolated periodic points.

Proof of Theorem 3.3.1. Suppose that Γ admits an exceptional minimal set Λ. By Denjoy's theorem, the set of periodic points of each element in Γ is nonempty. By hypothesis, there exists a generator $g \in G$ whose periodic points are isolated. Let us denote by $\text{Per}(g)$ the set of these points, and let $P(g) = \text{Per}(g) \cap \Lambda$. Notice that $P(g)$ is nonempty. Indeed, if the periodic points of g have order k and $p \in \Lambda$ is not a fixed point of g^k, then $\lim_{j\to\infty} g^{jk}(p)$ and $\lim_{j\to\infty} g^{-jk}(p)$ are fixed points of g^k contained in Λ.

We now claim that there exist $p \in P(g)$ and $f \in G$ such that $f(p) \in S^1 \setminus P(g)$. If not, the finite set $P(g)$ would be invariant under Γ, thus contradicting the minimality of Λ.

Let $G = g^k \in \Gamma$, and let us denote by u and v the periodic points of g immediately to the left and to the right of $f(p)$, respectively. The map $F = f \circ g^k \circ f^{-1}$ has a fixed point in $[u, v]$, namely, $f(p)$. Let a be the fixed point of this map to the left of v, and let q be the fixed point to the right of a. Replacing G by G^{-1} and/or F by F^{-1} if necessary, we may suppose that $G(x) < x$ and $F(x) > x$ for every $x \in \,]a, v[$ (see Figure 19).

We now claim that if V_0 is small enough, then the point $b = F(G^{-1}(a))$ belongs to the interval $]a, v[$. To show this, we first notice that

$$V(F^{-1}; [a, q]) = V(F; [a, q]) \leq \sum_{j=0}^{k-1} V(f \circ g \circ f^{-1}; f \circ g^j \circ f^{-1}([a, q]))$$

$$\leq V(f \circ g \circ f^{-1}) < 3V_0.$$

In the same way one obtains $V(G^{-1}; [u, v]) = V(G; [u, v]) < V_0$. Let $x_0 \in \,]a, v[$ and $y_0 \in \,]a, q[$ be such that

$$(F^{-1})'(x_0) = \frac{F^{-1}(v) - a}{v - a}, \qquad (F^{-1})'(y_0) = 1.$$

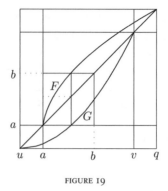

FIGURE 19

Clearly, we have $|\log(F^{-1})'(y_0) - \log(F^{-1})'(x_0)| \leq V(F^{-1}; [a, q])$, and hence

$$F^{-1}(v) - a > e^{-3V_0}(v - a). \qquad (3.33)$$

By an analogous argument one shows that

$$v - G^{-1}(a) > e^{-V_0}(v - a). \qquad (3.34)$$

If b were not contained in $]a, v[$, then $F^{-1}(v) \leq G^{-1}(a)$, and hence, by (3.33) and (3.34),

$$v - a \geq (F^{-1}(v) - a) + (v - G^{-1}(a)) > (e^{-V_0} + e^{-3V_0})(v - a).$$

Therefore, $e^{3V_0} - e^{2V_0} > 1$, which is impossible if $V_0 \leq \log(1.46557)$.

The elements F and G in Γ are thus as in Figure 17 over the interval $[a, b]$. Now notice that for every $x \in [a, q]$, one has

$$|\log(F')(x) - \log(F')(y_0)| \leq V(F; [a, q]) < 3V_0,$$

and hence $\sup_{x \in [a,q]} F'(x) < e^{3V_0}$. For $c = G^{-1}(a)$, inequality (3.32) applied to F and G yields

$$3V_0 + V_0 > V(F; [a, c]) + V(G; [c, b]) \geq \log\left(\dfrac{1}{1 - \dfrac{1}{\sup\limits_{x \in [a,c]} F'(x)}}\right)$$

$$> \log\left(\dfrac{1}{1 - e^{-3V_0}}\right).$$

Therefore, $e^{4V_0} - e^{V_0} > 1$, which is impossible if $V_0 \leq \log(1.22074)$. $\square$

Exercise 3.3.7.   Prove that for every group of circle homeomorphisms that is not semiconjugate to a group of rotations, and for every system of generators $\mathcal{G}$, there exists an element in the ball $B_{\mathcal{G}}(4)$ of radius 4 that is not semiconjugate to a rotation. Conclude that Duminy's theorem holds for groups of real-analytic diffeomorphisms generated by elements h satisfying $V(h) < V_0$ for some small $V_0 > 0$, even in the case where the hypothesis of the existence of a generator with isolated periodic points is not satisfied (see [185] for more details on this).

Exercise 3.3.8.   Prove that for every $g \in \mathrm{PSL}(2, \mathbb{R})$, one has

$$V(g) = 4\, dist(O, g(O)), \tag{3.35}$$

where O denotes the point $(0, 0)$ in the Poincaré disk, and $dist$ stands for the hyperbolic distance. Conclude that for subgroups of $\mathrm{PSL}(2, \mathbb{R})$, the claim of Duminy's theorem follows from either Marden's theorem or Margulis's inequality discussed at the beginning of § 3.3.1.

Remark.   By successive applications of Jørgensen's inequality, and using some results on the classification of Fuchsian groups, Yamada computed the value of the best constant for Marden's theorem. Thanks to (3.35), a careful reading of [255] then allows showing that the best constant V_0 of Duminy's theorem for subgroups of $\mathrm{PSL}(2, \mathbb{R})$ is $4\log(5/3)$, and this value is critical for the actions of $\mathrm{PSL}(2, \mathbb{Z})$ (with respect to its canonical system of generators).

3.4 Duminy's Second Theorem: On the Space of Semiexceptional Orbits

3.4.1 The statement of the result

The aim of this section is to present another great result by Duminy on the structure "to infinity" of some special orbits of pseudogroups of diffeomorphisms of one-dimensional manifolds. Roughly, in class $C^{1+\mathrm{Lip}}$, semiexceptional orbits are forced to have infinitely many ends. A proof of this result appears in [47]. In what follows, we will give the original (and unpublished) remarkable proof by Duminy.

We begin by recalling some basic notions. Given a compact subset K of M, we denote the set of connected components of $M \setminus K$ by $\mathcal{E}_K$, and we endow this set with the discrete topology. The **space of ends** of M is the inverse limit of the spaces $\mathcal{E}_K$, where K ranges over all compact subsets of M. For locally compact separable metric spaces, there is an equivalent but more concrete definition. Fix a sequence (K_n) of compact subsets of M such that $K_n \subset K_{n+1}$ for every $n \in \mathbb{N}$ and $M = \cup_{n \in \mathbb{N}} K_n$, and consider a sequence of connected components C_n of $M \setminus K_n$ such that $C_{n+1} \subset C_n$ for every $n \in \mathbb{N}$. If (K_n') and (C_n') satisfy the same properties, we say that (C_n) is equivalent to (C_n') if for every $n \in \mathbb{N}$ there exists $m \in \mathbb{N}$ such that $C_n \subset C_m'$ and $C_n' \subset C_m$. Then the space of ends of M identifies with the set of equivalence classes of this relation, and it is (compact and) metrizable: after one fixes the sequence (K_n), the distance between (the ends determined by) (C_n^1) and (C_n^2) is e^{-k}, where k is the minimal integer for which $C_k^1 \neq C_k^2$.

Let Γ be a finitely generated pseudogroup of homeomorphisms of some space, and let p be a point in this space. Fix a (symmetric) system of generators $\mathcal{G}$ of Γ. The **Cayley graph**[4] (with respect to $\mathcal{G}$) of the orbit $\Gamma(p)$ is the graph whose vertices are the points in $\Gamma(p)$ so that two vertices p' and p'' are connected by an edge (of length 1) if and only if there exists $g \in \mathcal{G}$ such that $g(p') = p''$. We define the **space of ends of the orbit of p** as the space of ends $\mathcal{E}(\Gamma(p))$ of the graph $Cay_{\mathcal{G}}(\Gamma(p))$. It is not difficult to check that although the Cayley graph depends on $\mathcal{G}$, the associated space of ends is independent of the choice of the system of generators. To simplify, in what follows we will omit the reference to $\mathcal{G}$ when this causes no confusion, and for each $n \in \mathbb{N}$, we will denote by $B(p, n)$ the ball of center p and radius n in $Cay(\Gamma(p))$ (compare § 2.2.5).

An orbit of a pseudogroup of homeomorphisms of a one-dimensional manifold is said to be **semiexceptional** if it corresponds to the orbit of a point p in a local exceptional set Λ such that p is an endpoint of one of the connected components of the complement Λ^c of Λ (see § 3.2.1). *Duminy's second theorem* may then be stated as follows:

Theorem 3.4.1. *Every semiexceptional orbit of a finitely generated pseudogroup of* $C^{1+\text{Lip}}$ *diffeomorphisms of a compact one-dimensional manifold has infinitely many ends.*

It is unknown whether this result holds for *all* the orbits of local exceptional sets Λ as above. (This is known in the real-analytic case, according to a result due to Hector.) To deal with semiexceptional orbits, the basic

---

4. Perhaps the appropriate terminology should be *Schreier graph*.

idea consists in associating to each point p in such an orbit the connected component of Λ^c having p as an endpoint. To be more concrete, we let $\mathcal{G}^* = \cup_{k \in \mathbb{N}} \mathcal{G}^k$, and to each $\hat{g} = (h_1, \ldots, h_k) \in \mathcal{G}^*$ we associate the element $g = h_k \cdots h_1$. For each connected component I of Λ^c on which g is defined, we set

$$S(\hat{g}; I) = |I| + \sum_{i=1}^{k-1} |h_i \cdots h_1(I)|.$$

For any two points p' and p'' belonging to some semiexceptional orbit $\Gamma(p)$, we denote the set of the elements $\hat{g} \in \mathcal{G}^*$ such that $g(p') = p''$ by $\mathcal{G}^*(p', p'')$. Then we consider the connected component I of Λ^c having p' as an endpoint, and we let

$$S(p', p'') = \inf\{S(\hat{g}; I) : \hat{g} \in \mathcal{G}^*(p', p''), I \subset dom(g)\}.$$

If there is no $\hat{g} \in \mathcal{G}^*(p', p'')$ for which $I \subset dom(g)$, we let $S(p', p'') = \infty$.

Lemma 3.4.2. *If p' and p'' are points of a semiexceptional orbit $\Gamma(p)$ for which there exists an element $f \in \Gamma$ such that $f(p') \neq p'$ and $f(p'') \neq p''$, then the sequences of points $p'_n = f^n(p')$ and $p''_n = f^n(p'')$ determine ends of $\Gamma(p)$, and these ends coincide if and only if $\lim_{n \to \infty} S(p'_n, p''_n) = 0$.*

Proof. Since $f(p') \neq p'$, the points $f^n(p')$ are two-by-two different. Therefore, the sequence $(f^n(p'))$ escapes from the compact sets of $Cay(\Gamma(p))$. Since the distance in $Cay(\Gamma(p))$ between p'_n and p'_{n+1} is independent of n, this easily shows that (p'_n) determines an end of the orbit of p; the same applies to (p''_n).

Now fix a constant $\delta > 0$ such that if $h \in \mathcal{G}$ and $x \in dom(h) \cap \Lambda$, then the 2δ-neighborhood of x is contained in the domain of h. Fix also $m_0 \in \mathbb{N}$ such that $B(p, m_0)$ contains all the points in $\Gamma(p)$ that are endpoints of some connected component Λ^c of length greater than or equal to δ. For each $n \in \mathbb{N}$, we denote by I'_n (resp. I''_n) the connected component of Λ^c whose closure contains p'_n (resp. p''_n). From the hypotheses $f(p') \neq p'$ and $f(p'') \neq p''$, one deduces that the intervals I'_n (resp. I''_n) are two-by-two disjoint, and hence the sum of their lengths is finite (in particular, $|I'_n|$ and $|I''_n|$ tend to zero as n goes to infinity). Since the space of ends does not depend on the choice of the system of generators, without loss of generality we may assume that f belongs to $\mathcal{G}$.

Suppose that for every large-enough $n \in \mathbb{N}$, every path $\hat{g}$ from p'_n to p''_n (i.e., such that $g(p'_n) = p''_n$) satisfies $I'_n \not\subset dom(g)$. If this is the case, these paths must intersect $B(p, m_0)$, from which one deduces that the ends determined by (p'_n) and (p''_n) are different, and

$$\lim_{n \to \infty} S(p'_n, p''_n) = \infty.$$

In what follows, suppose that for infinitely many $n \in \mathbb{N}$, there exists $\hat{g} \in \mathcal{G}^*(p'_n, p''_n)$ such that $I'_n \subset dom(g)$. One then easily checks that the same is true for *every* large-enough n. A concatenation-type argument then shows that for all very large $m \geq n$,

$$S(p'_m, p''_m) < S(p'_n, p''_n) + \sum_{j=n}^{m} (|I'_j| + |I''_j|).$$

This inequality obviously implies that $(S(p'_n, p''_n))$ is a Cauchy sequence and hence converges (to a finite limit).

Fix an arbitrary $\varepsilon > 0$, and denote the sum of the lengths of the connected components of Λ^c by S. Clearly, we may choose a finite family $\mathcal{I}_\varepsilon$ of these components such that the sum of their lengths is greater than $S - \varepsilon$. Let $m_\varepsilon \in \mathbb{N}$ be such that $B(p, m_\varepsilon)$ contains all the points in $\Gamma(p)$ situated in the closure of some interval in $\mathcal{I}_\varepsilon$. On the one hand, if the ends determined by (p'_n) and (p''_n) coincide, then for every large-enough n, there exists a path $\hat{g} \in \mathcal{G}^*$ from p'_n to p''_n avoiding $B(p, m_\varepsilon)$. This implies that $S(p'_n, p''_n) \leq S(\hat{g}; I'_n) \leq \varepsilon$, and since $\varepsilon > 0$ was arbitrary, this shows that $\lim_{n \to \infty} S(p'_n, p''_n) = 0$. On the other hand, if the ends we consider are different, then there exists an integer m'_0 such that for every large-enough n, each path from p'_n to p''_n passes through $B(p, m'_0)$. Denoting by ε_0 the minimum among the lengths of the connected components of Λ^c having a vertex in $B(p, m'_0)$ as an endpoint, we conclude that $S(p'_n, p''_n) > \varepsilon_0$ for all large-enough n. Therefore, $\lim_{n \to \infty} S(p'_n, p''_n) > 0$. $\square$

To show Theorem 3.4.1, we will argue by contradiction. Assuming that the ends determined by two sequences such as those in Lemma 3.4.2 coincide, we will use the equivalent condition $\lim_{n \to \infty} S(p'_n, p''_n) = 0$ to find paths $\hat{g}_n$ from p'_n to p''_n with arbitrarily small distortion. We will then conclude the proof by applying Corollary 3.3.4 to f^{-1} and g_n^{-1}. However, we immediately point out that throughout the proof, we will need to overcome several nontrivial technical difficulties.

Remark 3.4.3. The orbits associated to a free action of $\mathbb{Z}$ (resp. $\mathbb{Z}^d$, for $d \geq 2$) clearly have two ends (resp. one end). Since there exist free non-minimal actions of $\mathbb{Z}^d$ by $C^{1+\tau}$ circle diffeomorphisms for some positive τ

(see § 4.1.4), this shows that the C^{1+Lip} regularity hypothesis is necessary for Duminy's second theorem. However, we do not know whether the theorem still holds for lower differentiability if we assume a priori that there is no invariant probability measure.

3.4.2 A criterion for distinguishing two different ends

To fix notation, in what follows, we will consider a semiexceptional orbit $\Gamma(p)$ associated to a finitely generated pseudogroup Γ of C^{1+Lip} diffeomorphisms. We denote by Λ the local exceptional set, we choose a finite and symmetric system of generators $\mathcal{G}$ of Γ, and we denote by C a simultaneous Lipschitz constant for the logarithm of the derivative of each element $h \in \mathcal{G}$ on a δ-neighborhood of $\Lambda \cap dom(h)$ (where $\delta > 0$).

By Sacksteder's theorem, there exists $f \in \Gamma$ with a fixed point $a \in \Lambda$ such that $f'(a) < 1$. Since the orbit of p intersects every open interval containing points of Λ, it approaches a from at least one side. In what follows, we will assume that $\Gamma(p)$ approaches a from the right, since the other case is analogous. Let $b > a$ be very near to a such that

$$0 < \inf_{x \in [a,b]} f'(x) = m(f) \leq M(f) = \sup_{x \in [a,b]} f'(x) < 1.$$

Given $\varepsilon > 0$, we may choose $b_\varepsilon \in]a, b[$ such that $C(b_\varepsilon - a)/(1 - M(f)) \leq \varepsilon/2$; thus

$$V(f^n; [a, b_\varepsilon]) \leq C \sum_{k=0}^{n-1} |f^k([a, b_\varepsilon])| \leq C(b_\varepsilon - a) \sum_{k=0}^{n-1} M(f)^k \leq \frac{\varepsilon}{2}. \quad (3.36)$$

Proposition 3.4.4. *Let $\varepsilon > 0$ be such that $\varepsilon \leq \frac{1}{2} \log \left(\frac{1}{1-m(f)} \right)$. Suppose that there exist connected components $]x, y[$ and $]u, v[$ of the complementary set of Λ such that*

(i) $a < x < y < u < v < f(b_\varepsilon)$,
(ii) $\{x, u\}$ (resp. $\{y, v\}$) is contained in $\Gamma(p)$, and
(iii) one has the inequalities

$$\frac{f^{-1}(u) - u}{f^{-1}(u) - x} \exp(2\varepsilon) \leq \frac{v - u}{y - x} \leq \frac{u - a}{x - a} \exp(-2\varepsilon).$$

Then the sequences $(f^n(x))$ and $(f^n(u))$ (resp. $(f^n(y))$ and $(f^n(v))$) determine two different ends of $\Gamma(p)$.

To show this proposition, we will need to refine slightly for the $C^{1+\text{Lip}}$ case Schwartz's estimates from § 3.2.1.

Lemma 3.4.5. *Let I be a connected component of the complementary set of Λ, and let J be an interval containing I. Suppose that for some $\lambda > 1$ and $n \in \mathbb{N}$, there exists $\hat{g} = (h_1, \ldots, h_n) \in \mathcal{G}^n$ such that*

$$S(\hat{g}; I) \leq \inf \left\{ \frac{\log(\lambda)}{\lambda C}, \frac{\delta}{\lambda} \right\}, \qquad \frac{|J|}{|I|} \leq \lambda \exp(-\lambda C S(\hat{g}; I)).$$

Then one has

(i) $J \subset dom(g)$,

(ii) $V(g; J) \leq \lambda C S(\hat{g}; I)$, and

(iii) $|g(J)| \leq \lambda |g(I)|$.

Proof. The proof proceeds by induction. For $n = 1$, we have $\hat{g} = g \in \mathcal{G}$ and $S(\hat{g}; I) = |I|$. Property (i) follows from the fact that J contains I and $|J| \leq \lambda |I| = \lambda S(\hat{g}; I) \leq \delta$. For (ii), notice that $V(g; J) \leq C|J| \leq \lambda C S(\hat{g}; I)$. Concerning (iii), notice that by hypothesis and (3.23),

$$\frac{|g(J)|}{|g(I)|} \leq \frac{|J|}{|I|} e^{V(g;J)} \leq \lambda.$$

Suppose now that the claim holds up to $k \in \mathbb{N}$, and consider $\hat{g} = (h_1, \ldots, h_{k+1})$ in $\mathcal{G}^{k+1}$ satisfying the hypothesis. If (i), (ii), and (iii) hold for $(h_1, \ldots, h_k) \in \mathcal{G}^k$, then property (i) holds for $\hat{g}$, since $h_k \cdots h_1$ is defined on J, the interval $h_k \cdots h_1(J)$ intersects Λ, and

$$|h_k \cdots h_1(J)| \leq \lambda |h_k \cdots h_1(I)| < \lambda S(\hat{g}; I) \leq \delta;$$

property (ii) holds for $\hat{g}$, since by the inductive hypothesis we have

$$V(g; J) \leq V(h_{k+1}; h_k \cdots h_1(J)) + V(h_k \cdots h_1; J)$$
$$\leq C|h_k \cdots h_1(J)| + \lambda C S((h_k, \ldots, h_1); I) \leq \lambda C S(\hat{g}; I);$$

and property (iii) holds for $\hat{g}$, since from (ii) and the hypothesis of the lemma it follows that

$$\frac{|g(J)|}{|g(I)|} \leq \frac{|J|}{|I|} \exp(V(g; J)) \leq \lambda. \qquad \square$$

Notice that the parameters $S(\hat{g}; I)$ and λ occur simultaneously in the hypothesis of the preceding lemma 3.4.5. In particular, if $S(\hat{g}; I)$ is small, then the claim of the lemma asserts that the distortion is controlled in a neighborhood of I whose length is large with respect to that of $|I|$. This clever remark by Duminy will be fundamental in the sequel.

Proof of Proposition 3.4.4.   We will give the proof only for the case where $\{x, u\}$ is contained in $\Gamma(p)$, since the other case is analogous. Suppose for a contradiction that the ends determined by the sequences $(f^n(x))$ and $(f^n(u))$ coincide, and let $x_n = f^n(x)$, $y_n = f^n(y)$, $u_n = f^n(u)$, and $v_n = f^n(v)$. By Lemma 3.4.2, $S_n = S(x_n, u_n)$ converges to zero as n goes to infinity. Hence by definition, for each $n \in \mathbb{N}$, there exists $\hat{g}_n \in \mathcal{G}^*$ such that $g_n(x_n) = u_n$ and $S(\hat{g}_n; [x_n, y_n]) \leq 2S_n$. Letting $\lambda_n = 1/\sqrt{S_n}$, we see that for large-enough n, one has $\lambda_n > 1$ and

$$2S_n \leq \min\left\{ \frac{\log(\lambda_n)}{\lambda_n C}, \frac{\delta}{\lambda_n} \right\}.$$

For these integers n, let

$$\alpha_n = \lambda_n \exp(-2\lambda_n C S_n), \quad x'_n = x_n - \alpha_n(y_n - x_n), \quad y'_n = y_n + \alpha_n(y_n - x_n).$$

Remark that α_n goes to infinity together with n. By Lemma 3.4.5, g_n is defined on the whole interval $[x'_n, y'_n]$, and

$$V(g_n; [x'_n, y'_n]) \leq 2\lambda_n C S_n. \tag{3.37}$$

Claim (i). There exists an integer N_1 such that if $n \geq N_1$, then $[a, f^{-1}(u_n)]$ is contained in $[x'_n, y'_n]$.

Indeed, by (3.36),

$$\frac{x_n - a}{y_n - x_n} = \frac{f^n(x) - f^n(a)}{f^n(y) - f^n(x)} \leq \frac{x - a}{y - x} \exp(V(f^n; [a, b_\varepsilon])) \leq \frac{x - a}{y - x} \exp(\varepsilon/2).$$

Hence if we choose n such that $\alpha_n \geq e^{\varepsilon/2}(x - a)/(y - x)$, then $(x_n - a)/(y_n - x_n) \leq \alpha_n$, that is, $x'_n \leq a$. Analogously, if n is such that $\alpha_n \geq e^{\varepsilon/2}(f^{-1}(u) - y)/(y - x)$, then the inequality

$$\frac{f^{-1}(u_n) - y_n}{y_n - x_n} \leq \frac{f^{-1}(u) - y}{y - x} \exp(\varepsilon/2)$$

implies that $y'_n \geq f^{-1}(u_n)$.

Claim (ii). There exists an integer $N_2 \geq N_1$ such that if $n \geq N_2$, then $g_n(t) > t$ for all $t \in [a, f^{-1}(u_n)]$.

To show this claim, notice that from (3.36) and hypothesis (iii) of the proposition it follows that

$$\frac{f^{-1}(u_n) - u_n}{f^{-1}(u_n) - x_n} \exp(\varepsilon) \le \frac{v_n - u_n}{y_n - x_n} \le \frac{u_n - a}{x_n - a} \exp(-\varepsilon).$$

From (3.37) one concludes that $V(g_n; [x'_n, y'_n])$ converges to 0 as n goes to infinity. Take $N_2 \ge N_1$ such that $V(g_n; [x'_n, y'_n]) \le \varepsilon/2$ for every $n \ge N_2$. For such an n and every $t \in [x'_n, y'_n]$ we have

$$g'_n(t) \le \frac{v_n - u_n}{y_n - x_n} \exp(\varepsilon/2) < \frac{u_n - a}{x_n - a},$$

$$g'_n(t) \ge \frac{v_n - u_n}{y_n - x_n} \exp(-\varepsilon/2) > \frac{f^{-1}(u_n) - u_n}{f^{-1}(u_n) - x_n}.$$

By the mean value theorem, the first inequality shows that $g_n(t) > t$ for every $t \in [a, x_n]$, whereas the second one gives $g_n(t) > t$ for all $t \in [x_n, f^{-1}(u_n)]$ (see Figure 20).

Now fix $n \ge N_2$. From $g_n(a) < g_n(x_n) = u_n$ it follows that $f^{-1}g_n(a) < f^{-1}(u_n) < b_\varepsilon$. Let us consider the restrictions of f and g_n to the interval $[a, f^{-1}g_n(a)]$. We have

$$V(f; [a, f^{-1}g_n(a)]) + V(g_n; [a, g_n^{-1}f^{-1}g_n(a)])$$
$$\le V(f; [a, b_\varepsilon]) + V(g_n; [x'_n, y'_n])$$
$$\le \varepsilon$$
$$< \log\left(\frac{1}{1 - m(f)}\right).$$

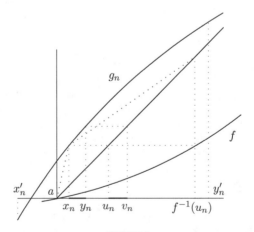

FIGURE 20

However, this contradicts Corollary 3.3.4 applied to the restrictions of f^{-1} and g_n^{-1} to the interval $[a, f^{-1}g_n(a)]$. $\square$

3.4.3 End of the proof

To complete the proof of Theorem 3.4.1, it suffices to find two intervals $]x, y[$ and $]u, v[$ satisfying the hypothesis of Proposition 3.4.4. However, we immediately point out that this is a nontrivial issue, and it is very instructive to follow the final steps (i.e., the proofs of claims (iii) and (iv)) in the following proof for the pseudogroup illustrated by Figure 16. We continue considering a positive constant $\varepsilon \leq \frac{1}{2} \log(\frac{1}{1-m(f)})$.

Claim (i). There exist $c_\varepsilon \in]a, f(b_\varepsilon)[$ and $g_\varepsilon \in \Gamma$ such that $[a, c_\varepsilon] \subset dom(g_\varepsilon)$, $g_\varepsilon(c_\varepsilon) = c_\varepsilon$, $g_\varepsilon(t) > t$ for all $t \in [a, c_\varepsilon[$, and $V(g_\varepsilon; [a, c_\varepsilon]) \leq \varepsilon$.

Indeed, since the orbit of $a \in \Lambda$ intersects the interval $]a, f(b_\varepsilon)[$, we may choose $g \in \Gamma$ and $c \in]a, f(b_\varepsilon)[$ such that $[g(a), g(c)] \subset]a, f(b_\varepsilon)[$ and $V(g; [a, c]) \leq \varepsilon/2$. For large-enough k, we have $f^k g(c) < c$, and since $f^k g(a) > a$, the map $g_\varepsilon = f^k g$ has fixed points in $]a, c[$. If we denote by c_ε the first of these points to the right of a, then from (3.36) one deduces that $V(g_\varepsilon; [a, c_\varepsilon]) \leq V(g; [a, c_\varepsilon]) + V(f^k; [a, b_\varepsilon]) \leq \varepsilon$.

Claim (ii). One has the inequality $f'(a) + g_\varepsilon'(c_\varepsilon) < e^\varepsilon$.

To show this claim, first notice that

$$V(f; [a, c_\varepsilon]) + V(g_\varepsilon; [a, c_\varepsilon]) \leq \frac{3\varepsilon}{2} < \log\left(\frac{1}{1-m(f)}\right).$$

Since the orbits of Λ are not dense in $\Lambda \cap]a, c_\varepsilon[$, by applying Corollary 3.3.4 to f^{-1} and g_ε^{-1}, we conclude that $f(c_\varepsilon) < g_\varepsilon(a)$, which obviously implies that $m(f) + m(g_\varepsilon) < 1$ (where $m(g_\varepsilon) = \inf_{x \in [a, c_\varepsilon]} g_\varepsilon'(x)$). We thus conclude that the value of $f'(a) + g_\varepsilon'(c_\varepsilon)$ is bounded from above by

$$m(f) \exp(V(f; [a, c_\varepsilon])) + m(g_\varepsilon) \exp(V(g_\varepsilon; [a, c_\varepsilon]))$$
$$\leq e^\varepsilon (m(f) + m(g_\varepsilon)) < e^\varepsilon.$$

In what follows, we will let $\alpha = f'(a)$ and $\beta = g_\varepsilon'(c_\varepsilon)$ (so that $\alpha + \beta < e^\varepsilon$), and we will impose on $\varepsilon > 0$ the extra condition (which holds for small-enough $\varepsilon > 0$)

$$1 - \alpha e^\varepsilon > (e^\varepsilon - \alpha)^2.$$

Claim (iii). For each $n \in \mathbb{N}$, one may choose $\kappa(\varepsilon, n) > 0$ such that $\lim_{\varepsilon \to 0} \kappa(\varepsilon, n) > 0$ and, for every $i \in \{1, \ldots, n\}$ and every $t \in [a, c_\varepsilon]$,

$$f^{n-i} g_\varepsilon f^i(t) - f^{n-i+1} g_\varepsilon f^{i-1}(t) \geq (c_\varepsilon - a)\kappa(\varepsilon, n). \tag{3.38}$$

Indeed, from $\alpha e^{-\varepsilon} \leq f'(t) \leq \alpha e^\varepsilon$ and $f(a) = a$, we conclude that for every $t \in [a, c_\varepsilon]$,

$$a + \alpha e^{-\varepsilon}(t - a) \leq f(t) \leq a + \alpha e^\varepsilon(t - a).$$

Analogously, from $\beta e^{-\varepsilon} \leq g_\varepsilon'(t) \leq \beta e^\varepsilon$ and $g_\varepsilon(c_\varepsilon) = c_\varepsilon$, we deduce that

$$c_\varepsilon - \beta e^\varepsilon(c_\varepsilon - t) \leq g_\varepsilon(t) \leq c_\varepsilon - \beta e^{-\varepsilon}(c_\varepsilon - t).$$

Hence

$$
\begin{aligned}
g_\varepsilon f(t) - f g_\varepsilon(t) &\geq g_\varepsilon(a + \alpha e^{-\varepsilon}(t - a)) - (a + \alpha e^\varepsilon(g_\varepsilon(t) - a)) \\
&\geq c_\varepsilon - \beta e^\varepsilon(c_\varepsilon - a - \alpha e^{-\varepsilon}(t - a)) \\
&\quad - (a + \alpha e^\varepsilon(c_\varepsilon - \beta e^{-\varepsilon}(c_\varepsilon - t) - a)) \\
&= (c_\varepsilon - a)[1 - \beta(e^\varepsilon - \alpha) - \alpha e^\varepsilon] \\
&> (c_\varepsilon - a)[1 - \alpha e^\varepsilon - (e^\varepsilon - \alpha)^2].
\end{aligned}
$$

Letting $\kappa(\varepsilon, n) = (\alpha e^{-\varepsilon})^n[1 - \alpha e^\varepsilon - (e^\varepsilon - \alpha)^2] > 0$, we then deduce that for every $i \in \{1, \ldots, n\}$ and every $t \in [a, c_\varepsilon]$,

$$
\begin{aligned}
f^{n-i} g_\varepsilon f^i(t) - f^{n-i+1} g_\varepsilon f^{i-1}(t) &\geq (\alpha e^{-\varepsilon})^{n-i}[g_\varepsilon f(f^{i-1}(t)) - f g_\varepsilon(f^{i-1}(t))] \\
&> (\alpha e^{-\varepsilon})^{n-i}(c_\varepsilon - a)[1 - \alpha e^\varepsilon - (e^\varepsilon - \alpha)^2] \\
&> (c_\varepsilon - a)\kappa(\varepsilon, n).
\end{aligned}
$$

Finally, notice that

$$\lim_{\varepsilon \to 0} \kappa(\varepsilon, n) = \alpha^n[1 - \alpha - (1 - \alpha)^2] > 0.$$

The preceding construction may be carried out for every very small $\varepsilon > 0$. Let us then fix $n \in \mathbb{N}$, and let us impose on $\varepsilon > 0$ the supplementary conditions

$$1 + \kappa(\varepsilon, n) \geq e^{2(n+2)\varepsilon}, \qquad 1 - \kappa(\varepsilon, n) \leq e^{-2(n+2)\varepsilon}. \tag{3.39}$$

Let I be a connected component of Λ^c contained in $]a, c_\varepsilon[$, and for each $i \in \{1, \ldots, n\}$, let $I_i = f^{n-i} g_\varepsilon f^i(I)$.

Claim (iv). For all $i < j$, the intervals $]x, y[= I_i$ and $]u, v[= I_j$ satisfy the conditions in Proposition 3.4.4.

Conditions (i) and (ii) easily follow from the construction. To check condition (iii), let us first notice that for every $k \in \{1, \ldots, n\}$,

$$(\alpha e^{-\varepsilon})^n \beta e^{-\varepsilon} \leq \frac{|I_k|}{|I|} \leq (\alpha e^{\varepsilon})^n \beta e^{\varepsilon},$$

and therefore

$$e^{-2(n+1)\varepsilon} \leq \frac{|I_i|}{|I_j|} \leq e^{2(n+1)\varepsilon}.$$

Hence by (3.38) and (3.39),

$$\begin{aligned}
\frac{u-a}{x-a} \cdot \frac{y-x}{v-u} &= \left(\frac{u-x}{x-a}+1\right)\frac{y-x}{v-u} \\
&\geq \left(\frac{(c_\varepsilon - a)\kappa(\varepsilon, n)}{x-a}+1\right)\frac{y-x}{v-u} \\
&\geq (1+\kappa(\varepsilon, n))2^{-2(n+1)\varepsilon} \\
&\geq e^{2\varepsilon},
\end{aligned}$$

that is,

$$\frac{u-a}{x-a}\exp(-2\varepsilon) \geq \frac{v-u}{y-x}.$$

Analogously,

$$\begin{aligned}
\frac{f^{-1}(u)-u}{f^{-1}(u)-x} \cdot \frac{y-x}{v-u} &= \left(1-\frac{u-x}{f^{-1}(u)-x}\right)\frac{y-x}{v-u} \\
&\leq \left(1-\frac{(c_\varepsilon - a)\kappa(\varepsilon, n)}{f^{-1}(u)-x}\right)\frac{y-x}{v-u} \\
&\leq (1-\kappa(\varepsilon, n))e^{2(n+1)\varepsilon} \leq e^{-2\varepsilon},
\end{aligned}$$

and hence

$$\frac{f^{-1}(u)-u}{f^{-1}(u)-x}\exp(2\varepsilon) \leq \frac{v-u}{y-x}.$$

This concludes the verification of condition (iii) of Proposition 3.4.4, thus completing the proof of Duminy's second theorem.

Exercise 3.4.6. Show that for the pseudogroup generated by the maps f and g in Figure 16, the Cayley graph associated to the semiexceptional orbit is a tree. Using this fact, show directly that the corresponding space of ends is infinite. For the orbits in Λ that are not semiexceptional, show

that despite the fact that they may not have the tree structure, they still
have infinitely many ends.

3.5 Two Open Problems

There exist two major open questions concerning the dynamics of groups of
$C^{1+\text{Lip}}$ circle diffeomorphisms, namely, the ergodicity of minimal actions
and the zero Lebesgue measure for exceptional minimal sets. In what
follows, we will present an overview of some partial results, and we will
explain some of the difficulties that lie behind those problems.

3.5.1 Minimal actions

Let us begin by making more precise the question of ergodicity of minimal
actions. To do this, recall that an action is **ergodic** if the measurable invariant sets have null or total measure. In what follows, the ergodicity will
be always considered with respect to the (normalized) Lebesgue measure.

Problem 1. Let Γ be a finitely generated subgroup of $\text{Diff}_+^{1+\text{Lip}}(S^1)$ all of
whose orbits are dense. Is the action of Γ on S^1 ergodic?

An analogous question for germs at the origin of analytic diffeomorphisms of the complex plane has an affirmative answer [100, 126]. This is
one of the reasons why one should expect that the answer to this problem
is positive. (See Exercise 3.5.12 for more evidence.) This is the case, for instance, when Γ has an element with irrational rotation number, according
to the following result obtained independently by Herman for the C^2 case
and by Katok for the general $C^{1+\text{bv}}$ case (see [115] and [133], respectively).
Here we reproduce Katok's proof, which strongly uses the combinatorial
structure of the dynamics of irrational rotations; see § 2.2.1.

Theorem 3.5.1. *If a diffeomorphism* $g \in \text{Diff}_+^{1+\text{bv}}(S^1)$ *has irrational rota-*
tion number, then the action of (the group generated by) g on the circle is
ergodic.

Proof. Recall that a **density point** of a measurable subset $A \subset S^1$ is a
point p such that

$$\frac{Leb(A \cap [p - \varepsilon, p + \varepsilon])}{Leb([p - \varepsilon, p + \varepsilon])}$$

converges to 1 as ε goes to zero. A classical theorem by Lebesgue asserts that in any measurable set, almost every point is a density point.

Let $A \subset S^1$ be a Γ-invariant measurable subset. Assuming that the Lebesgue measure of A is positive, we will show that this measure is actually total. To do this, let us consider a density point x_0 of A. Let φ be the conjugacy between g and the rotation of angle $\rho(g)$ such that $\varphi(x_0) = 0$. As in the proof of Denjoy's theorem, let $I_n(g) = \varphi^{-1}(I_n)$ and $J_n(g) = \varphi^{-1}(J_n)$. Notice that x_0 belongs to $J_n(g)$, and $|J_n(g)|$ converges to zero as n goes to infinity. Moreover, the intervals $g^k(J_n(g)), k \in \{0, \ldots, q_{n+1} - 1\}$, cover the circle, and each point in S^1 is contained in at most two of these intervals. If we use inequality (3.23), it is not difficult to conclude that for every $k \in \{0, \ldots, q_{n+1} - 1\}$,

$$\frac{Leb(g^k(J_n(g) \setminus A))}{Leb(g^k(J_n(g)))} \leq \exp(V(g^k; J_n(g))) \cdot \frac{Leb(J_n(g) \setminus A)}{Leb(J_n(g))}$$

$$\leq \exp(2V(g)) \cdot \frac{Leb(J_n(g) \setminus A)}{Leb(J_n(g))}.$$

Therefore, by the invariance of A,

$$Leb(S^1 \setminus A) \leq \sum_{k=0}^{q_{n+1}-1} Leb(g^k(J_n(g) \setminus A))$$

$$\leq \exp(2V(g)) \frac{Leb(J_n(g) \setminus A)}{Leb(J_n(g))} \sum_{k=0}^{q_{n+1}-1} Leb(g^k(J_n(g)))$$

$$\leq 2 \exp(2V(g)) \frac{Leb(J_n(g) \setminus A)}{Leb(J_n(g))}.$$

Since x_0 is a density point of A, the value of $Leb(J_n(g) \setminus A)/Leb(J_n(g))$ converges to 0 as n goes to infinity, from which one easily concludes that $Leb(S^1 \setminus A) = 0$. $\square$

Remark 3.5.2. Although every C^{1+bv} circle diffeomorphism with irrational rotation number is topologically conjugate to the corresponding rotation, in "many" cases the conjugating map is *singular*, even in the real-analytic case (see the discussion at the begining of § 3.6). Thus the preceding theorem does not follow from Denjoy's theorem.

According to Proposition 1.1.1, every group of circle homeomorphisms acting minimally and preserving a probability measure is topologically conjugate to a group of rotations. Moreover, if such a group is finitely generated, then at least one of its generators must have irrational rotation number. From Theorem 3.5.1 one then deduces the following:

Corollary 3.5.3. *If a finitely generated subgroup of* $\mathrm{Diff}_+^{1+\mathrm{bv}}(\mathrm{S}^1)$ *acts minimally and preserves a probability measure, then its action on* S^1 *is ergodic.*

The differentiability hypothesis is necessary for this result. (The hypothesis of finite generation is also necessary; see [64] for an example illustrating this.) Indeed, in [197] the reader may find examples of C^1 circle diffeomorphisms with irrational rotation number acting minimally but not ergodically. It is very plausible that by refining the construction method of [197], one may provide analogous examples in class $\mathrm{C}^{1+\tau}$ for every $\tau \in \,]0, 1[$.

Because of Theorem 3.5.1, for groups of $\mathrm{C}^{1+\mathrm{Lip}}$ circle diffeomorphisms acting minimally, the ergodicity question arises when the rotation number of each of its elements is rational. Under this hypothesis, the general answer to this question is unknown. Nevertheless, there exist very important cases where ergodicity is ensured.

Definition 3.5.4. Given a subgroup Γ of $\mathrm{Diff}_+^1(\mathrm{S}^1)$, a point $p \in \mathrm{S}^1$ is *expandable* if there exists $g \in \Gamma$ such that $g'(p) > 1$. The action is *differentiably expanding* if every point $p \in \mathrm{S}^1$ is expandable.

As we will see, if the action of a finitely generated subgroup of $\mathrm{Diff}_+^{1+\tau}(\mathrm{S}^1)$ is differentiably expanding and minimal, with $\tau > 0$, then it is necessarily ergodic. Notice, however, that the hypothesis of this claim is not invariant under smooth conjugacy, although the conclusion is. To state an a priori more general result that takes into account all of this, let us fix some notation. Given a finitely generated subgroup Γ of $\mathrm{Diff}_+^1(\mathrm{S}^1)$, let $\mathcal{G}$ be a finite symmetric system of generators. For each $n \geq 1$, let (compare Appendix B)

$$B_{\mathcal{G}}(n) = \{ h \in \Gamma : h = h_m \cdots h_1 \quad \text{for some } h_i \in \mathcal{G} \text{ and } m \leq n \},$$

and for each $x \in \mathrm{S}^1$, let

$$\lambda(x) = \limsup_{n \to \infty} \left(\max_{h \in B_{\mathcal{G}}(n)} \frac{\log(h'(x))}{n} \right).$$

Notice that this number is always finite, since it is bounded from above by

$$\sup_{h \in \mathcal{G}, y \in S^1} \log(h'(y)).$$

For each $\lambda > 0$, let $E_\lambda(\Gamma) = \{x \in S^1 : \lambda(x) \geq \lambda\}$. The **exponential set** $E(\Gamma)$ of the action is defined as the union of the sets $E_\lambda(\Gamma)$ with $\lambda > 0$; its complement $S(\Gamma)$ is called the **subexponential set**. Notice that every $E_\lambda(\Gamma)$, as well as $E(\Gamma)$ and $S(\Gamma)$, is a Borel set. Moreover, the function $x \mapsto \lambda(x)$ is invariant under the Γ-action. Therefore, the sets $E_\lambda(\Gamma)$, $E(\Gamma)$, and $S(\Gamma)$ are invariant under Γ. Finally, if (the action of) Γ is differentiably expanding, then $\lambda(x) > 0$ for every $x \in S^1$.

Theorem 3.5.5. *Let Γ be a finitely generated subgroup of $\mathrm{Diff}_+^{1+\tau}(S^1)$ whose action is minimal, where $\tau > 0$. If the exponential set of Γ has positive Lebesgue measure, then it has full measure, and the action is ergodic.*

Remark 3.5.6. Both hypotheses (that $\tau > 0$ and that the group is finitely generated) are necessary for the validity of the theorem. For a detailed discussion of this, see [64].

Example 3.5.7. Theorem 3.5.5 allows showing that if a subgroup of $\mathrm{Diff}_+^{1+\mathrm{Lip}}(S^1)$ satisfies the hypothesis of Duminy's first theorem and has no finite orbit, then its action is ergodic. Indeed, let Γ be a subgroup of $\mathrm{Diff}_+^{1+\mathrm{Lip}}(S^1)$ acting minimally and generated by a family $\mathcal{G}$ of diffeomorphisms satisfying $V(f) < V_0$ for every $f \in \mathcal{G}$. If we assume that the set $\mathrm{Per}(g)$ of periodic points is finite for at least one element $g \in \mathcal{G}$, then there must exist $p \in \mathrm{Per}(g)$ and $f \in \mathcal{G}$ such that $f(p)$ belongs to $S^1 \setminus \mathrm{Per}(g)$. Using some of the arguments in the proof of Theorem 3.3.1, starting with f and g, it is possible to create elements F and G in Γ that are as in Figure 17 over an interval $[a, b]$ of S^1 in such a way that $V(F; [a, c])$ and $V(G; [c, b])$ are small, where $c = G^{-1}(a)$. Fix a very small $\varepsilon > 0$, and let $\{I_1, \dots, I_n\}$ be a family of intervals covering the circle such that there exist $h_1, \dots, h_n$ in Γ satisfying $h_i(x) \in [c + \varepsilon, b]$ for every $x \in I_i$. Let C be the constant defined by $C^{-1} = \inf\{h_i'(x) : x \in I_1 \cup \dots \cup I_n\}$, and let N be a sufficiently large integer that each branch of the return map H^N induced by F and G is C-expanding (see Proposition 3.3.3). For $g_i = H^N h_i \in \Gamma$, one has $g_i'(x) > 1$ for every $x \in I_i$ (where we consider the right derivative in case of discontinuity). Thus the action of Γ is differentiably expanding, and hence its ergodicity follows from Theorem 3.5.5.

Unfortunately (or perhaps fortunately), there exist minimal actions for which the exponential set has zero Lebesgue measure. For instance, this is the case for the (standard action of the) modular group, as well as for the smooth, minimal actions of Thompson's group G; see [64].

Exercise 3.5.8. Give a precise statement and show a result in the following spirit: if Γ admits a continuous family of representations Φ_t in $\mathrm{Diff}_+^{1+\mathrm{bv}}(S^1)$ such that all the orbits by $\Phi_0(\Gamma)$ are dense and $\Phi_t(\Gamma)$ admits an exceptional minimal set for each $t > 0$, then the action of $\Phi_0(\Gamma)$ is not differentiably expanding.

Remark 3.5.9. Recall that if Γ is a non-Abelian countable group of circle homeomorphisms acting minimally, then its action is ergodic with respect to every stationary measure (this follows directly from Theorems 2.3.20 and 3.2.18). It is then natural to ask whether, in the case of groups of *diffeomorphisms*, there always exists a probability distribution on the group such that the corresponding stationary measure is absolutely continuous with respect to the Lebesgue measure (compare Remark 3.5.2). However, this relevant problem seems to be very hard. A partial result from [64] points in the negative direction: for the modular group and Thompson's group G, the stationary measure associated to any finitely supported probability distribution on the group is singular. Actually, for these cases the exponential set of the action has zero Lebesgue measure, but its mass with respect to the stationary measure is total.

Despite the preceding discussion, for the actions of the modular group and Thompson's group G already mentioned, as well as for most "interesting" actions in the literature, the set of points that are nonexpandable is finite and is made up of isolated fixed points of certain elements. Under this hypothesis we can give the following general result from [64], which covers Theorem 3.5.5 at least in the $C^{1+\mathrm{Lip}}$ case (for a complete proof of Theorem 3.5.5, we refer the reader to [180]).

Theorem 3.5.10. *Let Γ be a finitely generated subgroup of $C^{1+\mathrm{Lip}}$ circle diffeomorphisms whose action is minimal. Assume that for every nonexpandable point $x \in S^1$, there exist g_+ and g_- in Γ such that $g_+(x) = g_-(x) = x$, and such that g_+ (resp. g_-) has no fixed point in some interval $]x, x + \varepsilon[$ (resp. $]x - \epsilon, x[$). Then the action of Γ is ergodic.*

To show this result, we will use some slight modifications of Schwartz's estimates from § 3.2.1 that we leave as exercises for the reader.

Exercise 3.5.11. Given two intervals I and J and a C^1 map $f : I \to J$ that is a diffeomorphism onto its image, define the **distortion coefficient** of f on I as

$$\varkappa(f; I) = \log\left(\frac{\max_{x \in I} f'(x)}{\min_{y \in I} f'(y)}\right),$$

and its **distortion norm** as

$$\eta(f; I) = \sup_{\{x,y\} \subset I} \frac{\log\left(\frac{f'(x)}{f'(y)}\right)}{|f(x) - f(y)|} = \max_{z \in J} \left|(\log((f^{-1})'))'(z)\right|.$$

(i) Show that the distortion coefficient is subadditive under composition.

(ii) Show that $\varkappa(f, I) \leq C_f |I|$, where the constant C_f equals the maximum of the absolute value of the derivative of the function $\log(f')$. Conclude that if $\mathcal{G}$ is a subset of $\mathrm{Diff}_+^{1+\mathrm{Lip}}(S^1)$ such that the set $\{|(\log(f'(x)))'| : f \in \mathcal{G}, x \in S^1\}$ is bounded, then there exists a constant $C_{\mathcal{G}}$ (depending only on $\mathcal{G}$) such that for every interval I in the circle and every $f_1, \ldots, f_n$ in $\mathcal{G}$,

$$\varkappa(f_n \circ \cdots \circ f_1; I) \leq C_{\mathcal{G}} \sum_{i=0}^{n-1} |f_i \circ \cdots \circ f_1(I)|.$$

(iii) Under the assumptions in (ii), let I be an interval of S^1, and let x_0 be a point in I. Lettting $F_i = f_i \circ \cdots \circ f_1$, $I_i = F_i(I)$, and $x_i = F_i(x_0)$, show that

$$\exp\left(-C_{\mathcal{G}} \sum_{j=0}^{i-1} |I_j|\right) \cdot \frac{|I_i|}{|I|} \leq F_i'(x_0) \leq \exp\left(C_{\mathcal{G}} \sum_{j=0}^{i-1} |I_j|\right) \cdot \frac{|I_i|}{|I|}, \qquad (3.40)$$

$$\sum_{i=0}^{n} |I_i| \leq |I| \exp\left(C_{\mathcal{G}} \sum_{i=0}^{n-1} |I_i|\right) \sum_{i=0}^{n} F_i'(x_0). \qquad (3.41)$$

(iv) Still under the conditions in (ii), show that if for $x_0 \in S^1$ we set $S = \sum_{i=0}^{n-1} F_i'(x_0)$, then for every $\delta \leq \log(2)/2C_{\mathcal{G}}S$, one has

$$\varkappa(F_n;]x_0 - \delta/2, x_0 + \delta/2[) \leq 2C_{\mathcal{G}}S\delta. \qquad (3.42)$$

Exercise 3.5.12. The action of a group of circle homeomorphisms Γ is said to be **conservative** if for every measurable subset $A \subset S^1$ of positive Lebesgue measure there exists $g \neq id$ in Γ such that $Leb(A \cap g(A)) > 0$. Show that every ergodic action is conservative. Conversely, following the steps below, show that if the action of a finitely generated group of $C^{1+\text{Lip}}$ circle diffeomorphisms is minimal, then it is conservative.

(i) Assuming that the action is not conservative, show that the set $\sum(\Gamma)$ formed by the points $y \in S^1$ for which the series $\sum_{g \in \Gamma} g'(y)$ converges is nonempty.

(ii) Using (3.42), from (i) deduce that $\sum(\Gamma)$ is open and, using the minimality of the action, conclude that this set actually coincides with the whole circle.

(iii) Obtain a contradiction by choosing an infinite-order element $g \in \Gamma$ and using the equality

$$\sum_{i=0}^{n-1} \int_{S^1} (g^i)'(y)dy = 2\pi n.$$

From now on, we will assume that Γ is a subgroup of $\text{Diff}_+^{1+\text{Lip}}(S^1)$ satisfying the hypothesis of Theorem 3.5.10. In this case the set of non-expandable points is closed, since it coincides with $\cap_{g \in \Gamma}\{x : g'(x) \leq 1\}$. Together with the following lemma, this implies that this set is actually finite.

Lemma 3.5.13.   *The set of nonexpandable points is made up of isolated points.*

Proof.   For each nonexpandable point $y \in S^1$, we will find an interval of the form $]y, y + \delta[$ that does not contain any nonexpandable point. The reader will notice that a similar argument provides us with an interval of the form $]y - \delta', y[$ sharing this property.

By hypothesis, there exist $g_+ \in G$ and $\varepsilon > 0$ such that $g_+(y) = y$ and such that g_+ has no other fixed point in $]y, y + \varepsilon[$. Replacing g_+ by its inverse if necessary, we may assume that y is a right topologically repelling fixed point of g_+. Let us consider the point $\bar{y} = y + \varepsilon/2 \in]y, y + \varepsilon[$, and for each integer $k \geq 0$, let $\bar{y}_k = g_+^{-k}(\bar{y})$ and $J_k =]\bar{y}_{k+1}, \bar{y}_k[$. Taking $a = \bar{y}_1$ and $b = \bar{y}$ and applying (3.40), we conclude the existence of $k_0 \in \mathbb{N}$ such that $(g_+^k)'(x) \geq 2$ for all $k \geq k_0$ and all $x \in J_k$. Hence each J_k with $k \geq k_0$ contains no nonexpandable point, which clearly implies that the same holds for the interval $]y, \bar{y}_{k_0}[$. $\qquad\square$

According to the preceding proof, for each nonexpandable point $y \in S^1$, one can find an interval $I_y^+ =]y, y + \delta^+[$, a positive integer k_0^+, and an element $g_+ \in \Gamma$ having y as a right topologically repelling fixed point and with no other fixed point than y in the closure of I_y^+ such that if for $x \in I_y^+$ we take the smallest integer $n \geq 0$ satisfying $g_+^n(x) \notin I_y^+$, then $(g_+^{n+k_0^+})'(x) \geq 2$. Obviously, one can also find an interval $I_y^- =]y - \delta^-, y[$, a positive integer k_0^-, and an element $g_- \in \Gamma$ satisfying analogous properties. We then let

$$U_y = I_y^+ \cup I_y^- \cup \{y\}.$$

By definition (and continuity), for every expandable point y, there exist $g = g_y \in \Gamma$ and a neighborhood V_y of y such that $\inf_{x \in V_y} g'(x) > 1$. The sets $\{U_y : y \text{ nonexpandable}\}$ and $\{V_y : y \text{ expandable}\}$ form an open cover of the circle, from which we can extract a finite subcover

$$\{U_y : y \text{ nonexpandable}\} \cup \{V_{y_1}, \ldots, V_{y_k}\}.$$

Let

$$\lambda = \min \left\{ 2, \inf_{x \in V_{y_1}} g'_{y_1}(x), \ldots, \inf_{x \in V_{y_k}} g'_{y_k}(x) \right\}.$$

Since λ is the minimum among finitely many numbers greater than 1, we have $\lambda > 1$.

Lemma 3.5.14. *For every point $x \in S^1$, either the set of derivatives $\{g'(x) : g \in \Gamma\}$ is unbounded, or x belongs to the orbit of some nonexpandable point.*

Proof. If $x \in S^1$ is expandable, then it belongs to some of the sets $I_y^\pm$ or V_{y_j}, and there exists a map $g \in \Gamma$ such that $g'(x) \geq \lambda$. Similarly, either the image point $g(x)$ is nonexpandable, or there exists $h \in \Gamma$ such that $h'(g(x)) \geq \lambda$. By repeating this procedure we see that if we do not fall into a nonexpandable point by some composition, then we can always continue expanding by a factor at least equal to λ by some element of Γ. Therefore, for each point not belonging to the orbit of any nonexpandable point, the set of derivatives $\{g'(x) : g \in \Gamma\}$ is unbounded. Since for each x in the orbit of a nonexpandable point this set is obviously bounded, this proves the lemma. $\square$

For the proof of Theorem 3.5.10, we will use the so-called "expansion argument" (essentially due to Sullivan), which is one of the most important techniques for showing the ergodicity of dynamical systems having some hyperbolic behavior. Let $A \subset S^1$ be an invariant measurable set of positive Lebesgue measure, and let a be a density point of A not belonging to the orbit of any nonexpandable point (notice that since there are only finitely many nonexpandable points and Γ is countable, such a point exists).

The idea now consists in "blowing up" a very small neighborhood of a by using a well-chosen sequence of compositions so that the distortion is controlled and the length of the final interval is "macroscopic" (i.e., larger than some prescribed positive number). Since a is a density point of A, this final interval will consist mostly of points in A, and by the minimality of the action, this will imply that the measure of the set A is very close to 1. Finally, performing this procedure starting with smaller and smaller neighborhoods of a will yield the total measure of the invariant set A.

In our context the expansion procedure will work by applying the "exit-maps" $g_{\pm}^{n+k_0^{\pm}}$ to points in $I_y^+ \cup I_y^-$ and the maps g_y to points in the neighborhoods V_y. To simplify the control of distortion estimates, in what follows, we consider a symmetric generating system $\mathcal{G}$ of Γ containing the elements of the form g_+ and g_-.

Lemma 3.5.15. *There exists a constant $C_1 > 0$ such that for every expandable point $x \in S^1$, one can find $f_1, \ldots, f_n$ in $\mathcal{G}$ such that $(f_n \circ \cdots \circ f_1)'(x) \geq \lambda$ and*

$$\frac{\sum_{j=1}^{n}(f_j \circ \cdots \circ f_1)'(x)}{(f_n \circ \cdots \circ f_1)'(x)} \leq C_1. \tag{3.43}$$

Proof. A compactness-type argument reduces the general case to studying (arbitrarily small) neighborhoods of nonexpandable points. To find a small interval of the form $]y, y + \delta[$ formed by points satisfying the desired property, take the corresponding element g_+ having y as a right-repelling fixed point and no fixed point in $]y, \bar{y}]$, and for each $k \geq 0$, let $y_k = g_+^{-k}(\bar{y})$ and $J_k = g^{-k}(J_0)$, where $J_0 = [\bar{y}_1, \bar{y}[$. We know that for some k_0^+ we have $(g_+^n)'(x) \geq \lambda$ for all $n \geq k_0^+$ and all $x \in J_n$. For each x in the interval $I_y^+ =]y, \bar{y}_{k_0^+}[$, take $n \in \mathbb{N}$ such that $x \in J_n$. Then the estimates in Exercise 3.5.11 show that

$$\sum_{j=1}^{n}(g_+^j)'(x) \leq \exp(C_{\mathcal{G}}) \cdot \frac{\sum_{j=1}^{n}|g_+^j(J_n)|}{|J_n|} \leq \frac{\exp(C_{\mathcal{G}})}{|J_n|},$$

$$(g_+^n)'(x) \geq \exp(-C_{\mathcal{G}}) \cdot \frac{|g^n(J_n)|}{|J_n|} = \exp(-C_{\mathcal{G}}) \cdot \frac{|J_0|}{|J_n|}.$$

Therefore, if we let $f_1, \ldots, f_n$ all be equal to g_+,

$$\frac{\sum_{j=1}^{n}(f_j \circ \cdots \circ f_1)'(x)}{(f_n \circ \cdots \circ f_1)'(x)} = \frac{\sum_{j=1}^{n}(g_+^j)'(x)}{(g_+^n)'(x)} \leq \frac{\exp(2C_{\mathcal{G}})}{|J_0|}.$$

Moreover, for $n \geq k_0^+$, we have $(f_n \circ \cdots \circ f_1)'(x) = (g_+^n)'(x) \geq 2 \geq \lambda$. Analogous arguments may be applied to the interval $[y - \delta, y[$ associated to y. Since there are only finitely many nonexpandable points, this concludes the proof. $\qquad \square$

Lemma 3.5.16. *There exists a constant C_2 such that for every point x that does not belong to the orbit of any nonexpandable point and for every $M > 1$, one can find $f_1, \ldots, f_n$ in $\mathcal{G}$ such that $(f_n \circ \cdots \circ f_1)'(x) \geq M$ and*

$$\frac{\sum_{j=1}^{n}(f_j \circ \cdots \circ f_1)'(x)}{(f_n \circ \cdots \circ f_1)'(x)} \leq C_2. \tag{3.44}$$

Proof. Starting with $x_0 = x$, we let

$$\widetilde{F}_k = f_{k,n_k} \circ \cdots \circ f_{k,1}, \qquad x_k = \widetilde{F}_k(x_{k-1}),$$

where the elements $f_{k,j} \in \mathcal{G}$ satisfy

$$\frac{\sum_{j=1}^{n_k}(f_{k,j} \circ \cdots \circ f_{k,1})'(x_{k-1})}{(f_{k,n_k} \circ \cdots \circ f_{k,1})'(x_{k-1})} \leq C_1, \qquad (f_{k,n_k} \circ \cdots \circ f_{k,1})'(x_{k-1}) \geq \lambda.$$

If we perform this procedure $K \geq \log(M)/\log(\lambda)$ times, then for the compositions $F_k = \widetilde{F}_k \circ \cdots \circ \widetilde{F}_1$, we obtain

$$F_K'(x) = \prod_{k=1}^{K}(f_{k,n_k} \circ \cdots \circ f_{k,1})'(x_{k-1}) \geq \lambda^K \geq M.$$

To estimate the left-side expression of (3.44), notice that for $y = f_n \circ \cdots \circ f_1(x)$,

$$\frac{\sum_{j=1}^{n}(f_j \circ \cdots \circ f_1)'(x)}{(f_n \circ \cdots \circ f_1)'(x)} = \sum_{j=1}^{n}(f_{j+1}^{-1} \circ \cdots \circ f_n^{-1})'(y). \tag{3.45}$$

If we let $y = F_K(x)$, then, using (3.45), we see that the left-side expression of (3.44) is equal to

$$\sum_{k=1}^{K}\sum_{j=1}^{n_k}\left((f_{k,j+1}^{-1} \circ \cdots \circ f_{k,n_k}^{-1}) \circ (\widetilde{F}_{k+1}^{-1} \circ \cdots \circ \widetilde{F}_K^{-1})\right)'(y)$$

$$= \sum_{k=1}^{K}(\widetilde{F}_{k+1}^{-1} \circ \cdots \circ \widetilde{F}_K^{-1})'(y) \cdot \sum_{j=1}^{n_k}(f_{k,j+1}^{-1} \circ \cdots \circ f_{k,n_k}^{-1})'(x_k)$$

$$\leq \sum_{k=1}^{K}\frac{1}{\lambda^{K-k}} \cdot C_1 \leq \frac{C_1}{1-\lambda^{-1}}. \qquad \square$$

Lemma 3.5.17.   *For certain $\varepsilon > 0$, the following property holds: for every point x not belonging to the orbit of any nonexpandable point, there exists a sequence V_k of neighborhoods of x converging to x such that to each $k \in \mathbb{N}$ one may associate a sequence of elements $f_1, \ldots, f_{n_k}$ in $\mathcal{G}$ satisfying $|f_{n_k} \circ \cdots \circ f_1(V_k)| = \varepsilon$ and $\varkappa(f_{n_k} \circ \cdots \circ f_1; V_k) \le \log(2)$.*

Proof.   We will check the conclusion for $\varepsilon = \log(2)/2C_{\mathcal{G}}C_2$. To do this, fix $M > 1$, and consider the sequence of compositions $f_n \circ \cdots \circ f_1$ associated to x and M provided by the preceding lemma. Letting $\bar{F}_n = f_n \circ \cdots \circ f_1$ and $y = \bar{F}_n(x)$, for the neighborhood $V = \bar{F}_n^{-1}(]y - \varepsilon/2, y + \varepsilon/2[)$ of x, we have

$$\varkappa(\bar{F}_n; V) = \varkappa(\bar{F}_n^{-1};]y - \varepsilon/2, y + \varepsilon/2[)$$
$$= \varkappa(f_1^{-1} \circ \cdots \circ f_n^{-1};]y - \varepsilon/2, y + \varepsilon/2[).$$

The estimates in Exercise 3.5.11 then show that the distortion coefficient of the composition $f_1^{-1} \circ \cdots \circ f_n^{-1}$ is bounded from above by $\log(2)$ in a neighborhood of y of radius $r = \log(2)/4C_{\mathcal{G}}S$, where

$$S = \sum_{j=1}^{n}(f_{j+1}^{-1} \circ \cdots \circ f_n^{-1})'(\bar{F}_n(x_0)).$$

Now, according to (3.44) and (3.45), we have

$$S = \frac{\sum_{j=1}^{n}(f_j \circ \cdots \circ f_1)'(x)}{(f_n \circ \cdots \circ f_1)'(x)} \le C_2.$$

Thus $\varepsilon/2 \le r$, which yields the desired estimate for the distortion. Finally,

$$|V| = |\bar{F}_n^{-1}(]y - \varepsilon/2, y + \varepsilon/2[)| \le \frac{\varepsilon \exp(\varkappa(\bar{F}_n^{-1};]y - \varepsilon/2, y + \varepsilon/2[))}{(f_n \circ \cdots \circ f_1)'(x)} \le \frac{2\varepsilon}{M}$$

tends to zero as M goes to infinity. This concludes the proof of the lemma.   □

To complete the proof of Theorem 3.5.10, recall that for our invariant subset $A \subset S^1$ of positive Lebesgue measure, we found a density point a not belonging to the orbit of any nonexpandable point. Fix $\delta > 0$. By the preceding lemma, for some $\varepsilon > 0$, there exists a sequence V_k of neighborhoods of a converging to a such that for each $k \in \mathbb{N}$, there exist elements $f_1, \ldots, f_{n_k}$ in $\mathcal{G}$ satisfying $|f_{n_k} \circ \cdots \circ f_1(V_k)| = \varepsilon$ and $\varkappa(f_{n_k} \circ \cdots \circ f_1; V_k) \le \log(2)$. Passing to a subsequence if necessary, we may assume that the sequence of intervals $f_{n_k} \circ \cdots \circ f_1(V_k)$ converges to some interval V of length ε. Because of the minimality of the action, the proof will be completed when we show that $Leb((S^1 \setminus A) \cap V) = 0$. To do this, first observe that by the invariance of A,

$$\frac{Leb((S^1 \setminus A) \cap f_{n_k} \circ \cdots \circ f_1(V_k))}{\varepsilon}$$

$$= \frac{Leb((S^1 \setminus A) \cap f_{n_k} \circ \cdots \circ f_1(V_k))}{Leb(f_{n_k} \circ \cdots \circ f_1(V_k))}$$

$$\leq \exp(\varkappa(f_{n_k} \circ \cdots \circ f_1; V_k)) \cdot \frac{Leb((S^1 \setminus A) \cap V_k)}{Leb(V_k)}$$

$$\leq \frac{2 Leb((S^1 \setminus A) \cap V_k)}{Leb(V_k)}.$$

Now notice that the first expression in this inequality converges to $Leb((S^1 \setminus A) \cap V)/\varepsilon$, while the last expression converges to zero because a is a density point of A.

3.5.2 Actions with an exceptional minimal set

Another major open question in the theory concerns the Lebesgue measure of exceptional minimal sets.

Problem 2. If Γ is a finitely generated subgroup of $\mathrm{Diff}_+^{1+\mathrm{lip}}(S^1)$ having an exceptional minimal set Λ, is the Lebesgue measure of Λ necessarily equal to zero ?

A closely related problem concerns the finiteness of the number of orbits of connected components of $S^1 \setminus \Lambda$ (if this is always true, this could be considered a kind of generalization of the classical Ahlfors finiteness theorem [1]). In the same spirit, a conjecture by Dippolito suggests that the action of Γ on Λ should be topologically conjugate to the action of a group of piecewise affine homeomorphisms [66].

The Lebesgue measure of exceptional minimal sets is zero in the case of Fuchsian groups, but the techniques involved in the proof cannot be adapted to the case of general diffeomorphisms (see, for instance, [193]). In what follows, we will concentrate on a particular class of dynamics, namely that of *Markovian* minimal sets. Roughly speaking, over these sets the dynamics is conjugate to a *subshift*, and this information strongly simplifies the study of the combinatorics. Following [48, 163], we will see that for this case, the answer to Problem 2 is positive (the same holds for the finiteness of the connected components of the complement of exceptional minimal sets modulo the action [49]).

Let $P = (p_{ij})$ be a $k \times k$ **incidence matrix** (i.e., a matrix with entries 0 and 1). Let us consider the space $\Omega = \{1, \ldots, k\}^{\mathbb{N}}$ and the subspace Ω^* formed by the **admissible sequences**, that is, by the $\omega = (i_1, i_2, \ldots) \in \Omega$ such that $p_{i_j i_{j+1}} = 1$ for every $j \in \mathbb{N}$. We endow this subspace with the topology induced from the product topology on Ω, and we consider the dynamics of the shift $\sigma : \Omega^* \to \Omega^*$ on it.

Definition 3.5.18. Let $\mathcal{G} = \{g_1, \ldots, g_k\}$ be a family of homeomorphisms defined on open bounded intervals. If $\{I_1, \ldots, I_k\}$ is a family of closed intervals such that $I_j \subset ran(g_j)$ for every $j \subset \{1, \ldots, k\}$, then we say that $\mathcal{S} = (\{I_1, \ldots, I_k\}, \mathcal{G}, P)$ is a **Markov system** for the **Markov pseudogroup** $\Gamma_{\mathcal{S}}$ generated by the g_i's if the following properties are satisfied:

(i) The intersection $ran(g_i) \cap ran(g_j)$ is empty for every $i \neq j$.
(ii) If $p_{ij} = 1$ (resp. $p_{ij} = 0$), then $I_j \subset dom(g_i)$ and $g_i(I_j) \subset I_i$ (resp. $I_j \cap dom(g_i) = \emptyset$).

For each $\omega = (i_1, i_2, \ldots) \in \Omega^*$ and each $n \in \mathbb{N}$, the domain of definition of the map $\bar{h}_n(\omega) = g_{i_1} \cdots g_{i_n}$ contains the interval $g_{i_n}^{-1}(I_{i_n})$. Let $I_n(\omega) = \bar{h}_n(\omega)(g_{i_n}^{-1}(I_{i_n}))$ and

$$\Lambda_{\mathcal{S}} = \bigcap_{n \in \mathbb{N}} \bigcup_{\omega \in \Omega^*} I_n(\omega).$$

Notice that if $T : \cup_{i=1}^{k} ran(g_i) \to \cup_{i=1}^{k} dom(g_i)$ denotes the map whose restriction to each set $ran(g_i)$ coincides with g_i^{-1}, then the restriction of T to $\Lambda_{\mathcal{S}}$ is naturally **semiconjugate** to the shift $\sigma : \Omega^* \to \Omega^*$, in the sense that each $x \in \Lambda_{\mathcal{S}}$ uniquely determines an admissible sequence $\varphi(x) = \omega = (i_1, i_2, \ldots)$ such that x belongs to $I_n(\omega)$ for each n, and the equality $\varphi(T(x)) = \sigma(\varphi(x))$ holds.

Definition 3.5.19. A set Λ that is invariant under the action of a pseudogroup is said to be **Markovian** if there exists an open interval L intersecting Λ such that $L \cap \Lambda$ equals $\Lambda_{\mathcal{S}}$ for a Markov system $\mathcal{S}$ defined on L.

In this chapter we have already studied an example of a Markov pseudogroup, namely, the one illustrated by Figure 16. Indeed, the properties in the definition are satisfied for the elements g_1 and g_2 corresponding in

Figure 16 to f and g, respectively, where $I_1 = [a, b]$ and $I_2 = [c, d]$ (we consider an open interval slightly larger than $[a, d]$ as the domain of definition of g_1 and g_2). In this particular case, we have $p_{i,j} = 1$ for all i and j in $\{1, 2\}$, and thus $\Omega^* = \Omega$.

Theorem 3.5.20. *If Λ is a Markovian local exceptional set for a Markov pseudogroup of $C^{1+\mathrm{Lip}}$ diffeomorphisms of a one-dimensional compact manifold, then the Lebesgue measure of Λ is zero.*

In the case where the maps g_i may be chosen (uniformly) differentiably contracting, this result still holds in class $C^{1+\tau}$ (compare Theorem 3.5.22), but not in class C^1 (see Example 3.5.23). Actually, in the former case a stronger result holds [202]: the *Hausdorff dimension* of Λ is less than 1. Nevertheless, this is no longer true in the noncontracting Markovian case, even in the real-analytic context [87].

We will give the proof of Theorem 3.5.20 only for the case of the Markov system discussed above and illustrated by Figure 16; the reader should have no problem adapting the following arguments to the general case by taking care of some technical combinatorial details. First, notice that for every $x \in I = [b, c]$ and every element $g = g_{i_n} \cdots g_{i_1} \in \Gamma$ (where each g_{i_j} belongs to $\mathcal{G} = \{g_1, g_2\}$), one has

$$| \log(g'(b)) - \log(g'(x))| \leq C \sum_{j=0}^{n-1} |g_{i_j} \cdots g_{i_1}(b) - g_{i_j} \cdots g_{i_1}(x)|$$

$$\leq C \sum_{j=0}^{n-1} |g(I)| \leq C(d - a).$$

Choosing $x \in I$ such that $g'(x) = |g(I)|/|I|$, from this inequality we conclude that $g'(b) \leq |g(I)|e^{C(d-a)}/|I|$, and therefore

$$\sum_{g \in \Gamma} g'(b) \leq \frac{e^{C(d-a)}}{|I|} \sum_{g \in \Gamma} |g(I)| \leq \frac{e^{C(d-a)}(d-a)}{|I|} = S.$$

If we let $\ell = \log(2)/4CS$, from the estimates in Exercise 3.5.11 it follows that for every g and h in Γ and every $x \in [a, d]$ in the ℓ-neighborhood of $h(b)$, one has $g'(x)/2 \leq g'(b) \leq 2g'(x)$, and hence

$$\sum_{g \in \Gamma} |g([h(b) - \ell, h(b) + \ell])| \leq 2\ell \sum_{g \in \Gamma} \sup_{|x-h(b)| \leq \ell} g'(x) \leq 4\ell S. \qquad (3.46)$$

Fix $r \in \mathbb{N}$ such that for every $(i_1, \ldots, i_r) \in \{1, 2\}^r$, the length of the interval $g_{i_1} \cdots g_{i_r}([a, b])$ is less than or equal to ℓ. Notice that these intervals cover

Λ, and hence for every $n \in \mathbb{N}$, the same holds for the intervals of the form $g_{i_1} \cdots g_{i_n} h([a, d])$, where $(i_1, \ldots, i_n)$ ranges over $\{1, 2\}^n$, and h ranges over the elements of *length* r (that is, the elements of type $g_{i_1} \cdots g_{i_r}$, with $(i_1, \ldots, i_r) \in \{1, 2\}^r$). Therefore,

$$Leb(\Lambda) \leq \sum_{(i_1, \ldots, i_n) \in \{1,2\}^n} \sum_{(h_1, \ldots, h_r) \in \{g_1, g_2\}^r} |g_{i_1} \cdots g_{i_n} h_1 \cdots h_r([a, d])|,$$

and since for each $(h_1, \ldots, h_r) \in \{g_1, g_2\}^r$, the point $h_1 \cdots h_r(b)$ belongs to $h_1 \cdots h_r([a, d])$, from this we conclude that

$$Leb(\Lambda) \leq \sum_{(h_1, \ldots, h_r) \in \{g_1, g_2\}^r} \sum_{length(g)=n} |g([h_1 \cdots h_r(b) - \ell, h_1 \cdots h_r(b) + \ell])|.$$

Letting n go to infinity in this inequality, from the convergence of the series in (3.46) one concludes that $Leb(\Lambda) = 0$, thus finishing the proof.

We refer the reader to [47] and [128] for the realization of Markov pseudogroups as holonomy pseudogroups of codimension-one foliations. Let us point out, however, that it is not difficult to construct exceptional minimal sets that are not induced by Markovian systems; see [49].

Exercise 3.5.21. Completing Exercise 3.4.6, show that if an exceptional minimal set Λ is Markovian, then all the orbits in Λ have infinitely many ends (see [48] in case of problems with this).

To close this section, let us briefly discuss another approach to Problem 2 that is closely related to our approach to the ergodicity question. Let us begin by pointing out that Theorem 3.5.5 has a natural analog (due to Hurder [119]) in the present context.

Theorem 3.5.22. *Let Γ be a subgroup of $\mathrm{Diff}_+^{1+\tau}(\mathrm{S}^1)$ admitting an exceptional minimal set Λ. If $\tau > 0$, then the set $\Lambda \cap E(\Gamma)$ has null Lebesgue measure.*

The necessity of the hypothesis $\tau > 0$ for this result is illustrated by the following classical example due to Bowen [23].

Example 3.5.23. As the reader can easily check, the maps f and g in Figure 16 may be chosen so that the following supplementary conditions are verified (to put ourselves in the Markovian context, we again let $g_1 = f$, $g_2 = g$, and $I = [b, c]$, with $a = 0$ and $d = 1$).

(i) There exists a sequence of positive real numbers ℓ_n (where $n \geq 0$) such that $|I| = \ell_0$,

$$\sum_{n\geq 0} \ell_n < 1, \qquad \lim_{n\to\infty} \frac{\ell_{n+1}}{\ell_n} = 1,$$

and such that if for $n \in \mathbb{N}$ and $(g_{i_1}, \ldots, g_{i_n}) \in \{g_1, g_2\}^n$ we let $I_{i_1,\ldots,i_n} = g_{i_1} \cdots g_{i_n}(I)$, then $|I_{i_1,\ldots,i_n}| = \ell_n/2^n$.

(ii) Each g_i is smooth over I and over each $I_{i_1,\ldots,i_n}$, its derivative at the endpoints of these intervals equals $1/2$, and

$$\lim_{n\to\infty} \max_{(i_1,\ldots,i_n)\in\{1,2\}^n} \sup_{x\in I_{i_1,\ldots,i_n}} \left| g_i'(x) - \frac{1}{2} \right| = 0.$$

(iii) Each g_i is differentiable on a neighborhood of $a = 0$ and $d = 1$, with derivative identically equal to $1/2$ to the left (resp. to the right) of a (resp. d).

With these conditions, it is not difficult to show that g_1 and g_2 are of class C^1 over the whole interval $[a, d] = [0, 1]$, and their derivatives equal $1/2$ over the Markov minimal set Λ. Now for the Lebesgue measure of Λ we have

$$Leb(\Lambda) = 1 - |I| - \sum_{n\geq 1} \sum_{(i_1,\ldots,i_n)} |I_{i_1,\ldots,i_n}| = 1 - \sum_{n\geq 0} \ell_n > 0.$$

Notice that, by using this example, it is not difficult to create a subgroup of $\mathrm{Diff}_+^1(S^1)$ with two generators and having an exceptional minimal set of positive Lebesgue measure entirely contained in the exponential set of the action.

Exercise 3.5.24. After reading § 4.1.4, show that for every $\tau > 0$, there exist finitely generated Abelian subgroups of $\mathrm{Diff}_+^{1+\tau}(S^1)$ admitting an exceptional minimal set of positive measure. Prove directly (i.e., without using Theorem 3.5.22) that for these examples, the exceptional minimal set is contained (up to a null-measure set) in the subexponential set of the action.

Example 3.5.25. Following an idea that seems to go back to Mañé, given $\tau < 1$, let us consider a $C^{1+\tau}$ circle diffeomorphism with irrational rotation number and admitting a minimal invariant Cantor set Λ_0 of positive Lebesgue measure (see the preceding exercise). Fix a connected component I of the complement of Λ_0, and let $\bar{H} : S^1 \to S^1$ be a degree-2 map that coincides with f outside I (i.e., the map $\bar{H}$ makes an "extra turn" around the circle over I). If one uses $\bar{H}$, it is easy to construct a pseudogroup with two generators admitting an exceptional minimal set Λ containing Λ_0. Since the Lebesgue measure of Λ_0 is positive, this is also the case for Λ.

Exercise 3.5.26. Prove that if Γ is a finitely generated group of real-analytic circle diffeomorphisms admitting an exceptional minimal set Λ, then, with the exception of at most countably many points, $S^1 \setminus \Lambda$ is contained in the subexponential set $S(\Gamma)$.

Remark. Quite possibly, the same holds for subgroups of $\text{Diff}_+^{1+\text{bv}}(S^1)$.

Hint. Use the analyticity to show that, discarding at most countably many points, if x belongs to $S^1 \setminus \Lambda$, then there exists an open interval I_x containing x in its interior such that the images of I_x under the elements in the group are two-by-two disjoint. Then use the relations

$$1 \geq \sum_{g \in \Gamma} |g(I_x)| = \sum_{g \in \Gamma} \int_{I_x} g'(y)dy = \int_{I_x} \left(\sum_{g \in \Gamma} g'(y) \right) dy.$$

3.6 On the Smoothness of the Conjugacy between Groups of Diffeomorphisms

The problem of the smoothness of the conjugacy between groups of circle diffeomorphisms is technical and difficult. The relevant case of free actions is already extremely hard. Nevertheless, this case is very well understood thanks to the works of Siegel, Arnold, Herman, Moser, Yoccoz, Khanin, Katznelson and Orstein, among others. For $r > 2$, the main technical tool for obtaining the differentiability of the conjugacy[5] of a C^r circle diffeomorphism f having irrational rotation number $\rho(f)$ to the rotation of angle $\rho(f)$ corresponds to the Diophantine nature of $\rho(f)$. Roughly, the conjugacy is forced to be smooth if the approximations of $\rho(f)$ by rational numbers are "bad" in the sense that they are slow with respect to the denominator of the approximating rational. This

---

5. Notice that from Exercise 2.2.12 one easily concludes that two different conjugacies as above differ by a Euclidean rotation. Because of this, there is no ambiguity when speaking about the differentiability of the conjugacy.

corresponds to one of the main issues of the so-called small-denominators theory.

In what follows, we will study the case of nonfree actions by circle diffeomorphisms, which is essentially different. There is no systematic theory for this case, and perhaps the only definitive results correspond to Theorem 3.6.6, to be studied in detail in the next section, and Proposition 4.1.16 in the next chapter. However, we point out that these results are much simpler than those of the theory of small denominators. Slightly stronger results hold in some particular cases. For instance, Sullivan showed that every topological and absolutely continuous conjugacy between Fuchsian groups of first kind is real-analytic, and he further obtained analogous results for the smoothness over the exceptional minimal set for conjugacies between Fuchsian groups of second kind [234, 235]. On the other hand, using classical results on the existence and uniqueness of absolutely continuous invariant measures for expanding maps of the interval, the author showed in [185] that if two groups of C^r circle diffeomorphisms, with $r \geq 2$, satisfy the hypothesis of Duminy's first theorem (i.e., if they are generated by elements near rotations), and if their orbits are dense, then every topological and absolutely continuous conjugacy between them is a C^r diffeomorphism (see [212, 213] for a stronger result in the real-analytic context).

3.6.1 Sternberg's linearization theorem and C^1 conjugacies

Let f and g be C^r diffeomorphisms of a neighborhood of the origin in the line into their images. If f and g fix the origin, we say that they are **equivalent** if there exists $\varepsilon > 0$ such that $f|_{]-\varepsilon,\varepsilon[} = g|_{]-\varepsilon,\varepsilon[}$. Modulo this equivalence relation, the class of g will be denoted by $[g]$. The set of classes forms a group with respect to the composition of representatives, that is, $[f][g] = [f \circ g]$. This group, called the group of **germs** of C^r diffeomorphisms of the line that fix the origin and preserve orientation, will be denoted by $\mathcal{G}^r_+(\mathbb{R}, 0)$.

Notice that the derivative $g^{(i)}(0)$ of order $i \leq r$ at the origin is well defined for all $[g] \in \mathcal{G}^r_+(\mathbb{R}, 0)$. We will say that $[g]$ is **hyperbolic** if $g'(0) \neq 1$.

Lemma 3.6.1. *If $[g] \in \mathcal{G}^r_+(\mathbb{R}, 0)$ is hyperbolic, with $1 \leq r < \infty$, then there exists $[h]$ in $\mathcal{G}^r_+(\mathbb{R}, 0)$ such that $h'(0) = 1$ and $(hgh^{-1})^{(i)}(0) = 0$ for every $2 \leq i \leq r$.*

Proof. Let $g(x) = ax + a_2x^2 + \ldots + a_rx^r + o(x^r)$ be the Taylor series expansion of g about the origin. Let us formally write

$$\hat{h}(x) = x + b_2x^2 + \ldots + b_rx^r + \ldots,$$

and let us try to find the coefficients b_i so that

$$\hat{h} \circ g = M_a \circ \hat{h}, \tag{3.47}$$

where M_a denotes multiplication by $a = g'(0)$. If we identify the coefficients of x^2 in both sides of this equality, we obtain $a_2 + b_2a^2 = ab_2$, and therefore $b_2 = a_2/(a - a^2)$. In general, assuming that $b_2, \ldots, b_{i-1}$ are already known (where $i \leq r$), from (3.47) one obtains

$$Q_i(b_1, \ldots, b_{i-1}, a_1, \ldots, a_r) + b_ia^i = ab_i$$

for some polynomial Q_i in $(r + i - 1)$ variables. Thus

$$b_i = \frac{Q_i(b_1, \ldots, b_{i-1}, a_1, \ldots, a_r)}{a - a^i}.$$

Now let h be a C^r diffeomorphism defined on a neighborhood of the origin and such that $h(0) = 0, h'(0) = 1$, and $h^{(i)}(0) = i!b_i$ for every $2 \leq i \leq r$. By reversing the preceding computations, it is easy to see that $[h] \in \mathcal{G}^r_+(\mathbb{R}, 0)$ satisfies the desired properties. $\qquad\square$

Sternberg's theorem appears in the following form in [256] (see [233] for the original, weaker form). Let us point out that an analogous result holds for germs of real-analytic diffeomorphisms [52].

Theorem 3.6.2. Let $g \in \mathcal{G}^r_+(\mathbb{R}, 0)$ be a hyperbolic germ, where $2 \leq r \leq \infty$. If we let $a = g'(0)$, then there exists $[h] \in \mathcal{G}^r_+(\mathbb{R}, 0)$ such that $h'(0) = 1$ and $h(g(x)) = ah(x)$ for every x near the origin. Moreover, if $[h_1] \in \mathcal{G}^1_+(\mathbb{R}, 0)$ satisfies the last two properties, then $[h_1]$ belongs to $\mathcal{G}^r_+(\mathbb{R}, 0)$, and $[h_1] = [h]$.

Proof. Let us first consider the case $r < \infty$. By Lemma 3.6.1, to obtain a conjugacy, we may suppose that $g^{(i)}(0) = 0$ for each $i \in \{2, \ldots, r\}$. Replacing g by g^{-1} if necessary, we may suppose, moreover, that $g'(0) = a < 1$. Let $0 < \delta < 1$ be such that the domain of definition of g contains the interval $[-\delta, \delta]$. Let us define

$$C(\delta) = \sup\{|g^{(r)}(t)| : t \in [-\delta, \delta]\}.$$

A direct application of the mean value theorem shows that for all $t \in [-\delta, \delta]$, one has $|g'(t)| \leq a + C(\delta)$ and $|g^{(i)}(t)| \leq C(\delta)$ for every $i \in \{2, \ldots, r\}$.

Let E_δ be the space of real-valued functions ψ of class C^r on $[-\delta, \delta]$ satisfying $\psi(0) = \psi'(0) = \ldots = \psi^{(r)}(0) = 0$. Endowed with the norm

$$\|\psi\| = \sup\{|\psi^{(r)}(t)| : t \in [-\delta, \delta]\},$$

E_δ becomes a Banach space. For $\psi \in E_\delta$, $t \in [-\delta, \delta]$, and $i \in \{0, \ldots, r\}$, one has

$$|\psi^{(i)}(t)| \leq \frac{t^{r-i}}{(r-i)!} \|\psi\|.$$

Let us consider the linear operator $S_\delta : E_\delta \to E_\delta$ defined by $S_\delta(\psi) = (\psi \circ g)/a$. We claim that if $\delta > 0$ is small enough, then S_δ is a contraction (that is, $\|S_\delta\| < 1$). Indeed, it is easy to check that

$$(\psi \circ g)^{(r)} = \sum_{i=1}^{r} (\psi^{(i)} \circ g) \cdot Q_i(g', \ldots, g^{(r-i+1)}),$$

where Q_i is a polynomial with positive coefficients in $(r + 1 - i)$ variables and $Q_r(x) = x^r$. For $t \in [-\delta, \delta]$, this yields

$$|(\psi \circ g)^{(r)}(t)| \leq K(\delta) \|\psi\|,$$

where

$$K(\delta) = (a + C(\delta))^r + \sum_{i=1}^{r-1} \frac{\delta^{r-i}}{(r-i)!} Q_i(a + C(\delta), \ldots, a + C(\delta)).$$

Notice that $K(\delta)$ tends to a^r as δ goes to zero. Moreover, $\|S_\delta\| \leq K(\delta)/a$. Since $r \geq 2$ and $a < 1$, this shows that the value of $\|S_\delta\|$ is smaller than 1 for δ small enough.

Now fix $\delta > 0$ such that $\|S_\delta\| < 1$. Notice that the restriction of the map $x \mapsto g(x) - ax$ to the interval $[-\delta, \delta]$ defines an element ψ_1 in E_δ. Since S_δ is a contraction, the equation (in the variable ψ)

$$S_\delta(\psi) + a^{-1}\psi_1 = \psi$$

has a unique solution $\psi_0 \in E_\delta$. For $h = Id + \psi_0$, we have

$$\begin{aligned} h(g(x)) = g(x) + \psi_0 \circ g(x) &= \psi_1(x) + ax + \psi_0 \circ g(x) \\ &= \psi_1(x) + ax + aS_\delta(\psi_0)(x) \\ &= a\psi_0(x) + ax = ah(x). \end{aligned}$$

Since $\psi_0 \in E_\delta$, we have $h(0) = 0$ and $h'(0) = 1$. Therefore, if $r < \infty$, then $[h]$ is a germ in $\mathcal{G}_+^r(\mathbb{R}, 0)$ satisfying the required properties.

If $h_1 \in \mathcal{G}_+^1(\mathbb{R}, 0)$ satisfies the same properties, then

$$ahh_1^{-1}(t) = h \circ g \circ h_1^{-1}(t) = hh_1^{-1}(at)$$

for every t near the origin. From this one deduces that

$$hh_1^{-1}(t) = \lim_{n \to \infty} \frac{hh_1^{-1}(a^n t)}{a^n} = t \lim_{n \to \infty} \frac{hh_1^{-1}(a^n t)}{a^n t} = t(hh_1^{-1})'(0) = t,$$

which shows the uniqueness.

Finally, the case $r = \infty$ easily follows from the uniqueness already established for each finite $r \geq 2$. $\square$

A direct consequence of the preceding theorem is the following result.

Corollary 3.6.3. *Let g_1 and g_2 be elements in $\mathcal{G}_+^r(\mathbb{R}, 0)$, where $2 \leq r \leq \infty$. Suppose that $[\varphi] \in \mathcal{G}_+^1(\mathbb{R}, 0)$ conjugates g_1 to g_2; that is, on a neighborhood of the origin one has $\varphi \circ g_1 = g_2 \circ \varphi$. If g_1 is hyperbolic, then $[\varphi]$ is an element of $\mathcal{G}_+^r(\mathbb{R}, 0)$.*

Proof. Let us begin by noticing that $[g_2]$ is also hyperbolic, because $g_2'(0) = g_1'(0)$. If $h \in \mathcal{G}_+^r(\mathbb{R}, 0)$ conjugates g_2 to its linear part, then $h \circ \varphi$ conjugates g_1 to its linear part. If $b = (h\varphi)'(0)$, then $M_{1/b}h\varphi$ still conjugates g_1 to its linear part; moreover, it satisfies $(M_{1/b}h\varphi)'(0) = 1$. By the uniqueness of such a conjugacy, we have $[M_{1/b}h\varphi] \in \mathcal{G}_+^r(\mathbb{R}, 0)$, and therefore $[\varphi] \in \mathcal{G}_+^r(\mathbb{R}, 0)$. $\square$

Sternberg's linearization theorem still holds in class $C^{1+\tau}$ for every $\tau > 0$ (the proof consists in a straightforward extension of the preceding arguments; see also [54]). However, the theorem is no longer true in class C^1, as is illustrated by the next example due to Sternberg himself.

Example 3.6.4. Let us consider the map g defined on an open interval about the origin by $g(0) = 0$ and $g(x) = ax(1 - 1/\log(x))$ for $x \neq 0$, where $0 < a < 1$. It is easy to see that $[g]$ is the germ of a C^1 diffeomorphism satisfying $g'(0) = a$. Despite this fact, $[g]$ is not conjugate to the germ of the linear map M_a by any germ of a Lipschitz homeomorphism (in particular, $[g]$ is not C^1 conjugate to $[M_a]$). To show this, fix a constant $\bar{a} \in]0, 1[$ such that $g'(x) \geq \bar{a}$ for every x near the origin, and choose a local homeomorphism h fixing 0 and such that for these points x one has $g(x) = h M_a h^{-1}(x)$. For every $k \in \mathbb{N}$, one has $g^k(x) = h(a^k h^{-1}(x))$, and hence

$$\frac{g^k(x)}{a^k} = \frac{h(a^k h^{-1}(x))}{a^k}. \qquad (3.48)$$

If h were Lipschitz with constant $\bar{C}$, then, on the one hand, the right-hand member in this equality would be bounded from above by $\bar{C} h^{-1}(x)$ for every $k \in \mathbb{N}$. On the other hand, from the definition of g one easily concludes that

$$\frac{g^k(x)}{a^k} = x \prod_{i=0}^{k-1} \left(1 - \frac{1}{\log(g^i(x))} \right).$$

Since $g^i(x) \geq \bar{a}^i x$, there exists an integer $i_0 \geq 1$ such that for some positive constant C and every $i \geq i_0$,

$$\log(g^i(x)) \geq i \log(\bar{a}) + \log(x) \geq -\frac{i}{C}.$$

From this it follows that

$$\prod_{i=i_0}^{k-1} \left(1 - \frac{1}{\log(g^i(x))} \right) \geq \prod_{i=i_0}^{k-1} \left(1 + \frac{C}{i} \right).$$

Since the last product diverges as k goes to infinity, we conclude that the left-hand member in (3.48) is unbounded, thus obtaining a contradiction.

Exercise 3.6.5. Show that the germ at the origin in Example 3.5.23 is (hyperbolic and) non linearizable by the germ of any C^1 diffeomorphism.

As an application of Sternberg's theorem and Corollary 3.2.6, we now reproduce (with a much simpler proof than the original one) a result obtained by Ghys and Tsuboi in [97].

Theorem 3.6.6. *Let Φ_1 and Φ_2 be two representations of a group Γ in $\mathrm{Diff}_+^r(S^1)$, where $2 \leq r \leq \infty$. Suppose that their orbits are dense and that $\Phi_1(\Gamma)$ is not topologically conjugate to a group of rotations. If a C^1 circle diffeomorphism φ conjugates Φ_1 and Φ_2, then φ is a C^r diffeomorphism.*

Proof. By Corollary 3.2.6, there exist $g \in \Gamma$ and $x \in S^1$ such that $\Phi_1(g)(x) = x$ and $\Phi_1(g)'(x) \neq 1$. In local coordinates we obtain a hyperbolic germ of a diffeomorphism fixing the origin. The point $\varphi(x) \in S^1$ is fixed by $\Phi_2(g) = \varphi \circ \Phi_1(g) \circ \varphi^{-1} \in \Gamma_2$. Therefore, φ induces a conjugacy between hyperbolic germs. Corollary 3.6.3 then implies that φ is of class C^r in a neighborhood of x. Since $\varphi \circ \Phi_1(h) = \Phi_2(h) \circ \varphi$ holds for every $h \in \Gamma$, the set of points around which φ is of class C^r is invariant under $\Phi_1(\Gamma)$, and since the orbits under $\Phi_1(\Gamma)$ are dense, φ must be of class C^r on the whole circle. To show that φ^{-1} is of class C^r on S^1, it suffices to interchange the roles of Φ_1 and Φ_2. $\qquad\square$

Remark 3.6.7. The preceding theorem still holds for bi-Lipschitz conjugacies, provided that we assume a priori that the actions are ergodic (see Theorem 3.6.9). Moreover, the theorem is also true for conjugacies between groups of $C^{1+\tau}$ circle diffeomorphisms that are non-Abelian and act minimally.

Exercise 3.6.8. Prove that Theorem 3.6.6 still holds (in class C^2 or $C^{1+\tau}$) for groups admitting an exceptional minimal set and that are non-semiconjugate to groups of rotations.

Hint. Apply the same argument as before, bearing in mind that every orbit must accumulate on the minimal invariant Cantor set.

Remark. This result does not generalize to bi-Lipschitz conjugacies; see Theorem 3.6.14.

3.6.2 The case of bi-Lipschitz conjugacies

The main object of this section is to extend Theorem 3.6.6 to bi-Lipschitz conjugacies under the hypothesis of ergodicity for the action (see § 3.5.1 for a discussion of this hypothesis).

Theorem 3.6.9. *Let Φ_1 and Φ_2 be two representations of a finitely generated group Γ into $\mathrm{Diff}^r_+(S^1)$, where $2 \leq r \leq \infty$. Suppose that $\Phi_1(\Gamma)$ acts minimally and ergodically. If a bi-Lipschitz circle homeomorphism φ conjugates Φ_1 and Φ_2, then φ is a C^1 diffeomorphism. Moreover, if $\Phi_1(\Gamma)$ is not topologically conjugate to a group of rotations, then φ is a C^r diffeomorphism.*

The proof of this result uses an equivariant version of a classical lemma of cohomological flavor due to Gottschalk and Hedlund. Let M be a compact metric space and Γ a finitely generated group acting on it by homeomorphisms. A *cocycle* associated to this action (compare § 5.2.1) is a map $c : \Gamma \times M \to \mathbb{R}$ such that for each fixed $f \in \Gamma$, the map $x \mapsto c(f, x)$ is continuous, and such that for every f and g in Γ and every $x \in M$, one has

$$c(fg, x) = c(g, x) + c(f, g(x)). \tag{3.49}$$

With this notation, the equivariant Gottschalk-Hedlund lemma from [173] may be stated as follows:

Lemma 3.6.10. *Suppose that Γ is finitely generated and its action on M is minimal. Then the following are equivalent:*

(i) *There exist $x_0 \in M$ and a constant $C_0 > 0$ such that $|c(f, x_0)| \leq C_0$ for every $f \in \Gamma$.*

(ii) *There exists a continuous function $\phi : M \to \mathbb{R}$ such that $c(f, x) = \phi(f(x)) - \phi(x)$ for all $f \in \Gamma$ and all $x \in M$.*

Proof. If the second condition is satisfied, then (i) follows from

$$|c(f, x_0)| \leq |\phi(f(x_0))| + |\phi(x_0)| \leq 2\|\phi\|_{C^0}.$$

Conversely, let us suppose that the first condition holds. For each $f \in \Gamma$, consider the homeomorphism $\hat{f}$ of the space $M \times \mathbb{R}$ defined by $\hat{f}(x, t) = (f(x), t + c(f, x))$. It is easy to see that the cocycle relation (3.49) implies that this defines a group action of Γ on $M \times \mathbb{R}$ in the sense that $\hat{f}\hat{g} = \widehat{fg}$ for all f and g in Γ. Moreover, condition (i) implies that the orbit of the point $(x_0, 0)$ under this action is bounded; in particular, its closure is a (nonempty) compact invariant set. One may then apply Zorn's lemma to deduce the existence of a nonempty minimal invariant compact subset Λ of $M \times \mathbb{R}$. We claim that Λ is the graph of a continuous real-valued function on M.

First of all, since the action of Γ on M is minimal, the projection of Λ into M is the whole space. Moreover, if $(\bar{x}, t_1)$ and $(\bar{x}, t_2)$ belong to Λ for some $\bar{x} \in M$ and some $t_1 \neq t_2$, then this implies that $\Lambda \cap \Lambda_t \neq \emptyset$, where $t = t_2 - t_1$ and $\Lambda_t = \{(x, s + t) : (x, s) \in \Lambda\}$. Notice that the action of Γ on $M \times \mathbb{R}$ commutes with the map $(x, s) \mapsto (x, s + t)$; in particular, Λ_t is also invariant. Since Λ is minimal, this implies that $\Lambda = \Lambda_t$. One then concludes that $\Lambda = \Lambda_t = \Lambda_{2t} = \ldots$, which is impossible because Λ is compact and $t \neq 0$.

We have then proved that for every $x \in M$, the set Λ contains exactly one point of the form (x, t). Putting $\phi(x) = t$, one obtains a function from M into $\mathbb{R}$. This function must be continuous, since its graph (which coincides with Λ) is compact.

Finally, since the graph of ϕ is invariant under the action, for all $f \in \Gamma$ and all $x \in \mathrm{M}$, the point $\hat{f}(x, \phi(x)) = (f(x), \phi(x) + c(f, x))$ must be of the form $(f(x), \phi(f(x)))$, which implies that $c(f, x) = \phi(f(x)) - \phi(x)$. $\square$

The following corresponds to a "measurable version" of the preceding lemma.

Lemma 3.6.11.   *Let* M *be a compact metric space and* Γ *a finitely generated group acting on it by homeomorphisms. Suppose that the action of* Γ *on* M *is minimal and ergodic with respect to some probability measure* μ, *and let* c *be a cocycle associated to this action. If* ϕ *is a function in* $\mathcal{L}^{\infty}_{\mathbb{R}}(\mathrm{M}; \mu)$ *such that for all* $f \in \Gamma$ *and* μ-*almost every* $x \in \mathrm{M}$, *one has*

$$c(f, x) = \phi(f(x)) - \phi(x), \tag{3.50}$$

then there exists a continuous function $\tilde{\phi} : \mathrm{M} \to \mathbb{R}$ *that coincides* μ *almost everywhere with* ϕ *and such that for all* $f \in \Gamma$ *and all* $x \in \mathrm{M}$,

$$c(f, x) = \tilde{\phi}(f(x)) - \tilde{\phi}(x). \tag{3.51}$$

Proof.   Let M_0 be the set of points for which (3.50) does not hold for some $f \in \Gamma$. Since Γ is finitely generated, $\mu(\mathrm{M}_0) = 0$. Let M'_1 be the complementary set of the essential support of ϕ, and let $\mathrm{M}_1 = \cup_{f \in \Gamma} f(\mathrm{M}'_1)$. Take a point x_0 in the full-measure set $\mathrm{M} \setminus (\mathrm{M}_0 \cup \mathrm{M}_1)$. Equality (3.50) then gives $|c(f, x_0)| \le 2\|\phi\|_{\mathcal{L}^{\infty}}$ for all $f \in \Gamma$. By the preceding lemma, there exists a continuous function $\tilde{\phi} : \mathrm{M} \to \mathbb{R}$ such that (3.51) holds for *every* x. This implies that μ almost everywhere one has

$$\tilde{\phi} \circ f - \tilde{\phi} = \phi \circ f - \phi;$$

hence

$$\tilde{\phi} - \phi = (\tilde{\phi} - \phi) \circ f.$$

Since the action of Γ on M is assumed to be μ-ergodic, the difference $\tilde{\phi} - \phi$ must be μ-a.e. equal to a constant C. To conclude the proof, just replace $\tilde{\phi}$ by $\tilde{\phi} - C$. $\square$

We may now pass to the proof of Theorem 3.6.9. First, notice that if φ is a bi-Lipschitz circle homeomorphism conjugating the actions Φ_1 and

Φ_2 of our group Γ, then φ and φ^{-1} are almost everywhere differentiable, and their derivatives belong to $\mathcal{L}_{\mathbb{R}}^{\infty}(S^1, Leb)$. Therefore, the function $x \mapsto \log(\varphi'(x))$ is also in $\mathcal{L}_{\mathbb{R}}^{\infty}(S^1, Leb)$. The relation $\Phi_1(f) = \varphi^{-1} \circ \Phi_2(f) \circ \varphi$ gives almost everywhere the equality

$$\log(\Phi_1(f)'(x)) = \log(\varphi'(x)) - \log(\varphi'(\Phi_1(f)(x))) + \log(\Phi_2(f)'(\varphi(x))).$$

If we let $\phi = -\log(\varphi')$ and $c(f, x) = \log(\Phi_1(f)'(x)) - \log(\Phi_2(f)'(\varphi(x)))$, this yields, for all $f \in \Gamma$ and almost every $x \in S^1$,

$$c(f, x) = \phi(\Phi_1(f)(x)) - \phi(x).$$

Using the relation $\Phi_2(g) = \varphi \circ \Phi_1(g) \circ \varphi^{-1}$, one easily checks the cocycle identity

$$c(fg, x) = c(g, x) + c(f, \Phi_1(g)(x)).$$

Since the action Φ_1 is supposed to be ergodic, Lemma 3.6.11 ensures the existence of a continuous function $\tilde{\phi}$ that coincides almost everywhere with ϕ and such that (3.51) holds for every x and all f. By integration, one concludes that the derivative of φ is well defined everywhere and coincides with $\exp(-\tilde{\phi})$. In particular, φ is of class C^1. Exchanging the roles of Φ_1 and Φ_2, one concludes that φ is a C^1 diffeomorphism. Finally, to show that φ is a C^r diffeomorphism in the case where the actions are non-conjugate to actions by rotations, it suffices to apply Theorem 3.6.6. $\square$

Exercise 3.6.12. Let Γ be a group of C^1 circle diffeomorphisms and φ a bi-Lipschitz circle homeomorphism. Suppose that for every $f \in \Gamma$ and almost every $x \in S^1$, one has

$$\log(f'(x)) - \log(f'(\varphi(x))) = \log(\varphi'(x)) - \log(\varphi'(f(x))).$$

Show that φ *centralizes* Γ (that is, it commutes with every element of Γ).

Exercise 3.6.13. Given a circle diffeomorphism f, let F be the diffeomorphism of $\mathbb{R}^2 \setminus \{O\}$ defined in polar coordinates by

$$F(r, \theta) = \left(\frac{r}{\sqrt{f'(\theta)}}, f(\theta) \right).$$

(i) Show that if f is of class C^2 and its rotation number is irrational, then f is C^1 conjugate to $R_{\rho(f)}$ if and only if there exist positive r_1 and r_2 such that $F^n(B(O, r_1)) \subset B(O, r_2)$ for all $n \geq 0$.

(ii) Show that F preserves the Lebesgue measure, and that for every θ, the parallelogram generated by the vectors $F(1, \theta)$ and $\frac{d}{d\theta}F(1, \theta)$ has area 1. Conclude that there exists a real-valued function ψ defined on the circle such that for every $\theta \in S^1$,

$$F(1, \theta) + \psi \frac{d^2}{d\theta^2} F(1, \theta) = 0.$$

(iii) Endow the circle with the canonical projective structure (cf. Exercise 1.3.4), and identify each angle $\theta \in [0, 2\pi]$ with the corresponding point in the projective space (so that θ and $\theta + \pi$ identify with the same point for $\theta \in [0, \pi]$). Show that in the corresponding projective coordinates, the value of the function ψ in (ii) is given by

$$\psi = \frac{S(f)}{2} - 1.$$

Remark. This corresponds to the starting point of a proof using Sturm-Liouville theory of a beautiful theorem due to Ghys: for every diffeomorphism of the circle (endowed with the canonical projective structure), there exist at least four points at which the Schwarzian derivative vanishes (see [238]).

Remark that a conjugacy of an action to itself corresponds to a homeomorphism that centralizes the action. Moreover, if Φ_1 and Φ_2 are actions by C^r circle diffeomorphisms that are conjugate by some C^r diffeomorphism φ_0, and if φ is another bi-Lipschitz homeomorphism that conjugates them, then the (bi-Lipschitz) homeomorphism $\varphi_0^{-1}\varphi$ centralizes the action Φ_1. This is the reason that it is so important to deal with the study of centralizers before studying the general problem of conjugacies (see, however, Exercise 3.6.2). In this regard, one may show that Theorem 3.6.9 is far from being true for nonminimal actions.

Theorem 3.6.14. *Let Γ be a finitely generated group of $C^{1+\text{Lip}}$ circle diffeomorphisms whose action is not minimal. If the restrictions of the stabilizers of intervals are either trivial or infinite cyclic, then there exist bi-Lipschitz homeomorphisms that are not C^1 and that commute with all the elements in Γ. Moreover, these homeomorphisms may be chosen to be nondifferentiable on every nonempty open interval in S^1.*

This theorem is based on a very simple construction related to the techniques of the next chapter, and therefore we postpone its proof to § 4.1.1. For the moment, let us remark that the hypothesis about the stabilizers is not very strong. For instance, it is always satisfied in the real-analytic case (this result is due to Hector; a complete proof may be found in the appendix of [182]). Obviously, it is also satisfied by many other interesting non-real-analytic actions, for instance, those of Thompson's group G. Nevertheless, without this hypothesis, bi-Lipschitz conjugacies are forced to be smooth in many cases.

Exercise 3.6.15. Give examples of finitely generated groups of C^∞ circle diffeomorphisms that are conjugate by some bi-Lipschitz homeomorphism although there is no smooth conjugacy between them.

Remark. These examples may be constructed having either finite orbits or an exceptional minimal set. However, we do not know whether there exist groups of real-analytic diffeomorphisms having the desired property.

Structure and Rigidity via Dynamical Methods

4.1 Abelian Groups of Diffeomorphisms

4.1.1 Kopell's lemma

For a group of homeomorphisms of an interval, the circle, or the real line, we will say that $]a, b[$ is an ***irreducible component*** of the action if it is invariant and does not contain strictly any invariant interval. Let us denote by $\text{Diff}_+^{1+\text{bv}}([0, 1[)$ the group of $C^{1+\text{bv}}$ diffeomorphisms of $[0, 1[$, that is, the group of C^1 diffeomorphisms f of $[0, 1[$ such that the total variation of the logarithm of the derivative is finite on each compact interval $[a, b] \subset [0, 1[$. Recall that this variation is denoted by $V(f; [a, b])$, that is,

$$V(f; [a, b]) = \sup_{a=a_0 < a_1 < \cdots < a_n = b} \sum_{i=1}^{n} |\log(f')(a_i) - \log(f')(a_{i-1})|.$$

The group $\text{Diff}_+^{1+\text{bv}}(]0, 1])$ is defined in a similar way. Notice that every element f in $\text{Diff}_+^2([0, 1[)$ belongs to $\text{Diff}_+^{1+\text{bv}}([0, 1[)$. Indeed, for $0 \le a < b < 1$, one has

$$V(f; [a, b]) = \int_a^b |(\log(f'))'(s)| ds = \int_a^b \left| \frac{f''(s)}{f'(s)} \right| ds.$$

The following important result is stated as a theorem, although it is widely known (and we will refer to it) as Kopell's lemma, since it corresponds to the first lemma in the thesis of Kopell [144]. Notwithstanding, we provide a much simpler proof than the original one.

Theorem 4.1.1. *Let f and g be commuting diffeomorphisms of the interval $[0, 1[$ or $]0, 1]$. Suppose that f is of class C^{1+bv} and g of class C^1. If f has no fixed point in $]0, 1[$ and g has at least one fixed point in $]0, 1[$, then g is the identity.*

Proof. We will give the proof only for the case of the interval $[0, 1[$, since the case of $]0, 1]$ is analogous. Replacing f by f^{-1} if necessary, we may suppose that $f(x) < x$ for every $x \in]0, 1[$. Let $b \in]0, 1[$ be a fixed point of g. Let $a = f(b)$, and for every $n \in \mathbb{Z}$, let $b_n = f^n(b)$. Since g commutes with f, it must fix all the intervals $[b_{n+1}, b_n]$, and hence

$$g'(0) = \lim_{n \to \infty} \frac{g(b_n) - g(0)}{b_n - 0} = 1.$$

Let $\delta = V(f; [0, b])$. If u and v belong to $[a, b]$, then

$$\left| \log \left(\frac{(f^n)'(v)}{(f^n)'(u)} \right) \right| \le \sum_{i=1}^{n} |\log(f')(f^{i-1}(v)) - \log(f')(f^{i-1}(u))| \le \delta.$$

Let $u = x \in [a, b]$ and $v = f^{-n}gf^n(x) = g(x) \in [a, b]$. If we use the equality

$$g'(x) = \frac{(f^n)'(x)}{(f^n)'(f^{-n}gf^n(x))} g'(f^n(x)) = \frac{(f^n)'(x)}{(f^n)'(g(x))} g'(f^n(x))$$

and pass to the limit as n goes to infinity, it follows that $\sup_{x \in [a,b]} g'(x) \le e^\delta$. Now remark that this remains true if we replace g by g^j for any $j \in \mathbb{N}$ (this is due to the fact that the constant δ depends only on f). Therefore,

$$\sup_{x \in [a,b]} (g^j)'(x) \le e^\delta.$$

Since g fixes a and b, this is not possible unless the restriction of g to $[a, b]$ is the identity. Finally, since f and g commute, this implies that g is the identity on the whole interval $[0, 1[$. □

The preceding theorem allows us to conclude that for every $f \in$ $\mathrm{Diff}_+^{1+bv}([0, 1[)$ without fixed points in $]0, 1[$, its centralizer in $\mathrm{Diff}_+^1([0, 1[)$ acts freely on $]0, 1[$. Hölder's theorem then implies that this centralizer is semiconjugate to a group of translations. In fact, if this group of translations is dense, then the semiconjugacy is actually a conjugacy, as is stated in the next proposition.

Proposition 4.1.2. *Let Γ be a subgroup of $\mathrm{Diff}^1_+([0, 1[)$ that is semi-conjugate to a dense subgroup of the group of translations. If Γ contains an element of class C^{1+bv} without fixed points in $]0, 1[$, then the semiconjugacy is a topological conjugacy.*

Proof. Suppose that Γ is a subgroup of $\mathrm{Diff}^1_+([0, 1[)$ that is semiconjugate to a dense subgroup of the group of translations without being conjugate to it. Let $f \in \Gamma$ be the element given by the hypothesis. Without loss of generality, we may suppose that f is sent to the translation $T_{-1} : x \mapsto x - 1$ by the homomorphism induced by the semiconjugacy. In particular, one has $f(x) < x$ for every $x \in]0, 1[$. Choose an interval $[a, b]$ not reduced to a point that is sent to a single point by the semiconjugacy, and that is maximal for this property. By the choice of $[a, b]$, there exists an increasing sequence (n_i) of positive integers such that for every $i \in \mathbb{N}$, there exists $\bar{f}_i \in \Gamma$ satisfying, for every $n \in \mathbb{N}$,

$$\bar{f}_i^{n_i}(f^n(a)) \geq f^{n+1}(a), \qquad \bar{f}_i^{n_i+1}(f^n(a)) < f^{n+1}(a),$$
$$\bar{f}_i^{n_i}(f^n(b)) \geq f^{n+1}(b), \qquad \bar{f}_i^{n_i+1}(f^n(b)) < f^{n+1}(b).$$

Let $a_n = f^n(a)$ and $b_n = f^n(b)$. Passing to the limit as n goes to infinity in the inequalities

$$\frac{\bar{f}_i^{n_i+1}(a_n)}{a_n} < \frac{f(a_n)}{a_n} \leq \frac{\bar{f}_i^{n_i}(a_n)}{a_n},$$

we obtain

$$(\bar{f}_i'(0))^{n_i+1} \leq f'(0) \leq (\bar{f}_i'(0))^{n_i}. \tag{4.1}$$

For $n \geq 0$, the intervals $f^n(]\bar{f}_i(a), b[)$ are two-by-two disjoint. If we let $\delta = V(f; [0, b])$, then for every u and v in $]\bar{f}_i(a), b[$, one has

$$\left| \log\left(\frac{(f^n)'(v)}{(f^n)'(u)} \right) \right| \leq \sum_{i=1}^{n} |\log(f')(f^{i-1}(v)) - \log(f')(f^{i-1}(u))| \leq \delta. \tag{4.2}$$

Passing to the limit as n goes to infinity in the inequality

$$|\bar{f}_i([a, b])| = |f^{-n} \bar{f}_i f^n([a, b])|$$
$$\geq \frac{\inf_{u \in [\bar{f}_i(a), b]}(f^n)'(u)}{\sup_{v \in [\bar{f}_i(a), b]}(f^n)'(v)} \cdot \inf_{x \in [f^n(a), f^n(b)]} \bar{f}_i'(x) \cdot |[a, b]|,$$

and using the estimates (4.1) and (4.2), we obtain, for some positive constant C and every $i \in \mathbb{N}$,

$$|\bar{f}_i([a, b])| \geq e^{-\delta}(f'(0))^{1/n_i} \cdot |[a, b]| \geq C.$$

However, this inequality is absurd, since $|\bar{f}_i([a, b])|$ obviously converges to zero as i goes to infinity. □

Exercise 4.1.3. Using Denjoy's theorem, prove that Proposition 4.1.2 holds for every finitely generated subgroup of $\mathrm{Diff}_+^{1+\mathrm{bv}}(]0, 1[)$ whose center contains an element without fixed points (compare Example 3.1.7).

From Proposition 4.1.2 one immediately deduces the following:

Corollary 4.1.4. *If f is an element in $\mathrm{Diff}_+^{1+\mathrm{bv}}([0, 1[)$ without fixed points in $]0, 1[$, then its centralizer in $\mathrm{Diff}_+^1([0, 1[)$ is topologically conjugate to a group of translations of the line.*

Exercise 4.1.5. Prove the "strong version" of Kopell's lemma (in class $C^{1+\mathrm{Lip}}$) given by the following proposition (see either [51] or [68] in case of problems with this).

Proposition 4.1.6. *Let $f : [0, 1[\to [0, 1[$ be a $C^{1+\mathrm{Lip}}$ diffeomorphism such that $f(x) < x$ for all $x \in]0, 1[$. Fix a point $a \in]0, 1[$, and for each $n \in \mathbb{N}$, let $g_n : [f(a), a] \to [f(a), a]$ be a diffeomorphism tangent to the identity at the endpoints. If $g :]0, 1] \to]0, 1]$ is such that its restriction to $[f^{n+1}(a), f^n(a)]$ coincides with $f^n g_n f^{-n}$ for every $n \in \mathbb{N}$, then g extends to a C^1 diffeomorphism of $[0, 1[$ if and only if (g_n) converges to the identity in the C^1 topology.*

Exercise 4.1.7. Prove the following "real-analytic version" of Kopell's lemma (due to Hector): if f and g are real-analytic diffeomorphisms that may be written in the form $f(x) = x + a_i x^i + \cdots$ and $g(x) = x + b_j x^j + \cdots$ for $|x| \leq \varepsilon$, with $j > i$ and $f(x) < x$ for every small positive x, then the sequence of maps $f^{-n} g f^n$ converges uniformly to the identity on an interval $[0, \varepsilon']$ (see [182] for an application of this claim).

To close this section, we give a proof of Theorem 3.6.14. Let us begin by considering a diffeomorphism f of class $C^{1+\mathrm{bv}}$ of an interval $I = [a, b]$ such that $f(x) < x$ for every $x \in]a, b[$. Fix an arbitrary point $c \in]a, b[$, and consider any bi-Lipschitz homeomorphism h from $[f(c), c]$ into itself. Extending h to $]a, b[$ so that it commutes with f, and then putting $h(a) = a$ and $h(b) = b$, we obtain a well-defined homeomorphism of $[a, b]$ (which we still denote by h). We claim that if C is a bi-Lipschitz constant for h in $[f(c), c]$, then Ce^V is a bi-Lipschitz constant for h on $[a, b]$, where $V = V(f; [a, b])$. Indeed, let us consider, for instance, a point $x \in f^n([f(c), c])$ for some $n \geq 0$, and such that h is differentiable on $f^{-n}(x) \in [f(c), c]$, with derivative less than or equal to C (notice that this is the case for almost every point $x \in [f^{n+1}(c), f^n(c)])$. From the relation $h = f^n h f^{-n}$ we obtain

$$h'(x) = h'(f^{-n}(x)) \cdot \frac{(f^n)'(hf^{-n}(x))}{(f^n)'(f^{-n})(x)} \leq C \cdot \frac{(f^n)'(hf^{-n}(x))}{(f^n)'(f^{-n})(x)}. \qquad (4.3)$$

Letting $y = f^{-n}(x) \in [f(c), c]$ and $z = h(y) \in [f(c), c]$, and arguing as in the proof of Kopell's lemma, we obtain

$$\left| \log \left(\frac{(f^n)'(z)}{(f^n)'(y)} \right) \right| = \left| \log \left(\frac{\prod_{i=0}^{n-1} f'(f^i(z))}{\prod_{i=0}^{n-1} f'(f^i(y))} \right) \right|$$
$$\leq \sum_{i=0}^{n-1} |\log(f'(f^i(z))) - \log(f'(f^i(y)))| \leq V.$$

Introducing this inequality in (4.3), we deduce that $h'(x) \leq Ce^V$. Since x was a generic point, this shows that the Lipschitz constant of h is bounded from above by Ce^V. Obviously, a similar argument shows that this bound also holds for the Lipschitz constant of h^{-1}.

For the proof of Theorem 3.6.14 we will use a similar construction. To simplify, we will prove only the first claim of the theorem, leaving to the reader the task of proving the second claim concerning the existence of bi-Lipschitz centralizers that are nondifferentiable on any open set.

Let us begin by recalling that if Γ is a group of $C^{1+\text{Lip}}$ circle diffeomorphisms preserving an exceptional minimal set, then the stabilizer of every connected component $]a, b[$ of the complement of this set is nontrivial (cf. Exercise 3.2.7). By the hypothesis of the theorem, the restriction to $]a, b[$ of this stabilizer is either trivial or infinite cyclic. In the first case, we define h as being any bi-Lipschitz and nondifferentiable homeomorphism of $[a, b]$. In the second case, let $f \in \Gamma$ be such that its restriction to $]a, b[$ generates the restriction of the corresponding stabilizer. Let $[\bar{a}, \bar{b}] \subset [a, b]$ be such that $f(x) \neq x$ for every $x \in]\bar{a}, \bar{b}[$, and such that $f(\bar{a}) = \bar{a}$ and $f(\bar{b}) = \bar{b}$. Replacing f by f^{-1} if necessary, we may assume that $f^n(x)$ converges to $\bar{a}$ as n goes to infinity for every $x \in [\bar{a}, \bar{b}]$. Arguing as in the case of the interval, fix a point $\bar{c} \in]\bar{a}, \bar{b}[$, and consider any bi-Lipschitz and non-differentiable homeomorphism h of $[f(\bar{c}), \bar{c}]$. This homeomorphism extends in a unique way to $[a, b]$ so that it commutes with the restriction of f to $[\bar{a}, \bar{b}]$ and coincides with the identity on $I \setminus [\bar{a}, \bar{b}]$.

Because of the hypothesis about the stabilizers, it is not difficult to check that there exists a unique extension of h to a circle homeomorphism (which we still denote by h) that commutes with (every element of) Γ and coincides with the identity on the complement of $\cup_{g \in \Gamma} g(]a, b[)$. We claim that this extension is bi-Lipschitz. More precisely, fixing a finite system of generators $\mathcal{G} = \{g_1, \ldots, g_k\}$ for Γ, letting V be the maximum among the total variation of the logarithm of the derivatives of the g_i's, and denoting by C the bi-Lipschitz constant of h on $[a, b]$, we claim that the globally defined homeomorphism h has a bi-Lipschitz constant less than or equal to Ce^{kV}. The proof of this claim is similar to that of the case of the interval. Let us fix, for instance, $x \in \cup_{g \in \Gamma}(g(I) \setminus I)$, and let us try to estimate the value of $h'(x)$. To do this, let us consider the smallest $n \in \mathbb{N}$ for which there exists an element $g = g_{i_n} \ldots g_{i_1} \in \Gamma$, with g_{i_j} in $\mathcal{G}$, such that $g(x) \in I$. The minimality of n implies that the intervals $I, g_{i_n}^{-1}(I), g_{i_{n-1}}^{-1}g_{i_n}^{-1}(I), \ldots, g_{i_1}^{-1} \cdots g_{i_n}^{-1}(I)$ have disjoint interior. Using the relation $h = g^{-1}hg$, we obtain, for a generic $x \in g^{-1}(I)$,

$$h'(x) = h'(g(x)) \cdot \frac{g'(x)}{g'(h(x))} \leq C \cdot \frac{g'(x)}{g'(y)}, \tag{4.4}$$

where $y = h(x) \in g^{-1}(I)$. Now, using the fact that the total variation of the logarithm of the derivative of each g_i is bounded from above by V, we obtain

$$\left| \log\left(\frac{g'(x)}{g'(y)}\right) \right| \leq \sum_{j=0}^{n-1} |\log(g_{i_{j+1}}'(g_{i_j} \cdots g_{i_1})(x)) - \log(g_{i_{j+1}}'(g_{i_j} \cdots g_{i_1})(y))|$$

$$\leq \sum_{i=1}^{k} V(\log(g_i'); S^1) \leq kV.$$

From (4.4) we conclude that $h'(x) \leq Ce^{kV}$, as we wanted to check.

Finally, let us consider the case where Γ admits finite orbits. If Γ is finite, then we consider any bi-Lipschitz and nondifferentiable diffeomorphism that commutes with its generator. If Γ is infinite, Hölder's theorem implies that the action cannot be free. Let I be a connected component of the complementary set of the union of the finite orbits, and let $f \in \Gamma$ be such that its restriction to I generates the restriction of the corresponding stabilizer. If we proceed as before with I and f, we may construct a bi-Lipschitz and nondifferentiable homeomorphism centralizing Γ. We leave the details to the reader.

4.1.2 Classifying Abelian group actions in class C^2

We may now give a precise description of the Abelian groups of C^{1+bv} diffeomorphisms of one-dimensional manifolds. The case of the interval is quite simple. Indeed, by Corollary 4.1.4, the restriction of such a group to each irreducible component is conjugate to a group of translations. The case of the circle is slightly more complicated. We begin with a lemma that is interesting in itself (compare § 2.3.1 and Exercise 3.1.6).

Lemma 4.1.8. *If Γ is an amenable subgroup of* Homeo$_+(\mathrm{S}^1)$*, then Γ either is semiconjugate to a group of rotations or contains a finite-index subgroup having fixed points.*

Proof. Since Γ is amenable, it must preserve a probability measure μ on S^1. If the orbits under Γ are dense, then μ has total support, and by reparameterizing the circle one easily checks that Γ is topologically conjugate to a group of rotations. If there exists a minimal invariant Cantor set, then the support of μ coincides with this set, and this allows semiconjugating Γ to a group of rotations. Finally, if there is a finite orbit, then the elements in Γ preserve the cyclic order of the points in this orbit. The stabilizer of these points is then a finite-index subgroup of Γ having fixed points. $\qquad\square$

From the preceding lemma (and its proof) it follows that if Γ is an Abelian subgroup of $\mathrm{Diff}_+^{1+bv}(\mathrm{S}^1)$, then it is either semiconjugate to a group of rotations or a finite central extension of an at-most-countable product of Abelian groups acting on disjoint intervals. By Corollary 4.1.4, the factors of this product are conjugate to groups of translations on each irreducible component. Finally, recall that for finitely generated subgroups of $\mathrm{Diff}_+^{1+bv}(\mathrm{S}^1)$, every semiconjugacy to a group of rotations is necessarily a conjugacy; however, this is no longer true for non–finitely generated groups (cf. Example 3.1.7).

Exercise 4.1.9. Prove that every virtually Abelian group of C^{1+bv} diffeomorphisms of the interval or the circle is actually Abelian.

Exercise 4.1.10. Prove that every Abelian subgroup of $\mathrm{Diff}_+^{1+\mathrm{bv}}(\mathbb{R})$ preserves a Radon measure on the line (compare Proposition 2.2.48).

4.1.3 Szekeres's theorem

In class C^2, the homomorphism given by Corollary 4.1.4 is necessarily surjective. This follows from the result below, due to Szekeres [237] (see also [256]). To state it properly, given a nonempty nondegenerate interval $[a, b[$ and $2 \leq r \leq \infty$, let us denote by $\mathrm{Diff}_+^{r,\Delta}([a, b[)$ the subset of $\mathrm{Diff}_+^r([a, b[)$ formed by the elements f such that $f(x) \neq x$ for all $x \in]a, b[$. For simplicity, we let $\infty - 1 = \infty$.

Theorem 4.1.11. *For every $f \in \mathrm{Diff}_+^{r,\Delta}([0, 1[)$, there exists a unique vector field X_f on $[0, 1[$ without singularities in $]0, 1[$ satisfying the following conditions:*

  (i) *X_f is of class C^{r-1} on $]0, 1[$ and of class C^1 on $[0, 1[$.*
  (ii) *If $f^{\mathbb{R}} = \{f^t : t \in \mathbb{R}\}$ is the flow associated with this vector field, then $f^1 = f$.*
  (iii) *The centralizer of f in $\mathrm{Diff}_+^1([0, 1[)$ coincides with $f^{\mathbb{R}}$.*

This theorem is particularly interesting if the germ of f at the origin is nonhyperbolic, since otherwise we may use Sternberg's linearization theorem. Actually, in the hyperbolic case, the claim before Example 3.6.4 allows extending Theorem 4.1.11 to class $C^{1+\tau}$ for every $\tau > 0$.

Exercise 4.1.12. Given $\lambda < 0$, consider a C^1 vector field $X = \varrho \frac{\partial}{\partial x}$ on $[0, 1]$ such that for every small-enough x, one has $\varrho(x) = \lambda x(1 - 1/\log(x))$. Prove that X is not C^1 linearizable, that is, there is no C^1 diffeomorphism conjugating X to its linear part $\lambda x \frac{\partial}{\partial x}$ (compare Example 3.6.4).

Hint. Show that if f is the time 1 of the flow associated to X, then

$$f(x) = e^{\lambda} x \left(\frac{1 - \log(x)}{1 - \log(f(x))} \right)$$

for all x near the origin. Using this, show more generally that f is not conjugate to its linear part by any bi-Lipschitz homeomorphism.

We will give the proof of Theorem 4.1.11 only for the case $r = 2$ (the extension to larger r is straightforward). First, notice that we may assume that

$f(x) > x$ for all $x \in]0, 1[$. Indeed, the vector field associated to a diffeomorphism that is topologically contracting at the origin may be obtained by changing the sign of the vector field associated to its inverse.

Lemma 4.1.13. *Let* $f : [0, 1[\to [0, 1[$ *be a* C^2 *diffeomorphism such that* $f(x) > x$ *for every* $x \in]0, 1[$. *If* $X(x) = \varrho(x)\frac{\partial}{\partial x}$ *is a* C^1 *vector field on* $[0, 1[$ *and* $a \in]0, 1[$, *then X is associated to f if and only if the following conditions are satisfied:*

(i) *The function* ϱ *is strictly positive on* $]0, 1[$.
(ii) *For every* $x \in [0, 1[$, *one has* $\varrho(f(x)) = f'(x)\varrho(x)$.
(iii) *One has* $\int_a^{f(a)} \frac{ds}{\varrho(s)} = 1$.

Proof. The first two conditions are clearly necessary. Concerning the third one, notice that if X is associated to f, then for every $t \geq 0$, we have

$$\frac{d}{dt} \int_a^{f^t(a)} \frac{ds}{\varrho(s)} = \frac{df^t(a)}{dt} \cdot \frac{1}{\varrho(f^t(a))} = 1.$$

Thus

$$\int_a^{f^t(a)} \frac{ds}{\varrho(s)} = t,$$

which implies (iii) by letting $t = 1$. Conversely, if condition (ii) is satisfied, then one easily checks that for every $x \in]0, 1[$, the derivative of the function $x \mapsto \int_x^{f(x)} \frac{ds}{\varrho(s)}$ is identically zero. Hence, for every $x \in]0, 1[$,

$$\int_x^{f(x)} \frac{ds}{\varrho(s)} = \int_a^{f(a)} \frac{ds}{\varrho(s)} = 1.$$

Therefore, if we denote by $\hat{f}$ the diffeomorphism obtained by integrating X up to time 1, then $\hat{f}(0) = f(0) = 0$, and for all $x \in]0, 1[$, one has

$$1 = \int_x^{\hat{f}(x)} \frac{ds}{\varrho(s)} = \int_x^{f(x)} \frac{ds}{\varrho(s)}.$$

From this one easily concludes that $\hat{f} = f$, that is, X is associated to f. $\square$

To construct the vector field X, we begin by considering the "discrete difference" $\Delta = \Delta_f$ defined by $\Delta(x) = f(x) - x$. Although $\Delta(f(x))$ and $\Delta(x)f'(x)$ do not coincide, the "error" has a nice expression. Indeed, if we define

$$\Theta(x) = \Theta_f(x) = \log(f'(x)) - \log\left[\int_0^1 f'(x + t\Delta(x))dt\right], \qquad (4.5)$$

then from the equality

$$f^2(x) = f(x) + \Delta(x)\int_0^1 f'(x + t\Delta(x))dt$$

it follows that

$$\Delta(f(x))\exp(\Theta(x)) = \Delta(x)f'(x). \qquad (4.6)$$

For later estimates we will further use the second-order Taylor series expansion

$$f^2(x) = f(x) + \Delta(x)f'(x) + \Delta(x)^2\int_0^1 (1-t)f''(x + t\Delta(x))dt,$$

which allows checking the equality

$$\Theta(x) = \log\left(1 - \frac{\Delta(x)^2}{\Delta(f(x))}\int_0^1 (1-t)f''(x + t\Delta(x))dt\right). \qquad (4.7)$$

The function ϱ corresponding to X will be obtained by multiplying Δ by the sum of the successive errors under the iteration, modulo some normalization. To be more concrete, let Σ be the function formally defined by $\Sigma(0) = 0$ and, for $x \in]0, 1[$,

$$\Sigma(x) = \sum_{n>0}\Theta(f^{-n}(x)).$$

This function satisfies the formal equality

$$\Sigma(f(x)) = \Sigma(x) + \Theta(x). \qquad (4.8)$$

Therefore, if we define the field $Y(x) = \Delta(x)\exp(\Sigma(x))\frac{\partial}{\partial x}$, from (4.6) and (4.8) we conclude that

$$\begin{aligned}
Y(f(x)) &= \Delta(f(x))\exp(\Sigma(f(x)))\frac{\partial}{\partial x} \\
&= \frac{\Delta(x)f'(x)}{\exp(\Theta(x))}\exp(\Theta(x) + \Sigma(x))\frac{\partial}{\partial x} = f'(x)Y(x).
\end{aligned}$$

The vector field X will then be of the form $X = cY$ for some normalizing constant $c = c(f)$ for which

$$\int_a^{f(a)}\frac{ds}{X(s)} = 1.$$

Remark that the condition imposed by this equality corresponds to

$$c = \int_a^{f(a)} \frac{dx}{\Delta(x)\exp(\Sigma(x))}. \tag{4.9}$$

Suppose that $f'(0) > 1$, and let $\lambda = f'(0)$. For $a \sim 0$, one has $\Sigma(x) \sim 0$ for $x \le f(a)$, and

$$\Delta(x) = x\left(\frac{f(x)}{x} - 1\right) \sim x(\lambda - 1).$$

Hence, according to (4.9), we must have

$$c = \int_a^{f(a)} \frac{dx}{\Delta(x)\exp(\Sigma(x))} \sim \int_a^{f(a)} \frac{dx}{x(\lambda - 1)} = \frac{\log(f(a)/a)}{\lambda - 1} \sim \frac{\log(\lambda)}{\lambda - 1},$$

and therefore the appropriate choice is $c(f) = \log(\lambda)/(\lambda - 1)$. When λ tends to 1, this expression converges to 1, and thus in the case $f'(0) = 1$ the normalizing constant to be considered should be $c(f) = 1$. We will now verify that the definition we have just sketched is pertinent and provides a vector field associated to f.

Proposition 4.1.14. *Let f be a diffeomorphism satisfying the hypothesis of Lemma 4.1.13. If we define $X = \varrho\frac{\partial}{\partial x}$ by $\varrho(x) = c(f)\Delta(x)\exp(\Sigma(x))$, then X is a C^1 vector field associated to f.*

Proof. We need to show three claims: the vector field X is well defined (in the sense that the series defining the function Σ converges), it is of class C^1 on $[0, 1[$, and f is the time 1 of the flow associated to it.

First step. The convergence of $\Sigma(x)$.

Since f is of class C^1, for each $x \in]0, 1[$, there exists $y = y(x) \in [x, f(x)]$ such that

$$\int_0^1 f'(x + t\Delta(x))dt = f'(y).$$

From this equality and (4.5) one deduces that

$$|\Theta(x)| = |\log(f'(x)) - \log(f'(y))| \le C|y - x| \le C\Delta(x),$$

where C is the Lipschitz constant of the function $\log(f')$ on $[x, f(x)]$ (which equals the supremum of the function $|f''|/|f'|$ on this interval).

One then concludes that the series corresponding to $\Sigma(x)$ converges for every $x \in\ ]0, 1[$. Moreover, $|\Sigma(x)| \leq Cx$, which implies that the function Σ extends continuously to $[0, 1[$ by letting $\Sigma(0) = 0$.

Second step. The differentiability of ϱ.

To show that ϱ is differentiable at the origin, it suffices to notice that

$$\lim_{t \to 0} \frac{\varrho(t)}{t} = c \lim_{t \to 0} \exp(\Sigma(t)) \lim_{t \to 0} \frac{\Delta(t)}{t} = c\Delta'(0) = \log(\lambda).$$

To check the differentiability at the interior, we will use an indirect argument. If we denote by L_X the Lie derivative along the vector field X, from the relation $\varrho(f(x)) = \varrho(x)f'(x)$ it follows that $L_X(\Theta \circ f^{-1}) = (L_X\Theta) \circ f^{-1}$. Now from (4.7) it is easy to conclude that for some constant $C > 0$,

$$\varrho(x)\Theta'(x) \leq C\Delta(x).$$

Therefore, the series $\sum_{n>0}(L_X\Theta) \circ f^{-n}$ converges, and its value equals $L_X(\Sigma)(x)$. This shows that Σ, and hence X, are of class C^1 on $]0, 1[$. Moreover, for every $x \in\ ]0, 1[$,

$$\varrho'(x) = \varrho(x)\frac{\Delta'(x)}{\Delta(x)} + \sum_{n>0} L_X(\varrho) \circ f^{-n}(x).$$

Letting x go to the origin, one readily concludes from this relation that $\varrho'(x)$ converges to $c\Delta'(0) = \log(\lambda)$, which shows that X is of class C^1 on the whole interval $[0, 1[$.

Third step. The time 1 of the flow.

We need to check that condition (iii) holds for every $x \in\ ]0, 1[$. Now, as we have already seen in the proof of Lemma 4.1.13, the equality $\varrho(f(x)) = \varrho(x)f'(x)$ implies that the function $x \mapsto \int_x^{f(x)} \frac{ds}{\varrho(s)}$ is constant.

Suppose first that f is tangent to the identity at the origin, that is, $\lambda = 1$. In this case, one has $\Delta'(0) = 0$. Hence for every $\varepsilon > 0$, there exists $\delta > 0$ such that if $a < \delta$ and $t \in [a, f(a)]$, then

$$|\Delta(t) - \Delta(a)| < \varepsilon(t - a), \qquad 1 - \varepsilon < \frac{1}{\exp(\Sigma(t))} < 1 + \varepsilon.$$

The first of these inequalities implies that $|\Delta(t) - (f(a) - a)| < \varepsilon(t - a)$, whereas the second one gives

$$\left| \frac{\Delta(t)}{\varrho(t)} - 1 \right| < \varepsilon.$$

One then obtains, for $a < \delta$,

$$\int_a^{f(a)} \frac{du}{\varrho(u)} > (1 - \varepsilon) \int_a^{f(a)} \frac{du}{\Delta(u)} > \frac{1 - \varepsilon}{1 + \varepsilon},$$

$$\int_a^{f(a)} \frac{du}{\varrho(u)} < (1 + \varepsilon) \int_a^{f(a)} \frac{du}{\Delta(u)} < \frac{1 + \varepsilon}{1 - \varepsilon}.$$

Nevertheless, since the function $x \mapsto \int_x^{f(x)} \frac{ds}{\varrho(s)}$ is constant, letting a go to the origin as ε goes to zero one deduces that this constant equals 1.

Suppose now that $\lambda > 1$. In this case, $\Delta'(0) = c - 1$, and hence for small t one has

$$|\Delta(t) - t(c - 1)| < \varepsilon t, \qquad \left| \frac{\lambda \Delta(t)}{\varrho(f)(t)} - 1 \right| < \varepsilon.$$

From this it follows that

$$\int_a^{f(a)} \frac{ds}{\varrho(s)}$$

$$< (1 + \varepsilon) \int_a^{f(a)} \frac{ds}{\lambda \Delta(s)} = \frac{1 + \varepsilon}{\lambda} \int_a^{f(a)} \frac{ds}{\Delta(s)} < \frac{1 + \varepsilon}{\lambda} \int_a^{f(a)} \frac{ds}{(c - 1 - \varepsilon)s}$$

$$= \frac{1 + \varepsilon}{\lambda(c - 1 - \varepsilon)} \log \left(\frac{f(a)}{a} \right) = \frac{1 + \varepsilon}{\lambda(c - 1 - \varepsilon)} \log \left(1 + \frac{\Delta(a)}{a} \right)$$

$$< \frac{1 + \varepsilon}{\lambda} \cdot \frac{\log(c + \varepsilon)}{c - 1 - \varepsilon} = (1 + \varepsilon) \frac{\log(c + \varepsilon)}{\log(c)} \cdot \frac{c - 1}{c - 1 - \varepsilon}.$$

If we pass to the limit, this yields

$$\int_a^{f(a)} \frac{ds}{\varrho(s)} \leq 1.$$

A similar argument shows the opposite inequality, thus concluding the proof. $\square$

Exercise 4.1.15. Show that if the original diffeomorphism is of class $C^{1+\text{Lip}}$, then the preceding construction provides us with a Lipschitz vector field associated to it.

We leave to the reader the task of checking that if the diffeomorphism f is of class C^r for some $r \geq 2$, then the associated vector field X is of class C^{r-1} on $]0, 1[$. However, X may fail to be twice differentiable at the origin. Moreover, if f is a diffeomorphism of the interval $[0, 1]$, then the vector field X defined on $[0, 1[$ extends continuously to $[0, 1]$ by letting $X(1) = 0$, but this extension is nondifferentiable in "most cases" (see [256, Chapters IV and V] for more details on this; see also [75]). Anyway, the uniqueness and the smoothness of X allow establishing an interesting result of regularity for the conjugacy between diffeomorphisms.

Proposition 4.1.16. *Let $r \geq 2$, and let f_1 and f_2 be C^r diffeomorphisms of an interval that is closed from at least one side. If f_1 and f_2 have no fixed point in the interior of this interval, then the restriction to the interior of every (if any) C^1 diffeomorphism conjugating them is a C^r diffeomorphism.*

4.1.4 Denjoy counterexamples

The regularity $C^{1+\text{Lip}}$ (or $C^{1+\text{bv}}$) is necessary for many of the dynamical results of the preceding chapter, as well as for some of the algebraic results of this one. Before we proceed to the construction of "counterexamples" to some of them, we need to recall the notion of ***modulus of continuity***.

Definition 4.1.17. Given a homeomorphism $\omega : [0, 1] \to [0, \omega(1)]$, we say that a function $\psi : [0, 1] \to \mathbb{R}$ is ω-continuous if there exists $C \in \mathbb{R}$ such that for every $x \neq y$ in $[0, 1]$,

$$\left| \frac{\psi(x) - \psi(y)}{\omega(|x - y|)} \right| \leq C.$$

Let us denote the supremum of the left-side expression by $\|\psi\|_\omega$, and let us call it the C^ω-norm of ψ. The main interest in the notion of ω-continuity comes from the obvious fact that every sequence of functions ψ_n defined on $[0, 1]$ and such that

$$\sup_{n \in \mathbb{N}} \|\psi_n\|_\omega < \infty$$

is equicontinuous. Actually, having a uniformly bounded modulus of continuity for a sequence of functions is a kind of quantitative (and sometimes easy-to-handle) criterion for establishing equicontinuity.

Example 4.1.18. For $\omega(s) = s^\tau$, where $0 < \tau < 1$, the notions of ω-continuity and τ-Hölder continuity coincide.

Example 4.1.19. For $\omega(s) = s$, the notion of ω-continuity corresponds to that of Lipschitz continuity.

Example 4.1.20. Given $\varepsilon \geq 0$, let $\omega = \omega_\varepsilon$ be such that $\omega_\varepsilon(s) = s[\log(1/s)]^{1+\varepsilon}$ for small $s > 0$. If a map is ω_ε-continuous, then it is τ-continuous for every $0 < \tau < 1$. Indeed, one easily checks that

$$s \log\left(\frac{1}{s}\right)^{1+\varepsilon} \leq C_{\varepsilon,\tau} s^\tau, \qquad \text{where} \qquad C_{\varepsilon,\tau} - \frac{1}{e^{1+\varepsilon}} \left(\frac{1+\varepsilon}{1-\tau}\right)^{1+\varepsilon}.$$

Notice that the map $s \mapsto s \log(1/s)^{1+\varepsilon}$ is not Lipschitz. Therefore, ω_ε-continuity of a function does not imply that the function is Lipschitz.

Example 4.1.21. A modulus of continuity ω satisfying $\omega(s) = 1/\log(1/s)$ for every small $s > 0$ is weaker than any modulus $s \mapsto s^\tau$, where $\tau > 0$.

For our construction, one of the main problems will consist in controlling the modulus of continuity for the derivatives of maps obtained by fitting together infinitely many diffeomorphisms defined on small intervals. To do this, the following elementary lemma will be quite useful.

Lemma 4.1.22. *Let $\{I_n : n \in \mathbb{N}\}$ be a family of closed intervals in $[0, 1]$ (resp. in S^1) having disjoint interiors and such that the complement of their union has zero Lebesgue measure. Suppose that φ is a homeomorphism of $[0, 1]$ such that its restrictions to each interval I_n are $C^{1+\omega}$ diffeomorphisms that are C^1-tangent to the identity at the endpoints of I_n and whose derivatives have ω-norms bounded from above by a constant C. Then φ is a $C^{1+\omega}$ diffeomorphism of the whole interval $[0, 1]$ (resp. of S^1), and the ω-norm of its derivative is less than or equal to $2C$.*

Proof. We will consider only the case of the interval, since that of the circle is analogous. Let x and y be points in $\cup_{n \in \mathbb{N}} I_n$ such that $x < y$. If they belong to the same interval I_n, then, by hypothesis,

$$\left| \frac{\varphi'(y) - \varphi'(x)}{\omega(y - x)} \right| \leq C.$$

Now take $x \in I_i = [x_i, y_i]$ and $y \in I_j = [x_j, y_j]$, with $y_i \leq x_j$. In this case we have

$$
\begin{aligned}
\left| \frac{\varphi'(y) - \varphi'(x)}{\omega(y - x)} \right| &= \left| \frac{(\varphi'(y) - 1) + (1 - \varphi'(x))}{\omega(y - x)} \right| \\
&\leq \left| \frac{\varphi'(y) - \varphi'(x_j)}{\omega(y - x)} \right| + \left| \frac{\varphi'(y_i) - \varphi'(x)}{\omega(y - x)} \right| \\
&\leq C \left[\frac{\omega(y - x_j)}{\omega(y - x)} + \frac{\omega(y_i - x)}{\omega(y - x)} \right] \leq 2C.
\end{aligned}
$$

The map $x \mapsto \varphi'(x)$ is therefore uniformly continuous on the dense subset $\cup_{n \in \mathbb{N}} I_n$ and hence extends to a continuous function defined on $[0, 1]$ whose derivative has C^ω-norm bounded from above by $2C$. Since $I \setminus \cup_{n \in \mathbb{N}} I_n$ has zero measure, this function must coincide (at every point) with the derivative of φ. $\qquad\square$

Because of the preceding lemma, it would be nice to exhibit of a "good" family of diffeomorphisms between intervals.

Definition 4.1.23. A family $\{\varphi_{a,b} : [0, a] \to [0, b]; a > 0, b > 0\}$ of homeomorphisms is said to be **equivariant** if $\varphi_{b,c} \circ \varphi_{a,b} = \varphi_{a,c}$ for all positive a, b, and c.

Given an equivariant family and two intervals $I = [x_1, x_2]$ and $J = [y_1, y_2]$, let us denote by $\varphi(I, J) : I \to J$ the homeomorphism defined by

$$
\varphi(I, J)(x) = \varphi_{x_2 - x_1, y_2 - y_1}(x - x_1) + y_1.
$$

Notice that $\varphi(I, I)$ must coincide with the identity.

Perhaps the simplest equivariant family of diffeomorphisms is the one formed by the linear maps $\varphi_{a,b}(x) = bx/a$. Nevertheless, it is clear that this family is not appropriate for fitting maps together in a smooth way. To overcome this difficulty, let us introduce a general procedure for constructing equivariant families. Given a family of homeomorphisms $\{\varphi_a :]0, a[\to \mathbb{R}; a > 0\}$, let us define $\varphi_{a,b} = \varphi_b^{-1} \circ \varphi_a :]0, a[\to]0, b[$. We have

$$
\varphi_{b,c} \circ \varphi_{a,b} = (\varphi_c^{-1} \circ \varphi_b) \circ (\varphi_b^{-1} \circ \varphi_a) = \varphi_c^{-1} \circ \varphi_a = \varphi_{a,c}.
$$

Thus, extending $\varphi_{a,b}$ continuously to the whole interval $[0, a]$ by setting $\varphi_{a,b}(0) = 0$ and $\varphi_{a,b}(a) = b$, we obtain an equivariant family. For some of our purposes, a good equivariant family is obtained via this procedure using the maps $\varphi_a :]0, a[\to \mathbb{R}$ defined by

$$\varphi_a(x) = -\frac{1}{a} \cot\left(\frac{\pi x}{a}\right). \tag{4.10}$$

The associated equivariant family was introduced by Yoccoz. The elements of this family satisfy remarkable differentiability properties that we now discuss.

Letting $u = \varphi_a(x)$, we have

$$\varphi'_{a,b}(x) = (\varphi_b^{-1})'(\varphi_a(x)) \cdot \varphi'_a(x) = \frac{(\varphi_b^{-1})'(u)}{(\varphi_a^{-1})'(u)} = \frac{u^2 + 1/a^2}{u^2 + 1/b^2}.$$

Notice that if $x \to 0$ (resp. $x \to a$), then $u \to -\infty$ (resp. $u \to +\infty$) and $\varphi'_{a,b}(x) \to 1$. Therefore, the map $\varphi_{a,b}$ extends to a C^1 diffeomorphism from $[0, a]$ into $[0, b]$ that is tangent to the identity at the endpoints. Moreover, for $a \geq b$ (resp. $a \leq b$), the function $u \mapsto \frac{u^2 + 1/a^2}{u^2 + 1/b^2}$ attains its minimum (resp. maximum) value at $u = 0$; because this value is equal to b^2/a^2, we have

$$\sup_{x \in [0,a]} |\varphi'_{a,b}(x) - 1| = \left|\frac{b^2}{a^2} - 1\right|.$$

For the second derivative of $\varphi_{a,b}$, we have

$$|\varphi''_{a,b}(x)| = \frac{d\varphi'_{a,b}(x)}{du} \cdot \left|\frac{du}{dx}\right|$$

$$= \frac{|2u(u^2 + 1/b^2) - 2u(u^2 + 1/a^2)|}{(u^2 + 1/b^2)^2} \pi(u^2 + 1/a^2)$$

$$= \pi \frac{u^2 + 1/a^2}{(u^2 + 1/b^2)^2}\left|\left[2u\left(\frac{1}{b^2} - \frac{1}{a^2}\right)\right]\right|$$

$$= \pi \frac{u^2 + 1/a^2}{u^2 + 1/b^2} \cdot \frac{|2u(1/b^2 - 1/a^2)|}{u^2 + 1/b^2}.$$

From this it follows that $\varphi_{a,b}$ is a C^2 diffeomorphism satisfying $\varphi''_{a,b}(0) = \varphi''_{a,b}(a) = 0$. The inequality $\frac{2|u|}{u^2 + t^2} \leq \frac{1}{t}$ applied to $t = 1/b$ yields

$$|\varphi''_{a,b}(x)| \leq \pi \frac{u^2 + 1/a^2}{u^2 + 1/b^2}\left|\frac{1}{b^2} - \frac{1}{a^2}\right| b.$$

For $a \leq b$, this implies that

$$|\varphi''_{a,b}(x)| \leq \pi \frac{b^2}{a^2}\left(\frac{b^2 - a^2}{a^2 b^2}\right) b = \frac{\pi b}{a^2}\left(\frac{b^2}{a^2} - 1\right).$$

Hence, if $a \leq b \leq 2a$, then

$$\left|\varphi_{a,b}''(x)\right| \leq 6\pi \left|\frac{b}{a} - 1\right| \frac{1}{a}.$$

Analogously, if $2b \geq a \geq b$, then

$$\left|\varphi_{a,b}''(x)\right| \leq \frac{\pi}{b}\left(1 - \frac{b^2}{a^2}\right) \leq 2\pi \left|\frac{b}{a} - 1\right| \frac{1}{b} \leq 4\pi \left|\frac{b}{a} - 1\right| \frac{1}{a}.$$

Therefore, in both cases we have

$$\left|\varphi_{a,b}''(x)\right| \leq 6\pi \left|\frac{b}{a} - 1\right| \frac{1}{a}. \tag{4.11}$$

The last inequality, together with the following exercise, shows that the family of maps $\varphi_{a,b}$ is nearly optimal.

Exercise 4.1.24. Let $\varphi : [0, a] \to [0, b]$ be a C^2 diffeomorphism. Suppose that $\varphi'(0) = \varphi'(a) = 1$. Show that there exists a point $s \in \,]0, a[$ such that

$$|\varphi''(s)| \geq \frac{2}{a}\left|\frac{b}{a} - 1\right|.$$

With the exception of Exercise 4.1.29, in what follows, the notation $\varphi_{a,b}$ and $\varphi_{I,J}$ will be used only to denote the maps in Yoccoz's family. Without loss of generality, we will suppose that the function $s \mapsto \omega(s)/s$ is decreasing. (Notice that the moduli from Examples 4.1.18, 4.1.20, and 4.1.21 may be modified far from the origin so that they satisfy this property.) Under this assumption we have the following:

Lemma 4.1.25. *If $a > 0$ and $b > 0$ are such that $a/b \leq 2$, $b/a \leq 2$, and*

$$\left|\frac{b}{a} - 1\right| \frac{1}{\omega(a)} \leq C,$$

then the C^ω-norm of $\varphi_{a,b}'$ is less than or equal to $6\pi C$.

Proof. According to (4.11), for every $x \in [0, a]$, one has

$$|\varphi_{a,b}''(x)| \leq \frac{6\pi C\omega(a)}{a}.$$

For every $y < z$ in $[0, a]$, there exists $x \in [y, z]$ satisfying $\varphi'_{a,b}(z) - \varphi'(y) = \varphi''_{a,b}(x)(z - y)$. Since the function $s \mapsto \omega(s)/s$ is decreasing and $z - y \leq a$, this implies that

$$\left| \frac{\varphi'_{a,b}(z) - \varphi'_{a,b}(y)}{\omega(z - y)} \right| = |\varphi''_{a,b}(x)| \left| \frac{z - y}{\omega(z - y)} \right| \leq |\varphi''_{a,b}(x)| \left| \frac{a}{\omega(a)} \right| \leq 6\pi C,$$

which shows the lemma. $\square$

On the basis of the preceding discussion, we can now give a conceptual construction of the so-called *Denjoy counterexamples* (i.e., we can give a proof of Theorem 3.1.2). The method of proof will be used later to smooth many other group actions on the interval and the circle.

Proof of Theorem 3.1.2.   Slightly more generally, for all $\varepsilon > 0$, every irrational angle θ, and every $k \in \mathbb{N}$, we will exhibit a $C^{1+\omega_\varepsilon}$ circle diffeomorphism with rotation number θ whose derivative has a C^{ω_ε}-norm bounded from above by $2/[\log(k)]^{\varepsilon/2}$. To do this, let us fix $x_0 \in S^1$, and for each $n \in \mathbb{Z}$, let us replace each point $x_n = R_\theta^n(x_0)$ of its orbit by an interval I_n of length

$$\ell_n = \frac{1}{(|n| + k)[\log(|n| + k)]^{1+\varepsilon/2}}.$$

The original rotation induces a homeomorphism $h = \bar{R}_\theta$ of a circle of length $\bar{\ell}_k = \sum_{n \in \mathbb{Z}} \ell_n$ by letting $\bar{R}_\theta(x) = \varphi(I_n, I_{n+1})(x)$ for $x \in I_n$ and extending it continuously to S^1. We now check that $\bar{R}_\theta$ is a diffeomorphism satisfying the desired properties. To do this, we need to estimate the value of expressions of the type

$$\left| \frac{\ell_{n+1}}{\ell_n} - 1 \right| \frac{1}{\omega_\varepsilon(\ell_n)}. \tag{4.12}$$

We will make the explicit computations for $n \geq 0$, leaving the case $n \leq 0$ to the reader. Expression (4.12) equals

$$\left| \frac{(n + k)[\log(n + k)]^{1+\varepsilon/2}}{(n + k + 1)[\log(n + k + 1)]^{1+\varepsilon/2}} - 1 \right|$$

$$\cdot \frac{(n + k)[\log(n + k)]^{1+\varepsilon/2}}{[\log(n + k) + (1 + \varepsilon/2)\log\log(n + k)]^{1+\varepsilon}},$$

which is bounded from above by

$$\left| \frac{(n+k)[\log(n+k)]^{1+\varepsilon/2}}{(n+k+1)[\log(n+k+1)]^{1+\varepsilon/2}} - 1 \right| \frac{n+k}{[\log(n+k)]^{\varepsilon/2}}.$$

By applying the mean value theorem to the function $s \mapsto s[\log(s)]^{1+\varepsilon/2}$, we obtain the following upper bound for the latter expression:

$$\frac{[\log(n+k+1)]^{1+\varepsilon/2} + (1+\varepsilon/2)[\log(n+k+1)]^{\varepsilon/2}}{[\log(n+k+1)]^{1+\varepsilon/2}[\log(n+k)]^{\varepsilon/2}} \leq \frac{2}{[\log(n+k)]^{\varepsilon/2}}.$$

The diffeomorphism h that we constructed acts on a circle of length $\bar{\ell}_k \sim 2/\varepsilon[\log(k)]^{\varepsilon/2}$. Therefore, to obtain a diffeomorphism of the unit circle, we must renormalize the circle, say, by an affine map φ. Notice that this procedure does not increase the C^{ω_ε}-norm for the derivative. Indeed, from the equality $\bar{R}'_\theta = (\varphi \circ h \circ \varphi^{-1})' = \bar{\ell}_k^{-1} \cdot (h' \circ \varphi^{-1}) \cdot L_k$ one deduces that

$$\frac{|\bar{R}'_\theta(x) - \bar{R}'_\theta(y)|}{\omega_\varepsilon(|x-y|)} = \frac{|(\varphi \circ h \circ \varphi^{-1})'(x) - (\varphi \circ h \circ \varphi^{-1})'(y)|}{\omega_\varepsilon(|\varphi^{-1}(x) - \varphi^{-1}(y)|)}$$
$$\cdot \frac{\omega_\varepsilon(|\varphi^{-1}(x) - \varphi^{-1}(y)|)}{\omega_\varepsilon(|x-y|)}$$
$$\leq \frac{|h'(\varphi^{-1}(x)) - h'(\varphi^{-1}(y))|}{\omega_\varepsilon(|\varphi^{-1}(x) - \varphi^{-1}(y)|)} \leq \frac{2}{[\log(k)]^{\varepsilon/2}},$$

where the first inequality comes from the fact that ω_ε is increasing and $\bar{\ell}_k < 1$. $\qquad\square$

The preceding construction might suggest that there exist Denjoy counterexamples for any modulus of continuity for the derivative weaker than the Lipschitz one. However, there exist subtle obstructions related to the Diophantine nature of the rotation number that are not completely understood; see [115, Chapter X] for a partial result on this.

Exercise 4.1.26. Show that there is no Denjoy counterexample of class $C^{1+\omega_0}$ whose derivative is identically equal to 1 on the minimal invariant Cantor set, where $\omega_0(s) = s \log(1/s)$ (compare Lemma 3.1.3 and the argument before it).

Remark. It seems to be unknown whether this claim is still true without the hypothesis about the derivative over the invariant Cantor set. Actually, the general problem of finding the weakest modulus of continuity ensuring the validity of Denjoy's theorem seems interesting.

We would now like to use the preceding technique to construct smooth actions on the circle of higher-rank Abelian groups with an exceptional minimal set. However, Lemma 3.1.3 suggests that we should find obstructions in differentiability classes less than $C^{1+\mathrm{Lip}}$. Indeed, this will be the main object of the next section, where we will show, for instance, that for every $d \geq 2$ and all $\varepsilon > 0$, every free action of $\mathbb{Z}^d$ by $C^{1+1/d+\varepsilon}$ circle diffeomorphisms is minimal.[1] According to [241] (see also [242]), the hypothesis $\varepsilon > 0$ should be superfluous for this result. More generally, the theorem should be true for actions by diffeomorphisms with bounded d-variation (cf. Exercise 4.1.32). Here we will content ourselves with showing that in regularity less than $C^{1+1/d}$, this result is no longer true.

Proposition 4.1.27. *For every $\varepsilon > 0$ and every positive integer d, there exist free $\mathbb{Z}^d$-actions by $C^{1+1/d-\varepsilon}$ circle diffeomorphisms admitting an exceptional minimal set.*

Proof. Notice that the case $d = 1$ corresponds to Theorem 3.1.2. However, the construction involves all the cases simultaneously and produces examples of groups of circle diffeomorphisms of class $C^{1+\omega}$, where $\omega(s) = s^{1/d}[\log(1/s)]^{1/d+\varepsilon}$.

Let us begin by fixing a rank-d torsion-free subgroup of $SO(2, \mathbb{R})$, and let $\theta_1, \ldots, \theta_d$ be the angles of the generators. Let $m \geq d - 1$ be an integer number and p a point in S^1. For each $(i_1, \ldots, i_d) \in \mathbb{Z}^d$, let us replace the point $R_{\theta_1}^{i_1} \cdots R_{\theta_d}^{i_d}(p)$ by an interval $I_{i_1,\ldots,i_d}$ of length

$$\ell_{(i_1,\ldots,i_d)} = \frac{1}{(|i_1| + \cdots + |i_d| + m)^d \, [\log(|i_1| + \cdots + |i_d| + m)]^{1+\varepsilon}}.$$

This procedure induces a new circle (of length $\bar{\ell}_m \leq 2^d/\varepsilon[\log(m)]^\varepsilon(d-1)!$), on which the rotations R_{θ_j} induce homeomorphisms f_j satisfying, for every $x \in I_{i_1,\ldots,i_j,\ldots,i_d}$,

$$f_j(x) = \varphi(I_{i_1,\ldots,i_j,\ldots,i_d}, I_{i_1,\ldots,1+i_j,\ldots,i_d})(x).$$

---

1. Actually, a stronger result holds: every free $\mathbb{Z}^d$-action by circle diffeomorphisms is minimal provided that the generators f_i, $i \in \{1, \ldots, d\}$, are, respectively, of class $C^{1+\tau_i}$, and $\tau_1 + \cdots + \tau_d > 1$ (see [140]).

By the equivariance properties of the $\varphi(I, J)$'s, these homeomorphisms f_j commute between them. The verification that each f_i is of class $C^{1+\omega}$ is analogous to the proof of Theorem 3.1.2, and we leave it to the reader. Finally, notice that if we renormalize the circle and let m go to infinity, each f_i converges (in the $C^{1+\omega}$ topology) to the corresponding rotation R_{θ_i}. $\quad\square$

An analogous procedure leads to counterexamples to Kopell's lemma. Slightly more generally, for each integer $d \geq 2$ and each $\varepsilon > 0$, there exist $C^{1+1/(d-1)-\varepsilon}$ diffeomorphisms $f_1, \ldots, f_d$ of $[0, 1]$ and disjoint open intervals $I_{n_1,\ldots,n_d}$ disposed in $]0, 1[$ according to the lexicographic ordering so that for every $(n_1, \ldots, n_d) \in \mathbb{Z}^d$ and every $j \in \{1, \ldots, d\}$,

$$f_j(I_{n_1,\ldots,n_j,\ldots,n_d}) = I_{n_1,\ldots,n_j-1,\ldots,n_d}.$$

To construct these diffeomorphisms, a natural procedure goes as follows. Given an integer $m \geq d - 1$, for each $(i_1, \ldots, i_d) \in \mathbb{Z}^d$, let

$$\ell_{i_1,\ldots,i_d} = \frac{1}{(|i_1| + \cdots + |i_d| + m)^d [\log(|i_1| + \cdots + |i_d| + m)]^{1+\varepsilon}}.$$

Let us inductively define $\ell_{i_1,\ldots,i_{j-1}} = \sum_{i_j \in \mathbb{Z}} \ell_{i_1,\ldots,i_{j-1},i_j}$. Let $[x_{i_1,\ldots,i_j,\ldots,i_d}, y_{i_1,\ldots,i_j,\ldots,i_d}]$ be the interval whose endpoints are

$$x_{i_1,\ldots,i_j,\ldots,i_d} = \sum_{i_1'<i_1} \ell_{i_1'} + \sum_{i_2'<i_2} \ell_{i_1,i_2'} + \cdots + \sum_{i_d'<i_d} \ell_{i_1,\ldots,i_d,i_d'},$$

$$y_{i_1,\ldots,i_j,\ldots,i_d} = x_{i_1,\ldots,i_j,\ldots,i_d} + \ell_{i_1,\ldots,i_d},$$

and let f_j be the diffeomorphism of $[0, 1]$ that, restricted to each $[x_{i_1,\ldots,i_j,\ldots,i_d}, y_{i_1,\ldots,i_j,\ldots,i_d}]$, coincides with

$$\varphi([x_{i_1,\ldots,i_j,\ldots,i_d}, y_{i_1,\ldots,i_j,\ldots,i_d}], [x_{i_1,\ldots,i_j-1,\ldots,i_d}, y_{i_1,\ldots,i_j-1,\ldots,i_d}]).$$

Using some of the estimates in this section, one readily checks that the obtained f_j's are of class $C^{1+\omega}$, where $\omega(s) = s^{1/d}[\log(1/s)]^{1/d+\varepsilon}$. In particular, these maps are $C^{1+1/d-\varepsilon}$ diffeomorphisms. However, notice that by this method, we have not achieved the optimal regularity $C^{1+1/(d-1)-\varepsilon}$ that we claimed. Actually, a direct consequence of Lemma 3.1.3 is that by this procedure it is impossible to reach the class $C^{1+1/d}$. To reach the optimal regularity $C^{1+1/(d-1)-\varepsilon}$, it is necessary to avoid many of the tangencies to the identity of the maps. To do this, we will use the original technique by Pixton [204], following the brilliant presentation of Tsuboi [241].

Let $X = \varrho \frac{\partial}{\partial x}$ be a C^∞ vector field on $[0, 1]$ satisfying $|\varrho'(x)| \leq 1$ for all $x \in [0, 1]$, and such that $\varrho(x) = x$ for every $x \in [0, 1/3]$, and $\varrho(x) = 0$ for every $x \in [1/2, 1]$. Let $\varphi^t(x)$ be the flow associated to X, that is, the solution of the differential equation

$$\frac{d\varphi^t(x)}{dt}(x) = \varrho(\varphi^t(x)), \qquad \varphi^0(x) = x.$$

For every positive a, b and each $t \geq 0$, the diffeomorphism $x \mapsto b\varphi^t(x/a)$ sends the interval $[0, a]$ onto $[0, b]$. Moreover, its derivative is identically equal to b/a in $[a/2, a]$ and equals be^t/a at the origin. Given $a' < 0 < a$ and $b' < 0 < b$, let us define a diffeomorphism $\varphi_{b',b}^{a',a} : [0, a] \to [0, b]$ by

$$\varphi_{b',b}^{a',a}(x) = b\varphi^{\log(b'a/a'b)}\left(\frac{x}{a}\right).$$

Notice that for every $c' < 0 < c$, one has the equivariance property

$$\varphi_{c',c}^{b',b} \circ \varphi_{b',b}^{a',a} = \varphi_{c',c}^{a',a}, \tag{4.13}$$

which is analogous to that in Definition 4.1.23. Nevertheless, the fact that in the family $\{\varphi_{b',b}^{a',a}\}$ four (and not only two) parameters are involved allows obtaining better control of the modulus of continuity of the derivatives (at the mild cost of having to suppress the tangencies to the identity at the endpoints).

Lemma 4.1.28. *Letting $C = \sup_{y \in [0,1]} \varrho''(y)$, for every $x \in [0, a]$, one has the inequality*

$$\left| \frac{\partial}{\partial x} \log\left(\frac{\partial \varphi_{b',b}^{a',a}}{\partial x}\right)(x) \right| \leq \frac{C}{a}\left|\frac{b'a}{a'b} - 1\right|. \tag{4.14}$$

Proof. From $d\varphi^t(x)/dt = \varrho(\varphi^t(x))$ one concludes that

$$\frac{d}{dt}\frac{\partial\varphi^t}{\partial x}(x) = \frac{\partial}{\partial x}\varrho(\varphi^t(x)) = \frac{\partial\varrho}{\partial x}(\varphi^t(x)) \cdot \frac{\partial\varphi^t}{\partial x}(x),$$

and hence

$$\frac{d}{dt}\log\left(\frac{\partial\varphi^t}{\partial x}\right)(x) = \frac{\partial\varrho}{\partial x}(\varphi^t(x)). \tag{4.15}$$

On the other hand, from $|\varrho'(x)| \leq 1$ it follows that

$$\left|\log\left(\frac{\partial\varphi^t}{\partial x}\right)(x)\right| = \left|\int_0^t \frac{d}{ds}\log\left(\frac{\partial\varphi^s}{\partial x}\right)(x)ds\right|$$

$$= \left|\int_0^t \frac{\partial\varrho}{\partial x}(\varphi^s(x))ds\right| \leq \int_0^t ds = t,$$

and therefore

$$\frac{\partial \varphi^t}{\partial x}(x) \le e^t. \tag{4.16}$$

Since

$$\log\left(\frac{\partial \varphi_{b',b}^{a',a}}{\partial x}\right)(x) = \log\left(\frac{b}{a}\right) + \log\left(\frac{\partial}{\partial x}\varphi^{\log(b'a/a'b)}\right)\left(\frac{x}{a}\right),$$

using (4.15) and (4.16), we obtain

$$\left|\frac{\partial}{\partial x}\log\left(\frac{\partial \varphi_{b',b}^{a',a}}{\partial x}\right)(x)\right|$$

$$= \left|\frac{\partial}{\partial x}\log\left(\frac{\partial}{\partial x}\varphi^{\log(b'a/a'b)}\right)\left(\frac{x}{a}\right)\right|$$

$$= \left|\int_0^{\log(b'a/a'b)} \frac{d}{dt}\frac{\partial}{\partial x}\log\left(\frac{\partial \varphi^t}{\partial x}\right)\left(\frac{x}{a}\right)dt\right|$$

$$= \frac{1}{a}\left|\int_0^{\log(b'a/a'b)} \frac{\partial}{\partial x}\frac{d}{dt}\log\left(\frac{\partial \varphi^t}{\partial x}\right)\left(\frac{x}{a}\right)\cdot\frac{d\varphi^t}{dt}\left(\frac{x}{a}\right)dt\right|$$

$$= \frac{1}{a}\left|\int_0^{\log(b'a/a'b)} \frac{\partial}{\partial x}\frac{\partial \varrho}{\partial x}\left(\varphi^t\left(\frac{x}{a}\right)\right)\cdot\frac{d\varphi^t}{dt}\left(\frac{x}{a}\right)dt\right|,$$

and hence

$$\log\left(\frac{\partial \varphi_{b',b}^{a',a}}{\partial x}\right)(x) \le \frac{C}{a}\left|\int_0^{\log(b'a/a'b)} e^t dt\right| = \frac{C}{a}\left|\frac{b'a}{a'b} - 1\right|. \qquad \square$$

Notice that

$$\frac{\partial \varphi_{b',b}^{a',a}}{\partial x}(0) = \frac{b}{a}\cdot\frac{\partial}{\partial x}\varphi^{\log(b'a/a'b)}(0)$$

$$= \frac{b}{a}\cdot\exp(\log(b'a/a'b)) = \frac{b'}{a'}, \qquad \frac{\partial \varphi_{b',b}^{a',a}}{\partial x}(a) = \frac{b}{a}. \tag{4.17}$$

Therefore, the diffeomorphisms $\varphi_{b',b}^{a',a}$ are not necessarily tangent to the identity at the endpoints. This allows fitting them together with a certain freedom. A careful choice of the (infinitely many) parameters $a' < 0$ and $b' < 0$ will then lead to the optimal regularity. In what follows, we will sketch the (quite technical) construction of commuting diffeomorphisms of the interval that are $C^{2-\varepsilon}$ counterexamples to Kopell's lemma. For the general case of $\mathbb{Z}^d$-actions on the interval, we refer to [241] (see also [130]).

Let us fix $k \in \mathbb{N}$ and $\varepsilon > 0$, and for each pair of integers m and n let

$$\ell_{m,n} = \frac{1}{(|m|^2 + |n|^2 + k^2)^{1+\varepsilon}}.$$

Let $\{I_{m,n} = [x_{m,n}, y_{m,n}]\}$ be a family of intervals placed inside a closed interval I respecting the lexicographic ordering and such that their union has full measure in I, with $|I_{m,n}| = \ell_{m,n}$. Define two homeomorphisms f and g of I by letting, for each $x \in I_{m,n}$,

$$f(x) = \varphi_{\ell_{m-1,n-1}, \ell_{m-1,n}}^{\ell_{m,n-1}, \ell_{m,n}}(x - x_{m,n}) + x_{m-1,n},$$

$$g(x) = \varphi_{\ell_{m,n-2}, \ell_{m,n-1}}^{\ell_{m,n-1}, \ell_{m,n}}(x - x_{m,n}) + x_{m,n-1}.$$

Properties (4.13) and (4.17) imply that f and g are commuting C^1 diffeomorphisms of I. Using (4.14), the reader should be able to check that f and g are actually $C^{2-\varepsilon'}$ diffeomorphisms, where $\varepsilon' > 0$ depends only on $\varepsilon > 0$ and goes to zero together with ε. However, we must warn that this is not an easy task, mainly because of the absence of an analogue to Lemma 4.1.22 for this case; see [241] for the details.

Exercise 4.1.29. Slightly abusing of notation, define $\varphi_{a,b} : [0, a] \to [0, b]$ by letting $\varphi_{a,b}(x) = \varphi_{-1,b/2}^{-1,a/2}(x)$ and $\varphi_{a,b}(x) = b - \varphi_{-1,b/2}^{-1,a/2}(x)$ for x in $[a/2, a]$ and $[0, a/2]$, respectively. Check that $\{\varphi_{a,b}\}$ is an equivariant family of diffeomorphisms tangent to the identity at the endpoints that satisfy (compare inequality (4.11))

$$\left| \frac{\partial}{\partial x} \log \left(\frac{\partial \varphi_{a,b}}{\partial x} \right)(x) \right| \leq \frac{2C}{a} \left| \frac{a}{b} - 1 \right|.$$

4.1.5 On intermediate regularities

The next result (obtained by the author in collaboration with Deroin and Kleptsyn in [65]) may be considered an extension (up to some parameter $\varepsilon > 0$) of Denjoy's theorem for free actions of higher-rank Abelian groups on the circle. We do not know whether, under the same hypothesis, the corresponding actions are ergodic (compare Theorem 3.5.1).

Theorem 4.1.30. *If d is an integer greater than or equal to 2 and $\varepsilon > 0$, then every free action of $\mathbb{Z}^d$ by $C^{1+1/d+\varepsilon}$ circle diffeomorphisms is minimal.*

Before giving the proof of this result, we would like to explain the main idea, which is inspired by the famous *Erdös principle* and shows once again how fruitful the probabilistic methods are in the theory. Suppose for a contradiction that I is a wandering interval for the dynamics of a rank-2 Abelian group of $C^{1+\tau}$ circle diffeomorphisms generated by elements g_1 and g_2. If we let $\mathbb{N}_0 = \mathbb{N} \cup \{0\}$, the family

$$\{g_1^m g_2^n(I) : (m, n) \in \mathbb{N}_0 \times \mathbb{N}_0, m + n \leq k\}$$

consists of $(k+1)(k+2)/2$ intervals. Since these intervals are two-by-two disjoint, we should expect that, "typically", their length has order $1/k^2$. Hence for a "generic" random sequence $I, h_1(I), h_2(I)\ldots$, where either $h_{n+1} = g_1 h_n$ or $h_{n+1} = g_2 h_n$, for $\tau > 1/2$ we should have

$$\sum_{k \geq 1} |h_k(I)|^\tau \leq C \sum_{k \geq 1} \frac{1}{k^{2\tau}} < \infty.$$

Now notice that the left-side expression of this inequality corresponds to the series that allows controlling the distortion of the successive compositions. More precisely, if this sum is finite, then Lemma 3.2.3 should allow us to find elements with hyperbolic fixed points, thus contradicting the hypothesis of freeness of the action.

To formalize this idea, we need to make the random nature of the compositions more precise. To do this, let us consider the Markov process on $\mathbb{N}_0 \times \mathbb{N}_0$ with transition probabilities

$$p((m, n) \to (m + 1, n)) = \frac{m+1}{m+n+2},$$
$$p((m, n) \to (m, n + 1)) = \frac{n+1}{m+n+2}. \tag{4.18}$$

This Markov process induces a probability measure $\mathbb{P}$ on the corresponding space of paths Ω. One easily shows that if one starts from the origin, the probability of arriving at the point (m, n) in k steps equals $1/(k+1)$ (resp. 0) if $m + n = k$ (resp. $m + n \neq k$).

To prove Theorem 4.1.30 in the case $d = 2$, let g_1 and g_2 be two commuting $C^{1+\tau}$ circle diffeomorphisms generating a rank-2 Abelian group. The semigroup Γ^+ generated by g_1 and g_2 identifies with $\mathbb{N}_0 \times \mathbb{N}_0$; therefore, the Markov process already described induces a "random walk" on Γ^+. In what follows, we will identify Ω with the corresponding space of paths over Γ^+. For every $\omega \in \Omega$ and all $n \in \mathbb{N}$, we denote by $h_n(\omega) \in \Gamma^+$ the product of the first n entries of ω. In other words, for $\omega = (g_{i_1}, g_{i_2}, \ldots) \in \Omega$, we let $h_n(\omega) = g_{i_n} \cdots g_{i_1}$ (where g_{i_j} equals either g_1 or g_2), and we let $h_0(\omega) = id$.

If the action of $\Gamma = \langle g_1, g_2 \rangle \sim \mathbb{Z}^2$ is free, then, by Hölder's theorem and § 2.2.2, the restriction of the rotation-number function to Γ is a group homomorphism. This implies that the rotation numbers $\rho(g_1)$ and $\rho(g_2)$ are **independent** over the rationals, in the sense that for all $(r_0, r_1, r_2) \in \mathbb{Q}^3$ distinct from $(0, 0, 0)$, one has $r_1 \rho(g_1) + r_2 \rho(g_2) \neq r_0$. Indeed, otherwise one could find nontrivial elements with rational rotation number. These elements would then have periodic points, and since they are of infinite order, this would contradict the freeness of the action.

Now suppose for a contradiction that the action of Γ is not minimal. In this case there exists a minimal invariant Cantor set for the action. Moreover, every connected component I of the complement of this set is wandering for the dynamics.

Lemma 4.1.31. *If $\tau > 1/2$, then the series $\ell_\tau(\omega) = \sum_{n \geq 0} |h_n(\omega)(I)|^\tau$ converges for $\mathbb{P}$-almost every $\omega \in \Omega$.*

Proof. We will show that if $\tau > 1/2$, then the expectation (with respect to $\mathbb{P}$) of the function ℓ_τ is finite, which obviously implies the claim of the lemma. To do this, first notice that since the arrival probabilities in k steps are equally distributed over the points at simplicial distance k from the origin,

$$\mathbb{E}(\ell_\tau) = \mathbb{E}\left(\sum_{k \geq 0} |h_k(\omega)(I)|^\tau\right) = \sum_{k \geq 0} \mathbb{E}(|h_k(\omega)(I)|^\tau) = \sum_{k \geq 0} \sum_{m+n=k} \frac{|g_1^m g_2^n(I)|^\tau}{k+1}.$$

By Hölder's inequality,

$$\sum_{m+n=k} \frac{|g_1^m g_2^n(I)|^\tau}{k+1} \leq \left(\sum_{m+n=k} |g_1^m g_2^n(I)|\right)^\tau \left((k+1) \cdot \frac{1}{(k+1)^{1/(1-\tau)}}\right)^{1-\tau},$$

and hence

$$\mathbb{E}(\ell_\tau) \leq \sum_{k \geq 0} \frac{\left(\sum_{m+n=k} |g_1^m g_2^n(I)|\right)^\tau}{(k+1)^\tau}.$$

By applying Hölder's inequality once again, we obtain

$$\mathbb{E}(\ell_\tau) \leq \left[\sum_{(m,n)\in\mathbb{N}_0\times\mathbb{N}_0} |g_1^m g_2^n(I)| \right]^\tau \left[\sum_{k\geq 1} \left(\frac{1}{k^\tau}\right)^{\frac{1}{1-\tau}} \right]^{1-\tau}.$$

Since $\tau > 1/2$, the series

$$\sum_{k\geq 1} \left(\frac{1}{k^\tau}\right)^{\frac{1}{1-\tau}} = \sum_{k\geq 1} \frac{1}{k^{\tau/(1-\tau)}}$$

converges, and since the intervals of type $g_1^m g_2^n(I)$ are two-by-two disjoint, this shows the finiteness of $\mathbb{E}(\ell_\tau)$.   □

Exercise 4.1.32. Given $\alpha > 0$, say that a function ψ has bounded α-**variation** if

$$\sup_{a_0 < a_1 < \ldots < a_n} \sum_{i=1}^n |\psi(a_i) - \psi(a_{i-1})|^\alpha < \infty.$$

Suppose that g_1 and g_2 are commuting circle diffeomorphisms that generate a rank-2 Abelian group acting freely and nonminimally. Let x and y be points in a connected component of the complement of the minimal invariant Cantor set. Prove that if g_1' and g_2' have bounded α-variation for some $\alpha < 2$, then the function $V_{x,y} : \Omega \to \mathbb{R}$ defined by

$$V_{x,y}(\omega) = \sum_{k\geq 0} |g_{i_{k+1}}'(h_k(\omega)(x)) - g_{i_{k+1}}'(h_k(\omega)(y))|$$

has finite expectation.

From Lemma 4.1.31 one deduces that if $S > 0$ is large enough, then the probability of the set $\Omega(S) = \{\omega \in \Omega : \ell_\tau(\omega) \leq S\}$ is positive (actually, $\mathbb{P}[\Omega(S)]$ converges to 1 as S goes to infinity). Fix such an $S > 0$, and let $\ell = \ell(\tau, S, |I|; \{g_1, g_2\})$ be the constant of Lemma 3.2.3. Finally, let us consider the open interval L' of length $|L'| = \ell$ that is next to I on the right. We claim that

$$\mathbb{P}[\omega \in \Omega : h_n(\omega)(I) \not\subset L' \text{ for all } n \in \mathbb{N}] = 0. \tag{4.19}$$

To show this, recall that the action of Γ is semiconjugate, but nonconjugate, to an action by rotations. Therefore, if we "collapse" each connected component of the complement of the minimal invariant Cantor set Λ, then we obtain a topological circle S_Λ^1 on which g_1 and g_2 induce minimal homeomorphisms. On the other hand, the interval L' becomes an interval U with nonempty interior in S_Λ^1. Since the rotation numbers of g_1 and g_2 are irrational, there must exist $N \in \mathbb{N}$ such that, after collapsing, $g_1^{-1}(U), \ldots, g_1^{-N}(U)$ cover S_Λ^1, and the same holds for $g_2^{-1}(U), \ldots, g_2^{-N}(U)$. On the original circle S^1 this implies that for every connected component I_0 of $S^1 \setminus \Lambda$, there exist n_1 and n_2 in $\{1, \ldots, N\}$ such that $g_1^{n_1}(I_0) \subset L'$ and $g_2^{n_2}(I_0) \subset L'$.

From the definition of N it follows immediately that for every integer $k \geq 0$, the conditional probability of the event

$$[g_1^i h_k(\omega)(I) \not\subset L' \text{ and } g_2^i h_k(\omega)(I) \not\subset L' \text{ for all } i \in \{1, \ldots, N\}]$$

given that $h_j(\omega)(I) \not\subset L'$ for all $j \in \{1, \ldots, k\}$ is zero. Remark now the following elementary property that follows directly from (4.18): the probabilities of "jumping" to the right (resp. upward) of the Markov process are greater than or equal to $1/2$ under (resp. over) the diagonal (see Figure 21). Together with what precedes, this implies that

$$\mathbb{P}[h_{k+i}(\omega)(I) \not\subset L', i \in \{1, \ldots, N\} \mid h_j(\omega)(I) \not\subset L', j \in \{1, \ldots, k\}]$$

$$\leq 1 - \frac{1}{2^N}. \tag{4.20}$$

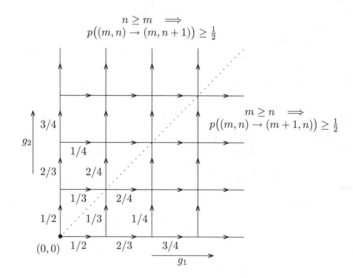

FIGURE 21

As a consequence, for every $r \in \mathbb{N}$,

$$\mathbb{P}[h_n(\omega)(I) \not\subset L', n \in \mathbb{N}]$$

$$\leq \mathbb{P}[h_n(\omega)(I) \not\subset L', i \in \{1, \ldots, rN\}] \leq \left(1 - \frac{1}{2^N}\right)^r,$$

from which (4.19) follows by letting r go to infinity.

To conclude the proof of Theorem 4.1.30 (in the case $d = 2$), remark that if $\omega \in \Omega(S)$ and $n \in \mathbb{N}$ satisfy $h_n(\omega)(I) \subset L'$, then Lemma 3.2.3 allows finding a hyperbolic fixed point of $h_n(\omega) \in \Gamma^+$, which contradicts the freeness of the action.

The proof of Theorem 4.1.30 for $d > 2$ is analogous to that given for the case $d = 2$. Assuming the existence of a wandering interval, one may consider the Markov process on $\mathbb{N}_0^d$ with transition probabilities

$$p((n_1, \ldots, n_i, \ldots, n_d) \longrightarrow (n_1, \ldots, 1 + n_i, \ldots, n_d)) = \frac{1 + n_i}{n_1 + \cdots + n_d + d}.$$

Once again, the arrival probabilities in k steps are equally distributed over the points at simplicial distance k from the origin. This allows controlling the distortion of almost every random sequence (that is, an analog of Lemma 4.1.31 holds for $\tau > 1/d$). Moreover, one easily checks that each $(n_1, \ldots, n_d) \in \mathbb{N}_0^d$ is the starting point of at least one half-line for which the transition probabilities between two adjacent vertices is greater than or equal to $1/d$ (it suffices to follow the direction of the ith coordinate for which n_i attains its maximum value). This allows obtaining an inequality that is analogous to (4.20) (whose right-hand member will be equal to $1 - 1/d^N$ for some large integer N). This inequality implies property (4.19), which, thanks to the control of distortion, allows using Lemma 3.2.3 to find elements with hyperbolic fixed points, thus contradicting the freeness of the action.

Exercise 4.1.33. Show the following generalization of Theorem 4.1.30: if $\tau > 1/d$ for some $d \in \mathbb{N}$ and Γ is a subgroup of $\text{Diff}_+^{1+\tau}(S^1)$ that is semiconjugate to a group of rotations and contains d elements with rotation numbers independent over the rationals, then the semiconjugacy is a conjugacy. In particular, if f is a Denjoy counterexample of class $C^{1+\tau}$, then its centralizer in $\text{Diff}_+^{1+\tau}(S^1)$ cannot contain elements $f_1, \ldots, f_{d-1}$ such that $\rho(f_1), \ldots, \rho(f_{d-1})$ and $\rho(f)$ are independent over the rationals.

Hint. A careful reading of the proof of Theorem 4.1.30 shows that the commutativity between the generators is needed only on the minimal invariant set.

Like Denjoy's theorem, Kopell's lemma may be extended to the case of Abelian groups of interval diffeomorphisms with intermediate regularity. The next result was also established in [65].

Theorem 4.1.34. *Let $d \geq 2$ be an integer and let $\varepsilon > 0$. Let $f_1, \ldots, f_{d+1}$ be C^1 commuting diffeomorphisms of $[0, 1[$ for which there exist disjoint open intervals $I_{n_1,\ldots,n_d}$ placed on $[0, 1]$ according to the lexicographic ordering and such that for all $(n_1, \ldots, n_d) \in \mathbb{Z}^d$ and all $i \in \{1, \ldots, d\}$,*

$$f_i(I_{n_1,\ldots,n_i,\ldots,n_d}) = I_{n_1,\ldots,n_i-1,\ldots,n_d}, \qquad f_{d+1}(I_{n_1,\ldots,n_d}) = I_{n_1,\ldots,n_d}.$$

If $f_1, \ldots, f_d$ are of class $C^{1+1/d+\varepsilon}$, then the action of f_{d+1} on the union of the intervals $I_{n_1,\ldots,n_d}$ is trivial.

Proof. We will deal only with the case $d = 2$, leaving to the reader the task of adapting the arguments to the general case. Let $\tau = 1/2 + \varepsilon$, let us identify with $\mathbb{N}_0 \times \mathbb{N}_0$ the semigroup Γ^+ generated by the elements f_1 and f_2 in $\mathrm{Diff}_+^{1+\tau}([0, 1])$, and let us consider again the Markov process given by (4.18). If we let $I = I_{0,0} = \,]a, b[$, and for each $\omega \in \Omega$ we let

$$\ell_\tau(\omega) = \sum_{i \geq 0} |h_i(\omega)(I)|^\tau,$$

then the argument of the proof of Lemma 4.1.31 allows showing that since $\varepsilon > 0$, the function $\ell_\tau : \Omega \to \mathbb{R}$ is almost surely finite.

Let us consider a τ-Hölder constant C for $\log(f_1')$ and $\log(f_2')$ over $[0, b]$. For each $\omega = (f_{j_1}, f_{j_2}, \ldots) \in \Omega$, each n, k in $\mathbb{N}$, and each $x \in I$, the equality $f_3^n = h_k(\omega)^{-1} f_3^n h_k(\omega)$ implies that $\log((f_3^n)'(x))$ coincides with

$$\log((f_3^n)'(h_k(\omega)(x)))$$

$$+ \sum_{i=1}^{k} [\log(f_{j_i}'(h_{i-1}(\omega)(x))) - \log(f_{j_i}'(f_3^n \circ h_{i-1}(\omega)(x)))],$$

and hence

$$|\log((f_3^n)'(x))| \leq |\log((f_3^n)'(h_k(\omega)(x)))|$$

$$+ C \sum_{i=1}^{k} |h_{i-1}(\omega)(x) - f_3^n \circ h_{i-1}(\omega)(x)|^\tau$$

$$\leq |\log((f_3^n)'(h_k(\omega)(x)))| + C \sum_{i=1}^{k} |h_{i-1}(\omega)(I)|^\tau.$$

If we choose $\omega \in \Omega$ such that $\ell_\tau(\omega) = S$ is finite, this inequality implies that

$$|\log((f_3^n)'(x))| \le |\log((f_3^n)'(h_k(\omega)(x)))| + CS.$$

Notice that the sequence $(h_k(\omega)(x))$ converges to a (necessarily parabolic) fixed point of f_3 (actually, for almost every $\omega \in \Omega$, this sequence converges to the origin). Letting k go to infinity, we conclude that $|\log((f_3^n)'(x))| \le CS$. Therefore, $(f_3^n)'(x) \le \exp(CS)$ for all $x \in I$ and all $n \in \mathbb{N}$, which implies that the restriction of f_3 to the interval I is the identity. By the commutativity hypothesis, the same is true over all the intervals I_{n_1, n_2}, which concludes the proof. $\square$

Exercise 4.1.35. Prove that Theorem 4.1.34 still holds if $f_1, \ldots, f_d, f_{d+1}$ are C^1 commuting diffeomorphisms of the interval and the $(d - \varepsilon)$-variations of the derivatives of $f_1, \ldots, f_d$ are finite for some $\varepsilon > 0$ (cf. Exercise 4.1.32).

Exercise 4.1.36. One may give a version of Theorem 4.1.34 for which the commutativity hypothesis between the f_i's is weaker (compare Exercise 4.1.5). More precisely, let $d \ge 2$ be an integer, and let $f_1, \ldots, f_{d+1}$ be C^1 diffeomorphisms of the interval $[0, 1[$ (which do not necessarily commute). Suppose that there exist disjoint open intervals $I_{n_1, \ldots, n_{d+1}}$ disposed on $]0, 1[$ respecting the lexicographic ordering and such that for every $(n_1, \ldots, n_{d+1}) \in \mathbb{Z}^{d+1}$,

$$f_i(I_{n_1, \ldots, n_i, \ldots, n_{d+1}}) = I_{n_1, \ldots, n_i-1, \ldots, n_{d+1}} \quad \text{for all} \quad i \in \{1, \ldots, d+1\}.$$

Prove that $f_1, \ldots, f_d$ cannot all simultaneously be of class $C^{1+1/d+\varepsilon}$ for any $\varepsilon > 0$.

Hint. Use similar arguments to those of the proof of Theorem 4.1.34. After establishing a (generic) control of the distortion, apply the technique of the proof of Lemma 4.2.12 to reach the conclusion.

The combinatorial hypothesis of Theorem 4.1.34 might seem strange. Nevertheless, according to [243], the actions of $\mathbb{Z}^{d+1}$ on $[0, 1]$ satisfying this hypothesis are very interesting from a cohomological viewpoint. Notice, however, that it is not difficult to construct actions of $\mathbb{Z}^{d+1}$ by $C^{2-\varepsilon}$ diffeomorphisms of the interval without global fixed points in the interior. To do this, let us consider, for instance, d diffeomorphisms $f_2, \ldots, f_{d+1}$ of an interval $[a, b] \subset]0, 1[$ whose open supports are disjoint, and let f_1 be a diffeomorphism of $[0, 1]$ without fixed point in the interior and sending $]a, b[$ into a disjoint interval. Extending $f_2, \ldots, f_{d+1}$ to the whole interval $[0, 1]$ so that they commute with f_1, one obtains a faithful action of $\mathbb{Z}^{d+1}$ by homeomorphisms of the interval and without global fixed point in $]0, 1[$. Clearly, the methods from § 4.1.4 allow smoothing this action up to class $C^{2-\varepsilon}$ for every $\varepsilon > 0$.

It is also interesting to remark the existence of actions by diffeomorphisms of the interval that are free but nonconjugate to actions by translations at the interior. According to Proposition 4.1.2, such a phenomenon cannot appear in class C^{1+bv}. However, if we use the methods from § 4.1.4, it is possible to construct examples of $\mathbb{Z}^2$-actions by $C^{3/2-\varepsilon}$ diffeomorphisms of $[0, 1]$ that are free on $]0, 1[$ but do admit wandering intervals. Once again, the regularity $C^{3/2-\varepsilon}$ is optimal for the existence of these examples. The next result from [65] should be compared with Proposition 4.1.2 and Corollary 4.1.4.

Theorem 4.1.37. *Let Γ be a subgroup of $\mathrm{Diff}_+^{1+\tau}([0, 1[)$ isomorphic to $\mathbb{Z}^d$, with $\tau > 1/d$ and $d \geq 2$. If the restriction to $]0, 1[$ of the action of Γ is free, then it is minimal and topologically conjugate to the action of a group of translations.*

Proof. Once again, we will give the complete proof only for the case $d = 2$. Let f_1 and f_2 be the generators of a group $\Gamma \sim \mathbb{Z}^2$ of $C^{1+\tau}$ diffeomorphisms of $[0, 1[$ acting freely on $]0, 1[$. Replacing some of these elements by their inverses if necessary, we may assume that they topologically contract toward the origin. Suppose that the action of Γ on $]0, 1[$ is not conjugate to an action by translations. In this case a simple control-of-distortion argument of hyperbolic type shows that all the elements in Γ are tangent to the identity at the origin. Indeed, suppose for a contradiction that there exist wandering intervals and an element $f \in \Gamma$ such that $f'(0) < 1$. Fix $\lambda < 1$ and $c \in]0, 1[$ such that $f'(x) \leq \lambda$ for all $x \in [0, c[$, and fix a maximal open wandering interval $I =]a, b[$ contained in $]0, c[$. If we denote by L the interval $[f(b), b]$, then

$$\sum_{n \geq 0} |f^n(L)|^\tau \leq |L|^\tau \sum_{n \geq 0} \lambda^{n\tau} = \frac{|L|^\tau}{1 - \lambda^\tau} = \bar{S}.$$

Consequently, $(f^n)'(x)/(f^n)'(y) \leq \exp(C\bar{S})$ for all x and y in L, where $C > 0$ is a τ-Hölder constant for $\log(f')$ over $[0, c]$. This estimate allows applying the arguments of the proof of Proposition 4.1.2, thus obtaining a contradiction.[2]

Let us now identify the semigroup Γ^+ generated by f_1 and f_2 with $\mathbb{N}_0 \times \mathbb{N}_0$, and let us consider once again our Markov process on it. If $\tau > 1/2$, then the proof of Lemma 4.1.31 shows the finiteness of the expectation of the function

$$\omega \mapsto \ell_\tau(\omega) = \sum_{k \geq 0} |h_k(\omega)(I)|^\tau.$$

Choose a large-enough $S > 0$ that the set $\Omega(S) = \{\omega \in \Omega : \ell_\tau(\omega) \leq S\}$ has positive probability, and let $\bar{\ell} = |I| / \exp(2^\tau C S)$. The first part of the proof of Lemma 3.2.3 shows that if I'' denotes the interval that is next to I on the left and has length $\bar{\ell}$, then for every x and y in $J = \bar{I}'' \cup \bar{I}$, every $\omega \in \Omega(S)$, and every $n \in \mathbb{N}$,

$$\frac{h_n(\omega)'(x)}{h_n(\omega)'(y)} \leq \exp(2^\tau C S). \tag{4.21}$$

Since the interval I is not strictly contained in any other open wandering interval, there must exist $h \in \Gamma$ such that $h(I) \subset I''$ (and hence $|h(I)| < |I| \exp(-2^\tau C S)$). Fix an arbitrary point $y \in I$. Since $h'(0) = 1$ and $h_n(\omega)(y)$ converges to the origin for all $\omega \in \Omega(S)$, from the equality

$$h'(y) = \frac{h_n(\omega)'(y)}{h_n(\omega)'(h(y))} h'(h_n(\omega)(y))$$

and (4.21) one deduces that $h'(y) \geq \exp(-2^\tau C S)$. By integrating this inequality, we obtain $|h(I)| \geq |I| \exp(-2^\tau C S)$, which contradicts the choice of h.   □

Remark 4.1.38. Let $f_1, \ldots, f_d$ be the generators of a group $\Gamma \sim \mathbb{Z}^d$ acting freely by homeomorphisms of $]0, 1[$ so that the action is not conjugate to an action by translations. If we identify the points in each orbit of f_1, the maps $f_2, \ldots, f_d$ become the generators of a rank-$(d - 1)$ Abelian group of circle homeomorphisms that is semiconjugate, but nonconjugate, to a

---

2. Remark that this argument uses only the hypothesis $\tau > 0$. This is related to the fact that Sternberg's linearization theorem still holds in class $C^{1+\tau}$ for any positive τ (see Remark 3.6.7). Indeed, since the centralizer of a nontrivial linear germ coincides with the 1-parameter group of linear germs, this prevents the existence of wandering intervals for the dynamics we are dealing with.

group of rotations. Theorem 4.1.30 then implies that the f_i's cannot all be of class $C^{1/(d-1)+\varepsilon}$. Notice that this argument uses only the regularity of the f_i's at the interior; in this context the obstruction appears in class $C^{1/(d-1)}$. However, according to Theorem 4.1.37, for interval diffeomorphisms of $[0, 1[$ the obstruction already appears in class $C^{1/d}$. Obviously, the difference lies in the differentiability of the maps at the origin. This actually plays an important role in the proof.

Exercise 4.1.39. Extend Theorem 4.1.37 to subgroups of $\mathrm{Diff}_+^{1+\tau}([0, 1[)$ that are semiconjugate (on $]0, 1[$) to groups of translations but do not act freely on the interior (compare Exercises 4.1.5 and 4.1.5).

4.2 Nilpotent Groups of Diffeomorphisms

4.2.1 The Plante-Thurston Theorems

Given a nilpotent group Γ, let us denote

$$\{id\} = \Gamma_k^{\mathrm{nil}} \lhd \Gamma_{k-1}^{\mathrm{nil}} \lhd \ldots \lhd \Gamma_1^{\mathrm{nil}} \lhd \Gamma_0^{\mathrm{nil}} = \Gamma$$

the **central series** of Γ; that is, $\Gamma_{i+1}^{\mathrm{nil}} = [\Gamma, \Gamma_i^{\mathrm{nil}}]$ and $\Gamma_{k-1}^{\mathrm{nil}} \neq \{id\}$. Remark that the subgroup $\Gamma_{k-1}^{\mathrm{nil}}$ is contained in the center of Γ. The next result corresponds to a (weak version of a) theorem due to Plante and Thurston [207, 208].

Theorem 4.2.1. *Every nilpotent subgroup of* $\mathrm{Diff}_+^{1+\mathrm{bv}}([0, 1[)$ *is Abelian.*

Proof. Without loss of generality, we may suppose that $]0, 1[$ is an irreducible component of the action of Γ. We will show that the restriction of Γ to $]0, 1[$ is free, which allows us to conclude the proof using Hölder's theorem. Suppose for a contradiction that there exists a nontrivial $f \in \Gamma$ whose set of fixed points in $]0, 1[$ is nonempty, and let $x_0 \in]0, 1[$ be a point in the boundary of this set. Fix a nontrivial element g in the center of Γ. We claim that $g(x_0) = x_0$. Otherwise, replacing g by its inverse if necessary, we may assume that $g(x_0) < x_0$. Let

$$a' = \lim_{n \to +\infty} g^n(x_0), \qquad b' = \lim_{n \to -\infty} g^n(x_0).$$

Notice that $[a', b'[\subset [0, 1[$. Moreover, the restriction of g to $]a', b'[$ has no fixed point. Since f and g commute, each $g^n(x_0)$ is a fixed point of f, and hence f fixes a' and b'. We then obtain a contradiction by applying Kopell's lemma to the restrictions of f and g to the interval $[a', b'[$. Therefore, $g(x_0) = x_0$.

Since g was nontrivial and $g(x_0) = x_0$, the boundary in $]0, 1[$ of the set of fixed points of g is nonempty. Fix a point $x_1 \in]0, 1[$ in this boundary, and let h be an arbitrary element in Γ. Since $g(x_1) = x_1$ and $gh = hg$, the same argument as before shows that $h(x_1) = x_1$. Therefore, the intervals $[a', x_1[$ and $[x_1, b'[$ are invariant under Γ, and this contradicts the fact that $]0, 1[$ was an irreducible component. $\square$

Exercise 4.2.2. Prove that Theorem 4.2.1 still holds for nilpotent groups of germs of C^{1+bv} diffeomorphisms of the line fixing the origin.

Let us now consider the case of the circle. Since nilpotent groups are amenable, if Γ is a nilpotent subgroup of $\mathrm{Diff}_+^{1+bv}(S^1)$, then it must preserve a probability measure μ on S^1. The rotation-number function $g \mapsto \rho(g)$ is then a group homomorphism from Γ into $\mathbb{T}^1$ (see § 2.2.2). If the rotation number of an element $g \in \Gamma$ is irrational, then μ is conjugate to the Lebesgue measure. The group Γ is therefore conjugate to a group of rotations; in particular, Γ is Abelian.

Suppose now that the rotation number of every element in Γ is rational. Then the support of μ is contained in the intersection of the set of periodic points of these elements. If Γ is not Abelian, then we can take f and g in $\Gamma_{k-2}^{\mathrm{nil}}$ such that $h = [f, g] \in \Gamma_{k-1}^{\mathrm{nil}}$ is nontrivial. Notice that h is contained in the center of Γ; moreover, $\rho(h) = 0$, and thus h has fixed points. From the equality $f^{-1}g^{-1}fg = h$ one obtains $g^{-1}fg = fh$. Hence $g^{-1}f^n g = f^n h^n$, and by means of successive conjugacies by g one concludes that $g^{-m}f^n g^m = f^n h^{mn}$ for all $m \in \mathbb{N}$. It follows that

$$h^{mn} = f^{-n} g^{-m} f^n g^m. \tag{4.22}$$

If x_0 belongs to the support of μ and m and n are positive integers such that x_0 is a fixed point of both f^n and g^m, then (4.22) shows that $h(x_0) = x_0$. Let us consider the restriction to $[x_0, x_0 + 1[$ of the group generated by f^n, g^m, and h. Since this group is nilpotent, this restriction must be Abelian, and hence from (4.22) one concludes that h^{mn} is the identity over $[x_0, x_0 + 1[$. Therefore, h itself is the identity, which is a contradiction. We have then proved the following result.

Theorem 4.2.3. *Every nilpotent subgroup of* $\mathrm{Diff}_+^{1+bv}(S^1)$ *is Abelian.*

Finally, in the case of the line we have the following result (recall that a group Γ is **metabelian** if its first derived subgroup $\Gamma' = [\Gamma, \Gamma]$ is Abelian).

Theorem 4.2.4. *Every nilpotent subgroup of* $\mathrm{Diff}_+^{1+\mathrm{bv}}(\mathbb{R})$ *is metabelian.*

This result is a direct consequence of Corollary 2.2.40 and the next proposition.

Proposition 4.2.5. *Let* Γ *be a nilpotent subgroup of* $\mathrm{Diff}_+^{1+\mathrm{bv}}(\mathbb{R})$. *If every element in* Γ *fixes at least one point in the line, then* Γ *is Abelian.*

Proof. Fix a nontrivial element g in the center of Γ, and let b be a point in the boundary of the set of fixed points of g. We claim that b is fixed by every element in Γ. Indeed, suppose for a contradiction that $f \in \Gamma$ is such that $f(b) \neq b$. Replacing f by f^{-1} if necessary, we may suppose that $f(b) < b$. By hypothesis, at least one of the sequences $(f^n(b))$ or $(f^{-n}(b))$ converges to a fixed point $a \in \mathbb{R}$ of f. Since both cases are analogous, let us deal only with the first one. Notice that $f^n(b)$ is a fixed point of g for every $n \in \mathbb{N}$, and hence $g(a) = a$. On the other hand, f has no fixed point in $]a, b[$. Therefore, letting $a' = \lim_{n \to \infty} f^{-n}(b)$, we obtain a contradiction by applying Kopell's lemma to the restrictions of f and g to the interval $[a, a'[$.

The preceding discussion implies that every element $f \in \Gamma$ fixes the intervals $]-\infty, b]$ and $[b, \infty[$. The conclusion of the proposition then follows from Theorem 4.2.1. $\square$

It is important to point out that Theorem 4.2.4 does not mean that the nilpotence degree of every nilpotent subgroup of $\mathrm{Diff}_+^{1+\mathrm{bv}}(\mathbb{R})$ is less than or equal to 2. Actually, $\mathrm{Diff}_+^{\infty}(\mathbb{R})$ contains nilpotent subgroups of arbitrary degree of nilpotence. The following example, taken from [76], uses ideas that are very similar to those of [206].

Example 4.2.6. Let us consider a C^{∞} diffeomorphism $g : [0, 1] \to [0, 1]$ that is infinitely tangent to the identity at the endpoints (that is, $g'(0) = g'(1) = 1$, and all the higher derivatives at 0 and 1 are zero). Let f be the translation $f(x) = x - 1$ on the line. For each integer $n \geq 0$, let $k(n, m) = \binom{m+n-1}{n}$, and let h_n be the diffeomorphism of the line defined by

$$h_n(x) = g^{k(n,m)}(x - m) + m \qquad \text{for } x \in [m, m+1[.$$

We leave to the reader the task of showing that the maps h_n commute, and that $[f, h_n] = f^{-1} h_n^{-1} f h_n = h_{n-1}$ for every $n \geq 1$, while $[f, h_0] = Id$. One then easily concludes that the group Γ_n generated by f and $h_0, \ldots, h_n$ is nilpotent with degree of nilpotence $n + 1$.

Exercise 4.2.7. Using Theorem 4.2.4, show directly the following weak version of Theorem 2.2.24: for $n \geq 3$, every action of a finite-index subgroup of $SL(n, \mathbb{Z})$ by C^{1+bv} diffeomorphisms of the line is trivial (compare Remark 5.2.33).

4.2.2 On growth of groups of diffeomorphisms

Another viewpoint of the Plante-Thurston theorem is related to the notion of *growth* of groups. Recall that for a group provided with a finite system of generators, the **growth function** is the one that assigns to each positive integer n the number of elements that may be written as a product of no more than n generators and their inverses. One says that the group has **polynomial growth** or **exponential growth** if its growth function has the corresponding asymptotic behavior. These notions do not depend on the choice of the system of generators (see Appendix B).

Exercise 4.2.8. Prove that if a group contains a free semigroup on two generators, then it has exponential growth.

Remark. There exist groups of exponential growth without free semi-groups on two generators; see, for instance, [198].

Example 4.2.9. The construction of groups with **intermediate growth** (that is, neither polynomial nor exponential growth) is a nontrivial and fascinating topic. The first examples were given by Grigorchuk in [103]. One of them, the so-called *first Grigorchuk group* $\hat{H}$, may be seen either as a group acting on the binary rooted tree $\mathcal{T}_2$ or as a group acting isometrically on the Cantor set $\{0, 1\}^{\mathbb{N}}$. (These points of view are essentially the same, since the boundary at infinity of $\mathcal{T}_2$ naturally identifies with $\{0, 1\}^{\mathbb{N}}$.) Here we record the explicit definition. For more details, we strongly recommend the lecture of the final chapter of [111].

If we use the convention $(x_1, (x_2, x_3, \ldots)) = (x_1, x_2, x_3, \ldots)$ for $x_i \in \{0, 1\}$, the generators of $\hat{H}$ are the elements $\hat{a}, \hat{b}, \hat{c}$, and $\hat{d}$ whose actions on sequences in $\{0, 1\}^{\mathbb{N}}$ are recursively defined by $\hat{a}(x_1, x_2, x_3, \ldots) = (1 - x_1, x_2, x_3, \ldots)$ and

$$\hat{b}(x_1, x_2, x_3, \ldots) = \begin{cases} (x_1, \hat{a}(x_2, x_3, \ldots)), x_1 = 0, \\ (x_1, \hat{c}(x_2, x_3, \ldots)), \; x_1 = 1, \end{cases}$$

$$\hat{c}(x_1, x_2, x_3, \ldots) = \begin{cases} (x_1, \hat{a}(x_2, x_3, \ldots)), x_1 = 0, \\ (x_1, \hat{d}(x_2, x_3, \ldots)), x_1 = 1, \end{cases}$$

$$\hat{d}(x_1, x_2, x_3, \ldots) = \begin{cases} (x_1, x_2, x_3, \ldots), \quad x_1 = 0, \\ (x_1, \hat{b}(x_2, x_3, \ldots)), \; x_1 = 1. \end{cases}$$

Given this definition, the action on $\mathcal{T}_2$ of the element $\hat{a} \in \hat{H}$ consists in permuting the first two edges (together with the trees that are rooted at the second-level vertices). The elements $\hat{b}$, $\hat{c}$, and $\hat{d}$ fix these edges, and their actions on the higher levels of $\mathcal{T}_2$ are illustrated in Figure 22.

A celebrated theorem of Gromov establishes that a group has polynomial growth if and only if it is almost nilpotent, that is, if it contains a finite-index nilpotent subgroup. Because of this and Exercise 4.1.9, the Plante-Thurston theorem may be reformulated by saying that every subgroup of polynomial growth of $\mathrm{Diff}_+^{1+bv}([0, 1[)$ is Abelian. Actually, this was the original statement of the theorem, which is prior to Gromov's theorem (see Exercise 4.2.2). Following [179], in what follows, we will generalize this statement to groups of subexponential growth, and even more generally to groups without free semigroups on two generators.

Theorem 4.2.10. *Every finitely generated subgroup of* $\mathrm{Diff}_+^{1+bv}([0, 1[)$ *without free semigroups on two generators is Abelian.*

Example 4.2.11. The **wreath product** $\Gamma = \mathbb{Z} \wr \mathbb{Z} = \mathbb{Z} \ltimes \oplus_{\mathbb{Z}} \mathbb{Z}$ naturally acts on the interval; it suffices to identify the generator of the $\mathbb{Z}$-factor in Γ

FIGURE 22

with a homeomorphism f of $[0, 1]$ satisfying $f(x) < x$ for all $x \in]0, 1[$, and the element $(\ldots, 0, 1, 0, \ldots)$ of the second factor in Γ with a homeomorphism g satisfying $g(x) \neq x$ for all $x \in]f(a), a[$ and $g(x) = x$ for all $x \in [0, 1] \setminus [f(a), a]$, where a is some point in $]0, 1[$. This action may be smoothed up to class C^∞ (see Example 4.4.1 for more details on this construction). Notice that Γ is a metabelian, non–virtually nilpotent group. According to a classical theorem by Rosenblatt [219] (which generalizes prior results of Milnor [167] and Wolf [254]; see also [24]), every solvable group that is non–virtually nilpotent contains free semigroups on two generators. Therefore, Γ contains such a semigroup. Actually, looking at the action on the line, one may easily verify that the semigroup generated by f and g is free.

To show Theorem 4.2.10, we will need a version of Kopell's lemma that does not use the commutativity hypothesis too strongly. This version will be very useful in § 4.4 for the study of solvable groups of diffeomorphisms of the interval. The following proof is taken from [46].

Lemma 4.2.12. *Let h_1 and h_2 be two diffeomorphisms from the interval $[0, 1[$ into their images that fix the origin. Suppose that h_1 is of class C^{1+bv} and h_2 is of class C^1. Suppose, moreover, that $h_1(x) < x$ for all $x \in]0, 1[$, that $h_2(x_0) = x_0$ for some $x_0 \in]0, 1[$, and that for each $n \in \mathbb{N}$, the point $x_n = h_1^n(x_0)$ is fixed by h_2. Suppose, finally, that $h_2(y) \geq z > y$ (resp. $h_2(y) \leq z < y$) for some y and z in $]x_1, x_0[$. Then there exists $N \in \mathbb{N}$ such that $h_2(h_1^n(y)) < h_1^n(z)$ (resp. $h_2(h_1^n(y)) > h_1^n(z)$) for all $n \geq N$.*

Proof. Notice that the sequence (x_n) tends to zero as $n \in \mathbb{N}$ goes to infinity. Let $\delta = V(h_1; [0, x_0])$. For all u and v in $[x_1, x_0]$ and all $n \in \mathbb{N}$, one has (compare inequality (4.2))

$$\left| \log \left(\frac{(h_1^n)'(u)}{(h_1^n)'(v)} \right) \right| \leq \delta. \tag{4.23}$$

Since h_2 fixes each point x_n, one necessarily has $h_2'(0) = 1$. Suppose that the claim of the lemma is not satisfied by some $y < z$ in $]x_1, x_0[$ (the other case may be reduced to this one by replacing h_2 by h_2^{-1}). Let κ be a constant such that

$$1 < \kappa < 1 + \frac{z - y}{e^\delta (y - x_1)}.$$

Fix $N \in \mathbb{N}$, large enough that

$$h_2'(w) \leq \kappa \text{ and } (h_2^{-1})'(w) \leq \kappa \text{ for all } w \in [x_{N+1}, x_N]. \tag{4.24}$$

For some $n \geq N$, we have $h_2(h_1^n(y)) \geq h_1^n(z)$. Let $y_n = h_1^n(y)$ and $z_n = h_1^n(z)$. By the mean value theorem, there must exist points u and v in $[x_1, x_0]$ such that

$$\frac{h_2(y_n) - y_n}{y_n - x_{n+1}} \geq \frac{z_n - y_n}{y_n - x_{n+1}} = \frac{(h_1^n)'(u)(z - y)}{(h_1^n)'(v)(y - x_1)}.$$

From inequality (4.23) it follows that

$$\frac{h_2(y_n) - y_n}{y_n - x_{n+1}} \geq \frac{z - y}{e^\delta (y - x_1)} > \kappa - 1.$$

One then deduces that for some $w \in [x_{n+1}, y_n]$,

$$h_2'(w) = \frac{h_2(y_n) - h_2(x_{n+1})}{y_n - x_{n+1}} = \frac{h_2(y_n) - x_{n+1}}{y_n - x_{n+1}} > \kappa,$$

which contradicts (4.24).   □

In § 2.2.5 we introduced an elementary criterion (cf. Lemma 2.2.44) for detecting free semigroups inside groups of interval homeomorphisms. We now give another criterion for the smooth case that will be fundamental for the proof of Theorem 4.2.10. Example 4.2.11 well illustrates the following the lemma:

Lemma 4.2.13. *Let f and g be elements in* $\mathrm{Homeo}_+([0, 1[)$ *such that $f(x) < x$ for all $x \in]0, 1[$. Suppose that there exists an interval $[a, b]$ contained in $]0, 1[$ such that $g(x) < x$ for every $x \in]a, b[$, $g(a) = a$, $g(b) = b$, and $f(b) \leq a$. Let $[y, z] \subset]a, b[$ be an interval such that either $g(y) \geq z$ or $g(z) \leq y$. Suppose that g fixes all the intervals $f^n([a, b])$, and that there exists $N_0 \in \mathbb{N}$ such that g has fixed points inside $f^n(]a, b[)$ for every $n \geq N_0$. If the group generated by f and g has no crossed elements, then the semigroup generated by these elements is free.*

Proof. Replacing $[a, b]$ by its image under some positive iterate of f, we may assume that g has no fixed point inside $]a, b[$, but it has fixed points in $f^n(]a, b[)$ for all $n \geq 1$. We need to show that any two different words W_1 and W_2 in positive powers of f and g represent different homeomorphisms. Up to conjugacy, we may suppose that these words are of the form

$W_1 = f^n g^{m_r} f^{n_r} \cdots g^{m_1} f^{n_1}$ and $W_2 = g^q f^{p_s} g^{q_s} \cdots f^{p_1} g^{q_1}$, where m_j, n_j, p_j, and q_j are positive integers, $n \geq 0$, and $q \geq 0$ (with $n > 0$ if $r = 0$, and $p > 0$ if $s = 0$). Since the group generated by f and g has no crossed elements, according to Proposition 2.2.45 this group must preserve a Radon measure v on $]0, 1[$. Notice that

$$\tau_v(W_1) = (n_1 + \cdots + n_r + n)\tau_v(f), \qquad \tau_v(W_2) = (p_1 + \cdots + p_s)\tau_v(f).$$

Since $\tau_v(f) \neq 0$, if the values of $(n_1 + \cdots + n_r + n)$ and $(p_1 + \cdots + p_s)$ are different, then $\tau_v(W_1) \neq \tau_v(W_2)$, and hence $W_1 \neq W_2$. Now, assuming that these values are both equal to some $N \in \mathbb{N}$, we will show that the elements $f^{-N} W_1$ and $f^{-N} W_2$ are different. To do this, notice that

$$
\begin{aligned}
f^{-N} W_1 &= f^{-N} f^n g^{m_r} f^{n_r} \cdots g^{m_1} f^{n_1} \\
&= f^{-N} f^n g^{m_r} f^{n_r} \cdots g^{m_2} f^{n_1+n_2}(f^{-n_1} g^{m_1} f^{n_1}) \\
&= f^{-N} f^n g^{m_r} f^{n_r} \cdots g^{m_3} f^{n_1+n_2+n_3}(f^{-(n_1+n_2)} g^{m_2} f^{n_1+n_2}) \\
&\quad \cdot (f^{-n_1} g^{m_1} f^{n_1}) \\
&\vdots \\
&= (f^{-(N-n)} g^{m_r} f^{N-n}) \cdots (f^{-(n_1+n_2)} g^{m_2} f^{n_1+n_2})(f^{-n_1} g^{m_1} f^{n_1}),
\end{aligned}
$$

and

$$
\begin{aligned}
f^{-N} W_2 &= f^{-N} g^q f^{p_s} g^{q_s} \cdots f^{p_1} g^{q_1} \\
&= f^{-N} g^q f^{p_s} g^{q_s} \cdots f^{p_3} g^{q_3} f^{p_1+p_2}(f^{-p_1} g^{q_2} f^{p_1}) g^{q_1} \\
&\vdots \\
&= (f^{-N} g^q f^N) \cdots (f^{-p_1} g^{q_2} f^{p_1}) g^{q_1}.
\end{aligned}
$$

Since the group generated by f and g does not contain crossed elements, and since all the maps $f^{-(N-n)} g^{m_r} f^{N-n}, \ldots, f^{-(n_1+n_2)} g^{m_2} f^{n_1+n_2}, f^{-n_1} g^{m_1} f^{n_1}$, and $f^{-N} g^q f^N, \ldots, f^{-p_1} g^{q_2} f^{p_1}$ have fixed points inside $]a, b[$, these maps must fix $]a, b[$. On the other hand, g^{q_1} fixes the interval $]a, b[$, but it does not have fixed points in its interior. Therefore, if $\bar{v}$ is a Radon measure on $]a, b[$ that is invariant under the group generated by (the restrictions to $]a, b[$ of) all these maps (including g^{q_1}), then $\tau_{\bar{v}}(f^{-N} W_1) = 0$ and $\tau_{\bar{v}}(f^{-N} W_2) = \tau_{\bar{v}}(g^{q_1}) \neq 0$. This shows that $f^{-N} W_1 \neq f^{-N} W_2$. $\square$

We may now proceed to the proof of Theorem 4.2.10. Let Γ be a finitely generated subgroup of $\mathrm{Diff}_+^{1+bv}([0, 1[)$ without free semigroups on two generators. To show that Γ is Abelian, without loss of generality we may suppose that Γ has no global fixed point in $]0, 1[$. By Proposition 2.2.45,

Γ preserves a Radon measure v on $]0, 1[$; moreover, Exercise 2.2.5 provides us with an element $f \in \Gamma$ such that $f(x) < x$ for all $x \in \mathbb{R}$. Let Λ be the set of points in $]0, 1[$ that are globally fixed by the action of the derived group $\Gamma' = [\Gamma, \Gamma]$. The set Λ is nonempty, since it contains the support of v. If Λ coincides with $]0, 1[$, then Γ is Abelian. Suppose now that Λ is strictly contained in $]0, 1[$ and that the restriction of Γ' to each connected component of $]0, 1[\setminus\Lambda$ is free. By Hölder's theorem, the restriction of Γ' to each of these connected components is Abelian, and hence Γ is metabelian. By Rosenblatt's theorem [24, 219], Γ is virtually nilpotent, and by the Plante-Thurston theorem, Γ is virtually Abelian. Finally, from Exercise 4.1.9 one concludes that in this case Γ is Abelian.

It remains the case where the action of Γ' on some connected component I of the complement of Λ is not free. The proof of Theorem 4.2.10 will then be concluded by showing that in this situation Γ contains free semigroups on two generators. To do this, let us consider an element $h \in \Gamma'$ and an interval $]y, z[$ strictly contained in I such that h fixes $]y, z[$, but no point in $]y, z[$ is fixed by h. We claim that there must exist an element $g \in \Gamma'$ sending $]y, z[$ into a disjoint interval (contained in I). Indeed, if this is not the case, then since Γ has no crossed elements, every element in Γ' must fix $]y, z[$, and hence the points y and z are contained in Λ, thus contradicting the fact that $]y, z[$ is strictly contained in the connected component I of $]0, 1[\setminus\Lambda$.

The element $g \in \Gamma'$ fixes the connected component $f^n(I)$ of $]0, 1[\setminus\Lambda$ for every $n \geq 0$. Moreover, since Γ has no crossed elements, for each $n \geq 0$, the intervals $f^n(]y, z[)$ and $gf^n(]y, z[)$ either coincide or are disjoint. Lemma 4.2.12 (applied to $h_1 = f$ and $h_2 = g$, with x_0 being the right endpoint of I) implies the existence of an integer $N_0 \in \mathbb{N}$ such that g fixes the interval $f^n(]y, z[)$ for every $n \geq N_0$ (see Figure 23). If we consider the dynamics of f and g on the open convex closure of $\cup_{n \in \mathbb{Z}} g^n(]y, z[)$, it follows from Lemma 4.2.13 that the semigroup generated by these elements is free. This concludes the proof of Theorem 4.2.10.

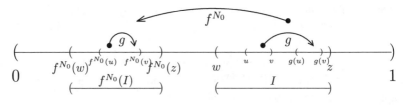

FIGURE 23

Exercise 4.2.14.   Show that subgroups of $\mathrm{Diff}_+^{1+\mathrm{bv}}(\mathbb{S}^1)$ (resp. $\mathrm{Diff}_+^{1+\mathrm{bv}}(\mathbb{R})$) without free semigroups on two generators are Abelian (resp. meta-belian) (compare Theorems 4.2.3 and 4.2.4).

Exercise 4.2.15.   Throughout this exercise, Γ will always be a finitely generated group having polynomial growth of degree k.

(i) Prove that every finitely generated subgroup of Γ has polynomial growth of degree less than or equal to k.

(ii) Prove that if $\Gamma_n \subset \Gamma_{n-1} \subset \ldots \subset \Gamma_0 = \Gamma$ is a series of subgroups such that for every $i \in \{0, \ldots, n-1\}$ there exists a nontrivial group homomorphism $\phi_i :$ $\Gamma_i \to (\mathbb{R}, +)$ satisfying $\phi_i(\Gamma_{i+1}) = \{0\}$, then $n \leq k$.

Hint. For each $i \in \{0, \ldots, n-1\}$, choose an element $g_i \in \Gamma_i$ such that $\phi_i(g_i) \neq 0$, and consider the words of the form $W = g_1^{m_1} g_2^{m_2} \cdots g_n^{m_n}$.

(iii) Suppose that Γ is a subgroup of $\mathrm{Homeo}_+([0, 1])$. Using (ii) and the translation-number homomorphism, show that Γ is solvable with degree of solvability less than or equal to k. Applying the Milnor-Wolf theorem [167, 254], conclude that Γ is virtually nilpotent.

(iv) Suppose now that Γ is a subgroup of $\mathrm{Diff}_+^{1+\mathrm{bv}}([0, 1[)$. Using Theorem 4.2.1 and Exercise 4.1.9, conclude that Γ is Abelian.

(v) Prove that the claim in (iv) still holds if Γ is contained in the group of germs $\mathcal{G}_+^{1+\mathrm{bv}}(\mathbb{R}, 0)$.

Hint. Instead of using the translation-number homomorphism, use Thurston's stability theorem from § 5.1.

Exercise 4.2.16.   The claim in item (ii) of this exercise may be proved either by a direct argument or by using the positive ping-pong lemma (see [24]).

(i) Under the hypothesis of Lemma 4.2.13, show that the group generated by the family of elements $\{f^n g f^{-n} : n \in \mathbb{Z}\}$ is not finitely generated.

(ii) Prove that if f and g are elements in a group such that the subgroup generated by $\{f^n g f^{-n} : n \in \mathbb{Z}\}$ is not finitely generated, then the semigroup generated by f and gfg^{-1} is free.

(iii) Use (ii) to give an alternative proof of Theorem 4.2.10.

Remark 4.2.17. We do not know whether Theorem 4.2.10 extends (at least in the case of subexponential growth) to groups of *germs* of C^{1+bv} diffeomorphisms of the line fixing the origin (compare Exercise 4.2.2).

4.2.3 Nilpotence, growth, and intermediate regularity

As in the cases of Denjoy's theorem and Kopell's lemma, the results of the preceding two sections are no longer true in regularity less than C^{1+Lip}. Indeed, using the first of the techniques in § 4.1.4, Farb and Franks provided in [76] a C^1 realization of each group N_n from Exercise 2.2.3 (notice that N_n is nilpotent for every n and non-Abelian for $n \geq 3$). On the other hand, according to a result by Malcev (see [211]), every finitely generated torsion-free nilpotent group embeds into some N_n. All of this shows the existence of a large variety of C^1 counterexamples to the Plante-Thurston theorem.

In fact, using the second method of construction from § 4.1.4, one may show that the canonical action of N_n on the interval can be smoothed until we reach any differentiability class less than $C^{1+2/(n-1)(n-2)}$ (see [130]). Notice that by Theorem 4.1.34, this action cannot exceed the class $C^{1+1/(n-2)}$. Filling the gap between these two differentiabilities remains an open problem. (It seems that the former differentiability is the critical one.)

Let us point out that N_n admits actions on the interval that are not semi-conjugate to the canonical one. In this direction, the following general question remains open: given a finitely generated torsion-free nilpotent group Γ, what is the best regularity for faithful actions of Γ by diffeomorphisms of the (closed) interval ? Concerning this, it is worth mentioning that the groups from Example 4.2.6 can be realized as groups of $C^{1+\tau}$ diffeomorphisms of $[0, 1]$ for any $\tau < 1$ (see [130]). Therefore, it is very likely that the right parameter giving the critical degree of obstruction should be associated to the order of solvability rather than to the order of nilpotence of the group.

Remark 4.2.18. For each pair of positive integers m and n such that $m < n$, the group N_m naturally embeds into N_n. If we denote by N the union of all these groups, then one easily checks that N is orderable, countable, and non–finitely generated. Moreover, N contains all torsion-free finitely generated nilpotent groups. It would be interesting to check whether the methods from § 4.1.4 allow showing that N is isomorphic to a group of C^1 diffeomorphisms of the interval.

To summarize the preceding remarks, every finitely generated torsion-free nilpotent group may be embedded into $\text{Diff}_+^{1+\tau}([0, 1])$ for some small-enough $\tau > 0$. It is then natural to ask whether some other groups of sub-exponential growth may appear as groups of $C^{1+\tau}$ diffeomorphisms of the interval. According to Exercise 4.2.8, the following result (due to the author [179]) provides a negative answer to this.

Theorem 4.2.19. *For all $\tau > 0$, every finitely generated group of $C^{1+\tau}$ diffeomorphisms of the interval $[0, 1[$ that does not contain free semigroups on two generators is virtually nilpotent.*

Remark 4.2.20. In recent years another notion has come to play an important role in the theory of growth of groups, namely, the notion of **uniformly exponential growth**. One says that a finitely generated group Γ has uniformly exponential growth if there exists a constant $\lambda > 1$ such that for every (symmetric) finite system of generators $\mathcal{G}$ of Γ, the number $L_{\mathcal{G}}(n)$ of elements in Γ that may be written as a product of at most n elements in $\mathcal{G}$ satisfies

$$\lim_{n \to \infty} \frac{\log(L_{\mathcal{G}}(n))}{n} \geq \lambda.$$

Although in many situations groups of exponential growth are forced to have uniform exponential growth (see, for instance, [24, 26, 74, 200]), there exist groups for which these properties are not equivalent [251]. We do not know whether this may happen for groups of $C^{1+\tau}$ diffeomorphisms of the interval.

The condition $\tau > 0$ in Theorem 4.2.19 is necessary. To show this, we sketch the construction of a subgroup of $\text{Diff}_+^1([0, 1])$ having intermediate growth. This subgroup is closely related to the group $\hat{H}$ from Example 4.2.9. Indeed, although $\hat{H}$ is a torsion group (the reader may easily check that the order of every element is a power of 2; see also [111]), starting with $\hat{H}$, one may create a *torsion-free* group having intermediate growth. This group was introduced in [102]. Geometrically, the idea consists in replacing $\mathcal{T}_2$ by a rooted tree whose vertices have (countable) infinite valence. More precisely, we consider the group $\bar{H}$ acting on the space $\Omega = \mathbb{Z}^{\mathbb{N}}$ and generated by the elements $\bar{a}, \bar{b}, \bar{c}$, and $\bar{d}$, recursively defined by $\bar{a}(x_1, x_2, x_3, \ldots) = (1 + x_1, x_2, x_3, \ldots)$ and

$$\bar{b}(x_1, x_2, x_3, \ldots) = \begin{cases} (x_1, \bar{a}(x_2, x_3, \ldots)), \, x_1 \text{ even}, \\ (x_1, \bar{c}(x_2, x_3, \ldots)), \, x_1 \text{ odd}, \end{cases}$$

$$\bar{c}(x_1, x_2, x_3, \ldots) = \begin{cases} (x_1, \bar{a}(x_2, x_3, \ldots)), \, x_1 \text{ even}, \\ (x_1, \bar{d}(x_2, x_3, \ldots)), \, x_1 \text{ odd}, \end{cases}$$

$$\bar{d}(x_1, x_2, x_3, \ldots) = \begin{cases} (x_1, x_2, x_3, \ldots), \quad x_1 \text{ even}, \\ (x_1, \bar{b}(x_2, x_3, \ldots)), \, x_1 \text{ odd}. \end{cases}$$

The group $\bar{H}$ preserves the lexicographic ordering on Ω, from which one concludes that it is orderable, and hence, according to § 2.2.3, it may be realized as a group of homeomorphisms of the interval (compare [104]). To explain this better, we give a concrete realization of $\bar{H}$ as a group of bi-Lipschitz homeomorphisms of $[0, 1]$ (compare Proposition 2.3.15).

Example 4.2.21. Given $C > 1$, let $(\ell_i)_{i \in \mathbb{Z}}$ be a sequence of positive real numbers such that

$$\sum_{i \in \mathbb{Z}} \ell_i = 1, \qquad \max\left\{\frac{\ell_{i+1}}{\ell_i}, \frac{\ell_i}{\ell_{i+1}}\right\} \leq C \quad \text{for all} \quad i \in \mathbb{Z}.$$

Let I_i be the interval $]\sum_{j < i} \ell_j, \sum_{j \leq i} \ell_j[$, and let $f : [0, 1] \to [0, 1]$ be the homeomorphism sending affinely each interval I_i onto I_{i+1}. Let us denote by g the affine homeomorphism sending $[0, 1]$ onto $\bar{I}_0$, and by $\lambda = 1/\ell_0$ the (constant) value of its derivative. Let a, b, c, and d be the maps recursively defined on a dense subset of $[0, 1]$ by $a(x) = f(x)$ and, for $x \in I_i$,

$$b(x) = \begin{cases} f^i gag^{-1} f^{-i}(x), i \text{ even}, \\ f^i gcg^{-1} f^{-i}(x), i \text{ odd}, \end{cases}$$

$$c(x) = \begin{cases} f^i gag^{-1} f^{-i}(x), i \text{ even}, \\ f^i gdg^{-1} f^{-i}(x), i \text{ odd}, \end{cases}$$

$$d(x) = \begin{cases} x, \qquad\qquad\quad i \text{ even}, \\ f^i gbg^{-1} f^{-i}(x), i \text{ odd}. \end{cases}$$

We claim that a, b, c, and d are bi-Lipschitz homeomorphisms with Lipschitz constant bounded from above by C (notice that C may be chosen as near to 1 as we want). Indeed, this is evident for a, while for b, c, and d, this may be easily checked by induction. For instance, if $x \in I_i$ for some odd integer i, then

$$b'(x) = \frac{(f^i)'(gag^{-1}f^{-i}(x))}{(f^i)'(f^{-i}(x))} \cdot \frac{g'(ag^{-1}f^{-i}(x))}{g'(g^{-1}f^{-i}(x))} \cdot a'(g^{-1}f^{-i}(x)),$$

and since $g'|_{[0,1]} \equiv \lambda$ and $(f^i)'|_{I_0} \equiv \ell_i/\ell_0$, we have $b'(x) = a'(g^{-1}f^{-i}(x)) \leq C$. It is geometrically clear that the group generated by $a, b, c,$ and d is isomorphic to $\bar{H}$.

Exercise 4.2.22.   Give an example of a finitely generated group Γ of homeomorphisms of the interval and/or the circle for which there exists a finite system of generators and a constant $\delta > 0$ such that for every topological conjugacy of Γ to a group of bi-Lipschitz homeomorphisms, at least one of these generators or its inverse has Lipschitz constant greater than or equal to $1 + \delta$.

Remark. It seems to be interesting to determine whether such a Γ may be a group of C^1 diffeomorphisms without crossed elements (cf. Definition 2.2.43).

The preceding idea is inappropriate to obtain an embedding of $\bar{H}$ into $\mathrm{Diff}^1_+([0,1])$, since the derivatives of the maps that are produced have discontinuities at each "level" of the action of H. Next we provide another realization for which it will be essential to "renormalize" the geometry at each step. We begin by fixing an equivariant family of homeomorphisms $\{\varphi_{u,v} : u > 0, v > 0\}$. For each $n \in \mathbb{N}$ and each $(x_1, \ldots, x_n)$ in $\mathbb{Z}^n$, let us consider a nondegenerate closed interval $I_{x_1,\ldots,x_n} = [u_{x_1,\ldots,x_n}, w_{x_1,\ldots,x_n}]$ and an interval $J_{x_1,\ldots,x_n} = [v_{x_1,\ldots,x_n}, w_{x_1,\ldots,x_n}]$, both contained in some interval $[0, T]$. Suppose that the following conditions hold (see Figure 24):

(i) $\sum_{x_1 \in \mathbb{Z}} |I_{x_1}| = T,$

(ii) $u_{x_1,\ldots,x_n} < v_{x_1,\ldots,x_n} \leq w_{x_1,\ldots,x_n}$, thus $J_{x_1,\ldots,x_n} \subset I_{x_1,\ldots,x_n},$

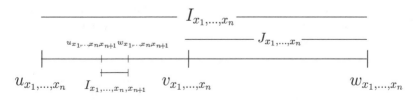

FIGURE 24

(iii) $w_{x_1,\ldots,x_{n-1},x_n} = u_{x_1,\ldots,x_{n-1},1+x_n}$,

(iv) $\lim_{x_n \to -\infty} u_{x_1,\ldots,x_{n-1},x_n} = u_{x_1,\ldots,x_{n-1}}$,

(v) $\lim_{x_n \to \infty} u_{x_1,\ldots,x_{n-1},x_n} = v_{x_1,\ldots,x_{n-1}}$, and

(vi) $\lim_{n \to \infty} \sup_{(x_1,\ldots,x_n) \in \mathbb{Z}^n} |I_{x_1,\ldots,x_n}| = 0$.

Notice that

$$|J_{x_1,\ldots,x_n}| + \sum_{x_{n+1} \in \mathbb{Z}} |I_{x_1,\ldots,x_n,x_{n+1}}| = |I_{x_1,\ldots,x_n}|. \tag{4.25}$$

Let us denote by $\bar{H}_n$ the stabilizer in $\bar{H}$ of the nth-level of the tree T_∞. For each $n \in \mathbb{N}$, we will define four homeomorphisms a_n, b_n, c_n, and d_n such that the group generated by them is isomorphic to $\bar{H}/\bar{H}_n$. To do this, let us consider the group homomorphisms ϕ_0 and ϕ_1 from the subgroup $\langle \bar{b}, \bar{c}, \bar{d} \rangle \subset \bar{H}$ into $\bar{H}$ defined by

$$\phi_0(\bar{b}) = \bar{a},\ \phi_0(\bar{c}) = \bar{a},\ \phi_0(\bar{d}) = id, \qquad \phi_1(\bar{b}) = \bar{c},\ \phi_1(\bar{c}) = \bar{d},\ \phi_1(\bar{d}) = \bar{b}.$$

Definition of a_n.

– If $p \in J_{x_1,\ldots,x_i}$ for some $i < n$, we let $a_n(p) = \varphi(J_{x_1,x_2,\ldots,x_i}, J_{1+x_1,x_2,\ldots,x_i})(p)$.

– For $p \in I_{x_1,\ldots,x_n}$, we let $a_n(p) = \varphi(I_{x_1,x_2,\ldots,x_n}, I_{1+x_1,x_2,\ldots,x_n})(p)$.

Definition of b_n. Suppose that $p \in]0, 1[$ belongs to $I_{x_1,\ldots,x_n}$, and denote by $(\bar{x}_1, \ldots, \bar{x}_n) \in \{0, 1\}^n$ the sequence obtained after reducing each entry modulo 2.

– If $\phi_{\bar{x}_1}(\bar{b})$, $\phi_{\bar{x}_2}\phi_{\bar{x}_1}(\bar{b})$, $\ldots$, $\phi_{\bar{x}_n} \ldots \phi_{\bar{x}_2}\phi_{\bar{x}_1}(\bar{b})$ are well defined, we let $b_n(p) = p$.

– In the other case we denote by $i = i(p) \le n$ the smallest integer for which $\phi_{\bar{x}_i} \ldots \phi_{\bar{x}_2}\phi_{\bar{x}_1}(\bar{b})$ is not defined, and we let

$$b_n(p) = \begin{cases} p, & p \in J_{x_1,\ldots,x_j},\ j < i, \\ \varphi(J_{x_1,\ldots,x_i,\ldots,x_j}, J_{x_1,\ldots,1+x_i,\ldots,x_j})(p), & p \in J_{x_1,\ldots,x_i,\ldots,x_j},\ i \le j < n, \\ \varphi(I_{x_1,\ldots,x_i,\ldots,x_n}, I_{x_1,\ldots,1+x_i,\ldots,x_n})(p), & p \in I_{x_1,\ldots,x_n}. \end{cases}$$

The definitions of c_n and d_n are similar to that of b_n. Clearly, the maps a_n, b_n, c_n, and d_n extend to homeomorphisms of $[0, T]$. The fact that they generate a group isomorphic to $\bar{H}/\bar{H}_n$ follows from the equivariance properties of the maps $\varphi(I, J)$. Condition (vi) implies, moreover, that the sequences of maps a_n, b_n, c_n, and d_n converge to homeomorphisms a, b, c, and d, respectively, which generate a group isomorphic to $\bar{H}$. We leave the details to the reader.

Example 4.2.23. Given a sequence $(\ell_i)_{i \in \mathbb{Z}}$ of positive real numbers such that $\sum \ell_i = 1$, let us define $|I_{x_1,\dots,x_n}|$ and $|J_{x_1,\dots,x_n}|$ by

$$|J_{x_1,\dots,x_n}| = 0, \qquad |I_{x_1,\dots,x_n}| = \ell_{x_1} \dots \ell_{x_n}.$$

If we carry out the preceding construction (for $T = 1$) using the equivariant family of affine maps $\varphi([0, u], [0, v])(x) = vx/u$, then we reobtain the inclusion of $\bar{\mathrm{H}}$ in the group of bi-Lipschitz homeomorphisms of the interval from Example 4.2.21 (under the same hypothesis $\ell_{i+1}/\ell_i \leq C$ and $\ell_i/\ell_{i+1} \leq C$ for all $i \in \mathbb{Z}$).

The details of the rest of the construction are straightforward, and we leave them to the reader (see [179] in case of problems). Let ω be a modulus of continuity satisfying $\omega(s) = 1/\log(1/s)$ for $s \leq 1/2e$, and such that the map $s \mapsto \omega(s)/s$ is decreasing. Fix a constant $C > 0$, and for each $k \in \mathbb{N}$, let $T_k = \sum_{i \in \mathbb{Z}} \frac{1}{(|i|+k)^2}$. Consider an increasing sequence (k_n) of positive integers, and for $n \in \mathbb{N}$ and $(x_1, \dots, x_n) \in \mathbb{Z}^n$, let

$$|I_{x_1,\dots,x_n}| = \frac{1}{(|x_1| + \dots |x_n| + k_n)^{2n}}.$$

If we use the equivariant family induced by (4.10), our general method provides us with subgroups of $\mathrm{Diff}_+^1([0, T_{k_1}])$ generated by elements a_n, b_n, c_n, and d_n and isomorphic to $\bar{\mathrm{H}}/\bar{\mathrm{H}}_n$. The goal then consists in controlling the C^ω-norm for the derivative of these diffeomorphisms. If the sequence (k_n) satisfies certain "rapid-growth" conditions (which depend only on C), it is not difficult to verify that these norms are bounded from above by C for all $n \in \mathbb{N}$ (the same holds for the derivatives of the inverses of these diffeomorphisms). The corresponding sequences are therefore equicontinuous, and it is easy to see that they converge to ω-continuous maps having C^ω-norm bounded from above by C. These maps correspond to the derivatives of $C^{1+\omega}$ diffeomorphisms a, b, c, and d (and their inverses), which generate a group isomorphic to $\bar{\mathrm{H}}$. Since this group acts on the interval $[0, T_{k_1}]$, to obtain an action on $[0, 1]$, we may conjugate by the affine map $g : [0, 1] \to [0, T_{k_1}]$. Since for large-enough k_1 one has $T_{k_1} \leq 1$, this conjugacy procedure does not increase the C^ω-norm for the derivatives. Finally, notice that since for every diffeomorphism f of $[0, 1]$ there exists a point at which the derivative equals 1, if the C^ω-norm of f' is bounded from above by C, then

$$\sup_{x \in [0,1]} |f'(x) - 1| \leq C\omega(1).$$

As a consequence, for small C, the preceding realization of $\bar{H}$ has its generators near the identity with respect to the $C^{1+\omega}$ topology.

Exercise 4.2.24. Inside the Grigorchuk-Maki group $\bar{H}$ there are many of commuting elements. More precisely, for each $d \in \mathbb{N}$, one may choose $d+1$ elements in $\bar{H}$ satisfying a combinatorial property similar to that of Theorem 4.1.34 (take, for instance, $f_1 = \bar{a}^{-2}$, $f_2 = \bar{b}^{-2}$, $f_3 = \bar{a}^{-1}\bar{b}^{-2}\bar{a}$, and so on). In this way, to verify that the natural action of $\bar{H}$ is not semiconjugate to an action by $C^{1+\tau}$ diffeomorphisms, it suffices to apply Theorem 4.1.34 for $d > 1/\tau$. The following items allow showing the same claim via more elementary methods (see [179] for more details).

(i) Prove that if h is a $C^{1+\tau}$ diffeomorphism of a closed interval $[u, v]$, and C denotes the Hölder constant of h', then for every $x \in [u, v]$, one has
$$|h(x) - x| \leq C|v - u|^{1+\tau}.$$

(ii) Using (i), prove directly (i.e., without using any probabilistic argument) that Theorem 4.1.34 holds for $d \geq d(\alpha)$, where $d(\alpha)$ is the minimal integer greater than or equal to 2 for which $\alpha(1 + \tau)^{d-2} \geq 1$.

(iii) Using (ii), conclude that the canonical action of $\bar{H}$ is not semiconjugate to an action by $C^{1+\tau}$ diffeomorphisms for any $\tau > 0$.

To conclude this section, we notice that the element $\bar{a}^2$ belongs to the center of $\bar{H}$. If we bear in mind the realization of $\bar{H}$ as a group of C^1 diffeomorphisms of the interval, the study of centralizers in $\text{Diff}^1_+([0, 1])$ of diffeomorphisms (or homeomorphisms) of the interval becomes natural. The next "weak Kopell lemma", due to Bonatti, Crovisier, and Wilkinson, is an interesting result in this direction.

Proposition 4.2.25. *If h is a homeomorphism of $[0, 1[$ without fixed points in $]0, 1[$, then the group of C^1 diffeomorphisms of $[0, 1[$ that commute with h has no crossed elements.*

Proof. Suppose for a contradiction that f and g are C^1 diffeomorphisms of $[0, 1[$ that commute with h, and that they are crossed on some interval $[x, y] \subset [0, 1]$. As in the proof of Lemma 2.2.44, we may restrict ourselves to the case where $f(x) = x$, $f(y) \in]x, y[$, $g(x) \in]x, y[$, and $g(y) = y$. Moreover, replacing f and/or g by some of their iterates, we may suppose that $g(x) > f(y)$. All of these properties are preserved under conjugacy, and hence f and g satisfy $f(h^n(x)) = h^n(x)$, $f(h^n(y)) \in]h^n(x), h^n(y)[$,

$g(h^n(x)) \in]h^n(x), h^n(y)[$, $g(h^n(y)) = h^n(y)$, and $g(h^n(x)) > f(h^n(y))$ for all $n \in \mathbb{Z}$. If $h(z) < z$ (resp. if $h(z) > z$) for all $z \in]0, 1[$, then the sequences $(h^n(x))$ and $(h^n(y))$ (resp. $(h^{-n}(x))$ and $(h^{-n}(y))$) converge to the origin. Since f and g are of class C^1, this implies that $f'(0) = g'(0) = 1$. On the other hand, since $g(h^n(x)) > f(h^n(y))$, there must exist a sequence of points $z_n \in]x_n, y_n[$ such that for each $n \in \mathbb{Z}$, either $f'(z_n) < 1/2$ or $g'(z_n) < 1/2$. However, this contradicts the continuity of the derivatives of f and g at the origin. $\qquad\square$

4.3 Polycyclic Groups of Diffeomorphisms

Unlike nilpotent groups, inside the group of C^2 diffeomorphisms of the interval there is a large variety of non-Abelian *polycyclic* groups. To see this, first notice that the affine group contains many polycyclic groups (see Exercise 4.3.1). On the other hand, the affine group is conjugate to a subgroup of $\mathrm{Diff}_+^\infty([0, 1])$. Indeed, inspired by Exercise 5.1.14, let us fix a constant $0 < \varepsilon < 1/2$, and let us consider two C^∞ diffeomorphisms $\varphi_1 :]0, 1[\to \mathbb{R}$ and $\varphi_2 :]0, 1[\to]0, 1[$ such that

$$\varphi_1(x) = -\frac{1}{x} \quad \text{and} \quad \varphi_2(x) = \exp\left(-\frac{1}{x}\right) \quad \text{for } x \in]0, \varepsilon],$$

$$\varphi_1(x) = \frac{1}{1-x} \quad \text{and} \quad \varphi_2(x) = 1 - \exp\left(\frac{1}{x-1}\right) \quad \text{for } x \in [1-\varepsilon, 1[.$$

Consider now the vector fields on the line $Y_1 = \frac{\partial}{\partial x}$ and $Y_2 = x\frac{\partial}{\partial x}$, which generate the Lie algebra of the affine group. One readily checks that the vector fields $X_j = \varphi^*(Y_j)$, where $j \in \{1, 2\}$ and $\varphi = \varphi_1 \circ \varphi_2^2$, extend to C^∞ vector fields on $[0, 1]$ that are (zero and) infinitely flat at the endpoints. Because of this, for each $g \in \mathrm{Aff}_+(\mathbb{R})$, the map $\varphi^{-1} \circ g \circ \varphi$ is a C^∞ diffeomorphism of $[0, 1]$ infinitely tangent to the identity at the endpoints.

Exercise 4.3.1. Show that every polycyclic subgroup of the affine group is isomorphic to $\mathbb{Z}^k \ltimes \mathbb{Z}^n$ for some nonnegative integers n and k.

The preceding construction was first given by Plante [205, 206], who showed that under a mild assumption, polycyclic groups of C^2 diffeomorphisms of the interval are forced to be conjugate to subgroups of the affine group; see Theorem 4.3.3, item (i). (Plante's results were used by

Matsumoto to classify certain codimension-one foliations; see [162].) It was Moriyama [172] who provided another kind of examples and gave a complete classification. To introduce his examples, given any integer $n \geq 2$ and a matrix A in $SL(n, \mathbb{Z})$ with an eigenvalue $\lambda \in]0, 1[$, let $(t_1, \ldots, t_n) \in \mathbb{R}^n$ be an eigenvector associated to it. Given an interval $[a, b] \subset]0, 1[$, let X be a nonzero vector field on $[a, b]$ that is infinitely tangent to zero at the endpoints. Consider a diffeomorphism f of $[0, 1[$ topologically contracting toward the origin such that $f(b) = a$ and such that if we extend X to $[0, 1[$ by imposing the condition $f_*(X) = \frac{1}{\lambda} X$, then the resulting X is of class C^∞ on $[0, 1[$ (to ensure the existence of such a vector field, one can either give an explicit construction or use Excrcise 4.3.2). Let $\{\varphi^t : t \in \mathbb{R}\}$ be the flow associated to X. For $i \in \{1, \ldots, n\}$, let f_i be the diffeomorphism φ^{t_i}. We claim that the group Γ generated by $f, f_1, \ldots, f_n$ is polycyclic. Indeed, from

$$f^{-1} f_i f = \varphi^{\lambda t_i} = \varphi^{\sum_j a_{i,j} t_j} = \prod_{j=1}^n (\varphi^{t_j})^{a_{i,j}} = \prod_{j=1}^n f_j^{a_{i,j}}$$

one easily concludes that the Abelian subgroup generated by $f_1, \ldots, f_n$ is normal. Moreover, Γ' coincides with the subgroup consisting of the elements of the form $f_1^{m_1} f_2^{m_2} \cdots f_n^{m_n}$, where $(m_1, \ldots, m_n)$ belongs to $(A - Id)(\mathbb{Z}^n)$. Therefore, Γ is polycyclic.

Exercise 4.3.2. Given $k \geq 1$, let $f \in \text{Diff}_+^{k+1}([0, 1[)$ be a diffeomorphism topologically contracting toward the origin and such that $f'(0) = 1$. Let X be a continuous vector field on $[0, 1[$ that is of class C^k on $]0, 1[$. Show that if $f_*(X) = \frac{1}{\lambda} X$ for some $\lambda \in]0, 1[$, then X is of class C^k on $[0, 1[$ (see [172] in case of problems).

To state Moriyama's theorem properly, recall that for every polycyclic group Γ, there is an associated **nilradical** $N(\Gamma)$ that corresponds to the maximal nilpotent normal subgroup of Γ (see Appendix A). For example, in the preceding example, $N(\Gamma)$ coincides with the Abelian group generated by $f_1, \ldots, f_n$. Moreover, although in general $N(\Gamma)$ does not necessarily contain the derived subgroup of Γ, there exists a finite-index subgroup Γ_0 of Γ such that $[\Gamma_0, \Gamma_0] \subset N(\Gamma)$.

Theorem 4.3.3. *Let Γ be a polycyclic subgroup of $\text{Diff}_+^{1+\text{bv}}([0, 1[)$ having no global fixed point in $]0, 1[$. The following two possibilities may occur:*

(i) *If $N(\Gamma)$ has no global fixed point in $]0, 1[$, then Γ is conjugate to a subgroup of the affine group.*

(ii) *If $N(\Gamma)$ has a global fixed point in $]0, 1[$, then Γ is a semidirect product $\mathbb{Z}^n \rtimes \mathbb{Z}$.*

Actually, in case (ii) the dynamics can be fully described, and it is very similar to that of the second example previously exhibited.

Here we show only that polycyclic subgroups of $\mathrm{Diff}_+^{1+\mathrm{bv}}([0, 1[)$ are metabelian. (Theorem 4.3.3 will then follow from the classification of solvable subgroups of $\mathrm{Diff}_+^{1+\mathrm{bv}}([0, 1[)$; see § 4.4.2.) Actually, this is quite easy. Let Γ be a non-Abelian polycyclic subgroup of $\mathrm{Diff}_+^{1+\mathrm{bv}}([0, 1[)$. By the Plante-Thurston theorem, the nilradical $N(\Gamma)$ is Abelian (and its restriction to each of its irreducible components is conjugate to a group of translations). Hence if Γ_0 is a finite-index subgroup of Γ such that $[\Gamma_0, \Gamma_0] \subset N(\Gamma)$, then Γ_0 is metabelian. The fact that Γ itself is metabelian then follows from Exercise 4.4.12.

We conclude this section with a clever result due to Matsuda [161]. It turns out that in most cases the algebraic properties of subgroups of $\mathrm{Diff}_+^2([0, 1[)$ and $\mathrm{Diff}_+^2([0, 1])$ are similar. However, the situation for polycyclic subgroups is rather special. We do not know whether the following result extends to the class $C^{1+\mathrm{bv}}$.

Theorem 4.3.4. *Every polycyclic subgroup of $\mathrm{Diff}_+^{1+\mathrm{Lip}}([0, 1])$ without global fixed points in $]0, 1[$ is topologically conjugate to a subgroup of the affine group.*

The proof of this theorem strongly uses Thurston's stability theorem. The discussion is thus postponed to § 5.1.

4.4 Solvable Groups of Diffeomorphisms

4.4.1 Some examples and statements of results

The discussion in the preceding section leads naturally to the question whether every solvable subgroup of $\mathrm{Diff}_+^{1+\mathrm{bv}}([0, 1[)$ is metabelian. However, the examples below (inspired by [98, Chapitre V]) show that there exist solvable groups of diffeomorphisms of the interval with arbitrary solvability degree having dynamics very different from that of the affine group. For this, it suffices to take successive extensions by $\mathbb{Z}$ in an appropriate way.

Example 4.4.1. Let f be a C^∞ diffeomorphism of $[0, 1]$ without a global fixed point in $]0, 1[$ and topologically contracting toward the origin. Suppose that f is the time 1 of the flow associated to a C^∞ vector field $[0, 1]$ that is infinitely flat at the endpoints. Fix $a \in]0, 1[$, and consider a vector field X on $[f(a), a]$ with zeros only at $f(a)$ and a, and that is infinitely flat at these points. Extend X by letting $X(x) = 0$ for $x \in [0, 1] \setminus [f(a), a]$, thus obtaining a C^∞ vector field on $[0, 1]$ that is infinitely flat at 0 and 1.

Let g be the C^∞ diffeomorphism obtained by integrating the vector field X (up to time 1), and let Γ be the group generated by f and g. Clearly, Γ has no fixed point in $]0, 1[$. We claim that Γ is solvable with order of solvability equal to 2. Indeed, let Γ^* be the Abelian subgroup of Γ formed by the elements fixing the points $f^n(a)$ (with $n \in \mathbb{Z}$) such that their restrictions to each interval $[f^{n+1}(a), f^n(a)]$ are contained in the group generated by the restriction of the element $f^n g f^{-n}$ to this interval. Clearly, Γ^* is an Abelian normal subgroup of Γ containing the derived group Γ'; moreover, the quotient Γ/Γ^* identifies with $(\mathbb{Z}, +)$, thus showing the claim.

In what follows, we will successively apply the preceding idea to obtain, for each $k \geq 2$, a solvable group $\bar{\Gamma}_k$ of C^∞ diffeomorphisms of $[2 - k, k - 1]$ with order of solvability equal to k and generated by elements $f_{1,k}, \ldots, f_{k,k}$, each of which is the time 1 of a flow associated to a C^∞ vector field that is infinitely flat at $2 - k$ and $k - 1$. To do this, we argue inductively. For $k = 2$, we let $\bar{\Gamma}_2 = \Gamma$, where $f_{1,2} = g$ and $f_{2,2} = f$. Suppose now that we have already constructed the group Γ_k, and let us consider the vector fields $X_{i,k} : [2 - k, k - 1] \to \mathbb{R}$ corresponding to the $f_{i,k}$'s. Let us consider a vector field $X_{k+1,k+1}$ on $[1 - k, k]$ that is of the form $\varrho \frac{\partial}{\partial x}$ for some C^∞ function $\varrho : [1 - k, k] \to \mathbb{R}$ that is negative in the interior and infinitely flat at the endpoints. After multiplying this vector field by a scalar factor if necessary, we may suppose that the time 1 of the associated flow is a diffeomorphism $f_{k+1,k+1}$ of $[1 - k, k]$ satisfying $f_{k+1,k+1}(k - 1) = 2 - k$.

For $i \in \{1, \ldots, k\}$, we extend $X_{i,k}$ to a vector field $X_{i,k+1}$ by letting $X_{i,k+1}(x) = 0$ for every point $x \in [1 - k, k] \setminus [2 - k, k - 1]$. The vector fields thus obtained are of class C^∞ on $[1 - k, k]$ and infinitely flat at $1 - k$ and k. Let $f_{i,k+1}$, $i \in \{1, \ldots, k\}$, be the time 1 of the associated flows. We claim that the group $\bar{\Gamma}_{k+1}$ generated by the $f_{i,k+1}$'s, $i \in \{1, \ldots, k+1\}$, is solvable with solvability degree $k + 1$. Indeed, the stabilizer in Γ of $[2 - k, k - 1]$ is a normal subgroup $\bar{\Gamma}^*$ that identifies with a direct sum of groups isomorphic to $\bar{\Gamma}_k$, and by the induction hypothesis the latter group is solvable with solvability degree k. On the other hand, $\bar{\Gamma}_{k+1}/\bar{\Gamma}^*$ identifies with $(\mathbb{Z}, +)$, and the derived group of Γ is contained in $\bar{\Gamma}^*$. Using this fact, one easily concludes the claim. Notice that $\bar{\Gamma}_{k+1}$ has no fixed point in the interior of $[1-k, k]$.

Example 4.4.2. We next improve the first step of the preceding example by using the construction of the beginning of § 4.3. We will thus obtain a more interesting family of solvable subgroups of $\text{Diff}_+^\infty([0, 1])$ with solvability degree 3. To do this, let us consider two C^∞ vector fields X_1 and X_2 defined on the interval $[1/3, 2/3]$ that are infinitely flat at the endpoints and whose flows induce a conjugate of the affine group. Let us denote by g and h the C^∞ diffeomorphisms obtained by integrating up to time 1 the vector fields X_1 and X_2, respectively.

Fix a C^∞ diffeomorphism $f : [0, 1] \to [0, 1]$ without fixed point in the interior and satisfying $f(2/3) = 1/3$. For $a = 2/3$, let us consider a sequence (to be fixed) of positive real numbers $(t_n)_{n\in\mathbb{Z}}$, and let us extend by induction the definition of the X_j's by letting

$$X_j(x) = t_n X_j(f^{-1}(x))/(f^{-1})'(x), \qquad x \in [f^{n+1}(a), f^n(a)], \quad n \geq 1,$$
$$X_j(x) = t_n X_j(f(x))/f'(x), \qquad x \in [f^{n+1}(a), f^n(a)], \quad n \leq -1.$$

Suppose that $\Pi_{i=1}^n t_i \to 0$ as $n \to +\infty$, that $\Pi_{i=n}^0 t_i \to 0$ as $n \to -\infty$, and that these convergences are very fast (to fix the ideas, let us suppose that the speed of convergence is superexponential; compare Exercise 4.1.5). In this case it is not difficult to check that the X_j's extend to C^∞ vector fields on $[0, 1]$ that are zero and infinitely flat at the endpoints. Let Γ be the group generated by f, g, and h. Obviously, the restriction of Γ to the interior of $[0, 1]$ is not semiconjugate to a subgroup of the affine group, and Γ has no fixed point in $]0, 1[$. We claim that Γ is solvable with degree of solvability 3. Indeed, let Γ^* be the metabelian subgroup of Γ formed by the elements fixing the points $f^n(a)$ and whose restrictions to the interior of each interval $[f^{n+1}(a), f^n(a)]$ are contained in the conjugate of the affine group generated by X_1 and X_2. If we denote by $g^t_{[f^{n+1}(a), f^n(a)]}$ and $h^t_{[f^{n+1}(a), f^n(a)]}$, respectively, the flows associated to the restrictions of X_1 and X_2 to $[f^{n+1}(a), f^n(a)]$ (with $t \in \mathbb{R}$), then for every $n \in \mathbb{N}$, one has

$$f^{-1} \circ g_{[f^{n+1}(a), f^n(a)]} \circ f = g^{t_n}_{[f^n(a), f^{n-1}(a)]},$$
$$f^{-1} \circ h_{[f^{n+1}(a), f^n(a)]} \circ f = h^{t_n}_{[f^n(a), f^{n-1}(a)]}.$$

The subgroup Γ^* is therefore normal in Γ. Moreover, the quotient Γ/Γ^* identifies with $(\mathbb{Z}, +)$, and Γ^* contains the derived group of Γ. Starting from this, one easily concludes our claim.

As in the preceding example, for each integer $k \geq 2$, one may take successive extensions of Γ by $(\mathbb{Z}, +)$ to obtain solvable subgroups of $\mathrm{Diff}_+^{\infty}([1-k, k])$ with solvability degree $k + 2$.

Exercise 4.4.3. Prove that in the preceding example, the convergences $\Pi_{i=1}^n t_i \to 0$ and $\Pi_{i=n}^0 t_i \to 0$ are necessary. More precisely, prove the following proposition:

Proposition 4.4.4. Let $g : [0, 1[\to [0, g(1)[$ be a C^2 diffeomorphism, and let $a < b$ be two fixed points of g in $]0, 1[$ such that g has no fixed point in $]a, b[$. Let $f : [0, b[\to [0, f(b)[$ be a C^2 diffeomorphism with no other fixed point in $[0, b]$ than 0, and such that $f(b) \leq a$. Suppose that there exists a sequence $(t_n)_{n \in \mathbb{N}}$ of positive real numbers such that for all $n \in \mathbb{N}$, one has $f^{-1} \circ g_{[f^n(a), f^n(b)[} \circ f = g_{[f^{n-1}(a), f^{n-1}(b)[}^{t_n}$. Then the value of $\Pi_{i=1}^n t_i$ converges to zero as n goes to infinity.

The method of construction of Examples 4.4.1 and 4.4.2 is essentially the only possible one for creating solvable groups of interval diffeomorphisms. Let us begin by giving a precise version of this fact in the metabelian case.

Theorem 4.4.5. If Γ is a metabelian subgroup of $\mathrm{Diff}_+^{1+bv}([0, 1[)$ without fixed points in $]0, 1[$, then Γ is either conjugate to a subgroup of the affine group or a semidirect product between $(\mathbb{Z}, +)$ and a subgroup of a product (at most countable) of groups that are conjugate to groups of translations.

We will give the proof of this theorem in the next section. A complete classification of solvable subgroups of $\mathrm{Diff}_+^{1+bv}([0, 1[)$ may be obtained by using the same ideas through a straightforward inductive argument [186]. To state the theorem in the general case, let us denote by $r(1)$ the family of groups that are conjugate to groups of translations, and by $r(2)$ the family of groups that are either conjugate to non-Abelian subgroups of the affine group or are semidirect products between $(\mathbb{Z}, +)$ and a subgroup of an at-most-countable product of nontrivial groups of translations. For $k > 2$, we define by induction the family $r(k)$ formed by the groups that are semidirect products between $(\mathbb{Z}, +)$ and a subgroup of an at-most-countable product of groups in $\mathcal{R}(k-1) = r(1) \cup \ldots \cup r(k-1)$, so that at least one of the factors does not belong to $\mathcal{R}(k-2)$.

Theorem 4.4.6. *Let* Γ *be a solvable subgroup of* $\mathrm{Diff}_+^{1+bv}([0, 1[)$ *without a fixed point in* $]0, 1[$. *If the solvability degree of* Γ *equals* $k \geq 2$, *then* Γ *belongs to the family* $r(k)$.

The preceding classification allows obtaining interesting rigidity results. For instance, the normalizer of a solvable group of diffeomorphisms of the interval is very similar to the original group, as is established in the next result from [184].

Theorem 4.4.7. *Let* Γ *be a solvable subgroup of* $\mathrm{Diff}_+^{1+bv}([0, 1[)$ *of solvability degree* $k \geq 1$ *and without a fixed point in* $]0, 1[$. *If* $\mathcal{N}(\Gamma)$ *denotes its normalizer in* $\mathrm{Diff}_+^1([0, 1[)$, *then one of the following possibilities occurs:*

(i) *If* $k > 1$, *then* $\mathcal{N}(\Gamma)$ *is solvable with solvability degree* k.
(ii) *If* $k = 1$ *and* Γ *is not infinite cyclic, then* $\mathcal{N}(\Gamma)$ *is topologically conjugate to a (perhaps non-Abelian) subgroup of the affine group.*
(iii) *If* $k = 1$ *and* Γ *is infinite cyclic, then* $\mathcal{N}(\Gamma)$ *is topologically conjugate to a subgroup of the group of translations.*

The classification of solvable groups of circle diffeomorphisms reduces, thanks to Lemma 4.1.8, to that of the case of the interval. Using Theorem 4.4.6, the reader should easily verify the following theorem:

Theorem 4.4.8. *Let* Γ *be a solvable subgroup of* $\mathrm{Diff}_+^{1+bv}(S^1)$. *If the degree of solvability of* Γ *equals* $k + 1$, *then one of the following possibilities occurs:*

(i) Γ *is topologically semiconjugate to a group of rotations.*
(ii) *There exists a nonempty finite subset* F *of* S^1 *that is invariant under* Γ *such that the derived group* Γ' *fixes each point of* F, *and the restriction of* Γ' *to each of the connected components of* $S^1 \setminus F$ *belongs to the family* $\mathcal{R}(k)$.

The case of the line involves some extra difficulties. However, using some of the results from § 2.2.5, one may obtain the following description:

Theorem 4.4.9. *Let* Γ *be a solvable subgroup of* $\mathrm{Diff}_+^{1+bv}(\mathbb{R})$. *If the solvability degree of* Γ *equals* $k \geq 1$, *then one of the following possibilities occurs:*

(i) Γ *is topologically semiconjugate to a subgroup of the affine group.*

(ii) Γ *is a subgroup of a product (at most countable) of groups in the family* $\mathcal{R}(k)$ *so that at least one of the factors does not belong to* $\mathcal{R}(k-1)$.

(iii) Γ *belongs either to* $r(k)$ *or* $r(k+1)$.

Notice that if Γ is solvable and semiconjugate to a group of affine transformations without being conjugate to it, then the second derived group Γ'' acts by fixing a countable family of disjoint open intervals whose union is dense. Theorems 4.4.5 and 4.4.6 then allow describing the dynamics of Γ''. On the other hand, the fact that Γ may belong to $r(k+1)$ when its solvability degree is k is quite natural, since unlike the case of the interval, in the case of the line it is possible to produce *central* extensions of nontrivial groups, even in class C^∞.

4.4.2 The metabelian case

The following elementary lemma will play a fundamental role in what follows.

Lemma 4.4.10. *The normalizer in* $\mathrm{Homeo}_+(\mathbb{R})$ *of every dense subgroup of the group of translations is contained in the affine group.*

Proof. Up to a scalar factor, the only Radon measure on the line that is invariant under a dense group of translations is the Lebesgue measure. The normalizer of such a group leaves this measure quasi-invariant, and hence the conclusion of the lemma follows from Proposition 1.2.2. $\qquad\square$

The next lemma is an improved version of the preceding one in class $C^{1+\mathrm{bv}}$.

Lemma 4.4.11. *If* Γ *is an Abelian subgroup of* $\mathrm{Diff}_+^{1+\mathrm{bv}}([0,1[)$ *without fixed points in* $]0,1[$*, then its normalizer in* $\mathrm{Diff}_+^1([0,1[)$ *is conjugate to a subgroup of the affine group.*

Proof. Let $g \in \Gamma$ be a nontrivial element. By Corollary 4.1.4, the hypothesis of the non-existence of fixed points in $]0,1[$ is equivalent to saying that Γ is contained in a 1-parameter group $g^{\mathbb{R}} = \{g^t : t \in \mathbb{R}\}$ that is conjugate to the group of translations and such that $g^1 = g$; moreover, the centralizer in $\mathrm{Diff}_+^1([0,1[)$ of every nontrivial element of Γ is contained in $g^{\mathbb{R}}$. If the image of Γ under the conjugacy is a dense subgroup of $(\mathbb{R},+)$, then

Lemma 4.4.10 implies that the image under this conjugacy of the normalizer $\mathcal{N}$ of Γ in $\mathrm{Homeo}_+([0, 1])$ is contained in the affine group. On the other hand, we claim that if $\{t \in \mathbb{R} : g^t \in \Gamma\}$ is infinite cyclic, then $\mathcal{N}$ equals $g^{\mathbb{R}}$. Indeed, if k is a positive integer such that $g^{1/k}$ is the generator of $\{t \in \mathbb{R} : g^t \in \Gamma\}$, then for every $h \in \mathcal{N}$, there exists positive integers n and m such that $hg^{1/k}h^{-1} = (g^{1/k})^n$ and $h^{-1}g^{1/k}h = (g^{1/k})^m$. One then has

$$(g^{1/k})^{mn} = ((g^{1/k})^m)^n = (h^{-1}g^{1/k}h)^n = h^{-1}(g^{1/k})^n h = g^{1/k},$$

from which one obtains $m = n = 1$. This implies that the elements in $\mathcal{N}$ commute with $g^{1/k}$, and hence $\mathcal{N}$ is contained in $g^{\mathbb{R}}$. Since $g^{\mathbb{R}}$ centralizes (and hence normalizes) Γ, this shows that $\mathcal{N} = g^{\mathbb{R}}$.   □

We may now pass to the proof of Theorem 4.4.5. By Corollary 4.1.4, if Γ is an Abelian subgroup of $\mathrm{Diff}_+^{1+\mathrm{bv}}([0, 1[)$ without fixed points in $]0, 1[$, then it is conjugate to a group of translations. Now let Γ be a metabelian and noncommutative subgroup of $\mathrm{Diff}_+^{1+\mathrm{bv}}([0, 1[)$ without fixed points in $]0, 1[$. If there exists $g \in \Gamma'$ such that the orbits under g accumulate at 0 and 1, then the Abelian group Γ' is contained in the topological flow associated to g. Therefore, Γ' acts without fixed point in $]0, 1[$, and the preceding lemma implies that Γ is conjugate to a (noncommutative) subgroup of the affine group.

In what follows, suppose that every $g \in \Gamma'$ has fixed points in $]0, 1[$. Kopell's lemma then easily implies that Γ' must have global fixed points in $]0, 1[$. Let $]a, b[$ be an irreducible component of Γ'. Notice that for every $h \in \Gamma$, the interval $h(]a, b[)$ is also an irreducible component of Γ'. In particular, if $h(]a, b[) \neq]a, b[$, then $h(]a, b[) \cap]a, b[= \emptyset$.

If $a = 0$ or $b = 1$, then every $f \in \Gamma$ fixes $]a, b[$, which contradicts the hypothesis of the nonexistence of fixed points in $]0, 1[$. Therefore, $[a, b]$ is contained in $]0, 1[$. If $f \in \Gamma$ fixes $]a, b[$, then Lemma 4.4.11 shows that the restriction of f to $]a, b[$ is affine in the coordinates induced by Γ'. The case of the elements that do not fix $]a, b[$ is more interesting.

Claim (i). If f is an element of Γ that does not fix $]a, b[$, and if u and v are the fixed points of f to the left and to the right of $]a, b[$, respectively, then the interval $]u, v[$ is an irreducible component of Γ (that is, $u = 0$ and $v = 1$).

Suppose not, and let $\bar{f} \in \Gamma$ be an element that does not fix $]u, v[$. Replacing f by f^{-1} if necessary, we may suppose that $f(x) > x$ for all $x \in]u, v[$. One then has $f(a) \geq b$. For $n \in \mathbb{N}$, the element $f^{-1}\bar{f}^{-n}f\bar{f}^n$ belongs to Γ'

and, therefore, fixes the points a and u. As a consequence, for every $n \in \mathbb{N}$,

$$f\bar{f}^n(u) = \bar{f}^n f(u) = \bar{f}^n(u), \qquad f\bar{f}^n(a) = \bar{f}^n f(a) \geq \bar{f}^n(b). \qquad (4.26)$$

One has $\bar{f} f \bar{f}^{-1} = f\bar{g}$ for some $\bar{g} \in \Gamma'$. On the other hand, $f\bar{g}$ has no fixed point in $]u, v[$ and fixes u and v. Therefore, f does not have fixed points in $]\bar{f}(u), \bar{f}(v)[$ and fixes $\bar{f}(u)$ and $\bar{f}(v)$. This shows that $]\bar{f}(u), \bar{f}(v)[\cap]u, v[= \emptyset$. Replace $\bar{f}$ by $\bar{f}^{-1}$ if necessary so that $\bar{f}(u) < u$. Notice that the sequence $(\bar{f}^n(u))$ tends to a fixed point of $\bar{f}$. One then easily checks that relations (4.26) contradict Lemma 4.2.12 (applied to the elements $h_1 = \bar{f}$ and $h_2 = f$ with respect to the points $x_0 = \bar{f}^{-1}(u)$, $y = a$, and $z = b \leq f(a)$). This concludes the proof of the claim.

Denote by Γ^* the normal subgroup of Γ formed by the elements fixing the irreducible components of Γ'. Since Claim (i) holds for every irreducible component $]a, b[$ of Γ', an element of Γ belongs to Γ^* if and only if it fixes at least one of the irreducible components of Γ'. The restriction of Γ^* to each irreducible component of Γ' is affine in the induced coordinates. Remark that Γ^* may admit irreducible components contained in the complement of the union of the irreducible components of Γ'. However, the restriction of Γ^* to such a component is Abelian and hence is conjugate to a subgroup of the group of translations. We then conclude that Γ^* is a subgroup of a product (at most countable) of groups conjugate to groups of affine transformations. Moreover, the quotient group Γ/Γ^* acts freely on the set $\text{Fix}(\Gamma')$. Fix an irreducible component $]a, b[$ of Γ', and define an order relation $\preceq$ on Γ/Γ^* by $f_1\Gamma^* \prec f_2\Gamma^*$ when $f_1(]a, b[)$ is to the left of $f_2(]a, b[)$. This relation is total, bi-invariant, and Archimedean. The argument of the proof of Hölder's theorem then shows that the group $H = \Gamma/\Gamma^*$ is naturally isomorphic to a subgroup of $(\mathbb{R}, +)$. Notice that H is nontrivial, since Γ does not fix $]a, b[$. Proposition 4.1.2 then implies the following claim.

Claim (ii). H is an infinite cyclic group.

The proof of Theorem 4.4.5 is then completed by the following claim:

Claim (iii). Γ^* is a subgroup of a product of groups that are conjugate to groups of translations.

To show this, fix an element $g \in \Gamma'$ whose restriction to an irreducible component $]a, b[$ of Γ' has no fixed point. Assume for a contradiction that there exists $h \in \Gamma$ fixing $]a, b[$ such that the restrictions of g and h to this interval generate a non-Abelian group. Without loss of generality, we may suppose that the fixed point a of h is topologically repelling on the right. Let $f \in \Gamma$ be an element such that $f\Gamma^*$ generates Γ/Γ^* and $f(b) \leq a$. For every $n \in \mathbb{N}$, the element $f^{-n}hf^n$ fixes the interval $]a, b[$, and hence its restriction therein equals the restriction of hg^{t_n} for some $t_n \in \mathbb{R}$, where $g^{\mathbb{R}}$ stands for the flow associated to the restriction of g to $[a, b[$. For $\delta = V(f; [0, b]) > 0$, inequality (4.2) allows showing that for every $x \in]a, b[$,

$$(hg^{t_n})'(x) = (f^{-n}hf^n)'(x) \leq \frac{(f^n)'(x)}{(f^n)'(f^{-n}hf^n(x))} \sup_{y \in]0, f^n(b)[} h'(y)$$

$$\leq e^\delta \sup_{y \in]0, f^n(b)[} h'(y), \qquad (4.27)$$

and

$$(hg^{t_n})'(x) \geq e^{-\delta} \inf_{y \in]0, f^n(b)[} h'(y). \qquad (4.28)$$

From (4.27) one concludes that

$$\sup_{x \in]a, b[} (g^{t_n})'(x) \leq e^\delta \cdot \frac{\sup_{y \in]0, b[} h'(y)}{\inf_{y \in]a, b[} h'(y)}.$$

This clearly implies that $(|t_n|)$ is a bounded sequence. Take a subsequence (t_{n_k}) converging to a limit $T \in \mathbb{R}$. Since h fixes each interval $[f^n(a), f^n(b)]$, one has $h'(0) = 1$. By integrating (4.27) and (4.28), we obtain, for every large-enough k and every $x \in]a, b[$,

$$(x - a)/2e^\delta \leq hg^{t_{n_k}}(x) - a \leq 2e^\delta(x - a),$$

and passing to the limit as k goes to infinity, we conclude that

$$(x - a)/2e^\delta \leq hg^T(x) - a \leq 2e^\delta(x - a).$$

Now notice that this argument still holds if we replace h by h^j for any $j \in \mathbb{N}$ (indeed, the constant δ depends only on f). Therefore, for every $x \in]a, b[$ and all $j \in \mathbb{N}$,

$$(x - a)/2e^\delta \leq (hg^T)^j(x) - a \leq 2e^\delta(x - a),$$

which is impossible for x near a since the latter point is fixed by hg^T and topologically repelling on the right. This concludes the proof.

Exercise 4.4.12. Show that if Γ is a subgroup of $\mathrm{Diff}_+^{1+\mathrm{bv}}([0, 1[)$ containing a finite-index subgroup Γ_0 that is metabelian (resp. solvable with solvability degree k), then Γ itself is metabelian (resp. solvable with solvability degree k) (compare Exercise 4.1.9).

Exercise 4.4.13. Show that every polycyclic subgroup of $\mathrm{Diff}_+^{1+\mathrm{bv}}([0, 1[)$ is strongly polycyclic (cf. Appendix A).

It is worth noticing that if a solvable subgroup of $\mathrm{Diff}_+^{1+\mathrm{bv}}([0, 1[)$ has solvability degree greater than 2, then it necessarily contains nontrivial elements having infinitely many fixed points in every neighborhood of the origin; in particular, these elements cannot be real-analytic. As a consequence, every solvable group of real-analytic diffeomorphisms of the interval is topologically conjugate to a subgroup of the affine group (provided that there is no global fixed point in the interior). The reader may find this and many other related results in [175] (see also [38, 186]). Finally, let us point out that the results already described for solvable subgroups of $\mathrm{Diff}_+^{1+\mathrm{bv}}([0, 1[)$ still hold –and may be refined– for subgroups of the group $\mathrm{PAff}_+([0, 1[)$ of piecewise affine homeomorphisms of the interval.

Exercise 4.4.14. Prove that if Γ is a nontrivial subgroup of $\mathrm{PAff}_+([0, 1[)$ acting freely on $]0, 1[$, then Γ is infinite cyclic. Using this, prove that every finitely generated metabelian subgroup of $\mathrm{PAff}_+([0, 1])$ is isomorphic to a semidirect product between $(\mathbb{Z}, +)$ and a *direct sum* (at most countable) of groups isomorphic to $(\mathbb{Z}, +)$ acting on two-by-two disjoint intervals. State and show a general result of classification for finitely generated solvable subgroups of $\mathrm{PAff}_+([0, 1])$ (see [184] for more details; see also [18]).

4.4.3 The case of the real line

For a solvable group of solvability degree k, we denote by $\{id\} = \Gamma_k^{\mathrm{sol}}$ $\lhd \ldots \lhd \Gamma_0^{\mathrm{sol}} = \Gamma$ its derived series; that is, $\Gamma_i^{\mathrm{sol}} = [\Gamma_{i-1}^{\mathrm{sol}}, \Gamma_{i-1}^{\mathrm{sol}}]$ for every $i \in \{1, \ldots, k\}$, with $\Gamma_{k-1}^{\mathrm{sol}} \neq \{id\}$. The following result should be compared with Theorem 2.2.56.

Proposition 4.4.15. *Let Γ be a solvable subgroup of $\mathrm{Homeo}_+(\mathbb{R})$ of solvability degree k. If there exists an index $i \leq k$ for which Γ_i^{sol} preserves a Radon measure υ_i such that the associated translation-number* *homomorphism* τ_{v_i} *satisfies* $\tau_{v_i}(\Gamma_i^{sol}) \neq \{0\}$, *then there exists a Radon measure on the line that is quasi-invariant under* Γ.

Proof. Let $j < k$ be the smallest index for which Γ_j^{sol} preserves a Radon measure, and let v_j be this measure. Using the hypothesis and (2.6), it is not difficult to check that $\tau_{v_j}(\Gamma_j^{sol}) \neq \{0\}$. By Lemma 2.2.53 one has $\kappa(\Gamma_{j-1}^{sol}) \neq \{1\}$, and by Lemma 2.2.54 this implies that v_j is quasi-invariant under Γ_{j-1}^{sol}. Lemma 2.2.55 then shows that v_j is quasi-invariant under $\Gamma_{j-2}^{sol}, \Gamma_{j-3}^{sol}, \ldots$, and hence under Γ. $\square$

We now give the proof of Theorem 4.4.9 (assuming Theorem 4.4.6). Fix a solvable subgroup Γ of $\mathrm{Diff}_+^{1+bv}(\mathbb{R})$ with solvability degree $k \geq 2$. If Γ has global fixed points, then Theorems 4.4.5 and 4.4.6 imply that Γ is a subgroup of a product (at most countable) of groups in the family $\mathcal{R}(k)$ such that at least one of the factors does not belong to $\mathcal{R}(k-1)$. Assume throughout that Γ has no global fixed point. We begin with a proposition that should be compared with the end of § 2.2.5.

Proposition 4.4.16. *Every solvable subgroup of* $\mathrm{Diff}_+^{1+bv}(\mathbb{R})$ *leaves quasi-invariant a Radon measure on the line.*

Proof. Let us consider the smallest index j for which Γ_j^{sol} has a global fixed point. Notice that since Γ has no fixed point, j must be positive. There are two distinct cases.

First case. The index j equals k.

We claim that Γ_{k-1}^{sol} preserves a Radon measure such that the translation-number homomorphism is not identically zero. Indeed, since Γ_{k-1}^{sol} is Abelian, using Kopell's lemma, one easily shows the existence of elements in Γ_{k-1}^{sol} without fixed points, and hence Proposition 2.2.48 implies the existence of a Radon measure invariant under Γ_{k-1} such that the translation-number homomorphism is nontrivial. The existence of a Radon measure v on the line that is quasi-invariant under Γ then follows from Proposition 4.4.15.

Second case. The index j is less than k.

Fix an irreducible component $]p_j, q_j[$ of Γ_j^{sol}.

Claim (i). If $\bar{f}_1$ and $\bar{f}_2$ are elements of $\Gamma^{\mathrm{sol}}_{j-1}$ that have fixed points but do not fix $]p_j, q_j[$, then the fixed points in $\mathbb{R} \cup \{-\infty, +\infty\}$ of these elements that are next to $]p_j, q_j[$ on the left coincide. The same holds for the fixed points that are next on the right.

To show this, let p and q be the fixed points of $\bar{f}_1$ to the left and to the right of $[p_j, q_j]$, respectively. Suppose that $\bar{f}_2$ does not fix $[p, q]$. For each $n \in \mathbb{Z}$, there exists $\bar{g} \in \Gamma^{\mathrm{sol}}_j$ such that $\bar{f}_2^n \bar{f}_1 \bar{f}_2^{-n} = \bar{f}_1 \bar{g}$. Since $\bar{f}_1 \bar{g}$ has no fixed point in $]p, q[$ and fixes p and q, the element $\bar{f}_1$ fixes the interval $\bar{f}_2^n(]p, q[)$ and has no fixed point inside it. One then deduces that the intervals $\bar{f}_2^n(]p, q[)$ are two-by-two disjoint for $n \in \mathbb{Z}$. Replacing $\bar{f}_2$ by its inverse if necessary, we may suppose that $\bar{f}_2^n(p_j)$ tends to a fixed point (in $\mathbb{R}$) of $\bar{f}_2$ as n goes to infinity. We then obtain a contradiction by applying the arguments of the proof of Claim (i) from § 4.4.2.

Let us define the interval $[p^*_{j-1}, q^*_{j-1}]$ as being equal to $[p_j, q_j]$ if every element in $\Gamma^{\mathrm{sol}}_{j-1}$ that does not fix $[p_j, q_j]$ has no fixed point. Otherwise, let us choose an element $f_j \in \Gamma^{\mathrm{sol}}_{j-1}$ that has fixed points but does not fix $[p_j, q_j]$, and define p^*_{j-1} and q^*_{j-1} as the fixed points of f_j to the left and to the right of $[p_j, q_j]$, respectively. Finally, denote by Γ^*_{j-1} the stabilizer of $[p^*_{j-1}, q^*_{j-1}]$ in Γ_{j-1}. The group Γ^*_{j-1} is normal in Γ_{j-1} because it is formed by the elements in Γ_{j-1} having fixed points in the line.

Claim (ii). The group $\Gamma^{\mathrm{sol}}_{j-1}$ preserves a Radon measure and has no fixed point.

Indeed, since Claim (i) holds for every irreducible component of Γ_j, the action of $\Gamma^{\mathrm{sol}}_{j-1} / \Gamma^*_{j-1}$ on $\mathrm{Fix}(\Gamma^*_{j-1})$ is free. The argument of the proof of Hölder's theorem then shows that $\Gamma^{\mathrm{sol}}_{j-1}$ fixes a Radon measure whose support is contained in $\mathrm{Fix}(\Gamma^*_{j-1})$.

Recall that by the definition of j, the subgroup $\Gamma^{\mathrm{sol}}_{j-1}$ has no fixed point (which, together with (i), implies the existence of elements therein without fixed points). The translation-number function with respect to the $\Gamma^{\mathrm{sol}}_{j-1}$-invariant Radon measure is therefore not identically zero. Proposition 4.4.15 then allows concluding the proof of the proposition in the second case.   □

We can now conclude the proof of Theorem 4.4.9. To do this, fix a Radon measure v that is quasi-invariant under Γ. Suppose first that v has no atom. Consider the equivalence relation $\sim$ that identifies two points if they belong to the closure of the same connected component of the complement of the support of v. The quotient space $\mathbb{R}/\sim$ is a topological line on which the group Γ naturally acts by homeomorphisms. Since v has no atom, it induces a Γ-quasi-invariant nonatomic Radon measure $\bar{v}$ on $\mathbb{R}/\sim$ whose support is total. Proposition 1.2.2 then implies that the latter action is topologically conjugate to an action by affine transformations. Therefore, the original action of Γ on the line is semiconjugate to an action by affine transformations.

It remains the case where v has atoms. First notice that Γ' preserves v, and hence the second derived group Γ'' fixes each atom of v. This argument shows in particular that in this case the index j considered in the proof of Proposition 4.4.16 equals either 1 or 2. We will see that it is actually equal to 1.

Denote by Γ_v the subgroup of Γ formed by the elements fixing v, and denote by Γ'_v its derived group. The elements in Γ'_v fix the atoms in v. Denote by Γ_v^* the normal subgroup of Γ formed by the elements fixing the irreducible components of Γ'_v. The arguments of the proofs of Claims (i) and (ii) in Proposition 4.4.16 show that Γ_v/Γ_v^* is isomorphic to a nontrivial subgroup H of $(\mathbb{R}, +)$. Notice that H cannot be dense, for otherwise there would be atoms of v of the same mass accumulating on some points in the line, thus contradicting the fact that v is a Radon measure. Therefore, H is infinite cyclic, and since Γ acts by automorphisms of H, the latter group must preserve v. Therefore, $\Gamma = \Gamma_v$ (this shows that $j = 1$). We then conclude that Γ is an extension (actually, a semidirect product) by $(\mathbb{Z}, +)$ of a solvable subgroup of a product of groups of diffeomorphisms of closed intervals. This concludes the proof of Theorem 4.4.9.

4.5 On the Smooth Actions of Amenable Groups

On the basis of the previous sections, it would be natural to continue the classification of groups of interval and circle diffeomorphisms by trying to describe the amenable ones. However, this problem seems to be extremely difficult. Actually, the only relevant result in this direction is Theorem 2.2.58, which, because of Thurston's stability theorem from § 5.1, does not provide any information in the case of groups of diffeomorphisms. As a matter of example, let us recall that according to § 1.5.2, Thompson's group F embeds into $\text{Diff}_+^\infty([0, 1])$. However, the problem of knowing whether F is amenable has been open for more than 30 years! In what follows, we will give some partial results on the classification of a particular but relevant family of amenable subgroups of $\mathrm{Diff}^2_+([0, 1])$.

Recall that amenability is stable under *elementary operations*, that is, it is preserved by passing to subgroups or quotients and by taking extensions or unions (see Exercise B.7). Following [201], we denote by SA the smallest family of amenable groups that is closed with respect to these operations and contains the groups of subexponential growth (see Exercise B.9).

A result in [186] asserts that every group of real-analytic diffeomorphisms of $[0, 1[$ contained in SA is metabelian. However, $\mathrm{Diff}^\infty_+([0, 1])$ contains interesting subgroups from SA. The following construction should be compared with those in [206].

Example 4.5.1. Given points $a < c < d < b$ in $]0, 1[$, let f be a C^∞ diffeomorphism of $[a, b]$ that is infinitely tangent to the identity at the endpoints and such that $f(c) = d$. Extend f to $[0, 1]$ by letting $f(x) = x$ for $x \notin]a, b[$. Let g be a C^∞ diffeomorphism of $[0, 1]$ that is infinitely tangent to the identity at the endpoints, with a single fixed point in the interior, and such that $g(c) = a$ and $g(d) = b$. Denote by Γ the subgroup of $\mathrm{Diff}^\infty_+([0, 1])$ generated by f and g. For each $n \in \mathbb{N}$, the subgroup Γ_n of Γ generated by $\{f_i = g^{-i} f g^i : |i| \le n\}$ is solvable with solvability degree $2n + 1$. Moreover, the subgroup $\Gamma^* = \cup_{n\in\mathbb{N}}\Gamma_n$ is normal in Γ, and the quotient Γ/Γ^* is isomorphic to $(\mathbb{Z}, +)$ (a generator is $g\Gamma^*$). The group Γ is therefore finitely generated and nonsolvable and belongs to the family SA (see [73, 105] for an interesting property concerning its Følner sequences).

Despite the preceding example, the subgroups of $\mathrm{Diff}^{1+\mathrm{bv}}_+([0, 1[)$ that belong to the family SA may be partially classified. To do this, let us define by transfinite induction the subfamilies SA_α of SA as follows:

(i) SA_1 is the family of countable groups all of whose finitely generated subgroups have subexponential growth.

(ii) If α is not a limit ordinal, then SA_α is the family of groups Γ obtained either as a quotient, as a subgroup, or as an extension of groups in $\mathrm{SA}_{\alpha-1}$. The latter means that Γ contains a normal subgroup G in $\mathrm{SA}_{\alpha-1}$ such that the quotient Γ/G also belongs to $\mathrm{SA}_{\alpha-1}$.

(iii) If α is a limit ordinal, then SA_α is the family of groups obtained as a union of groups in the union $\cup \mathrm{SA}_\beta$, with $\beta < \alpha$.

A group Γ belongs to SA if and only if it belongs to SA_α for some ordinal α [201]. For instance, the group from Example 4.5.1 belongs to $SG_{\alpha+1}$, where α is the first infinite ordinal. The next result (whose proof is based on those of § 4.4) appears in [184] (see also [29] for a more accurate version in the piecewise affine case).

Theorem 4.5.2. *Every subgroup of* $\mathrm{Diff}_+^{1+bv}([0, 1[)$ *that belongs to* SA_α *for some finite* α *is solvable with solvability degree less than or equal to* α.

Theorem 4.5.2 applies only to particular families of amenable groups. Indeed, from [11] it follows that the family SA is smaller than that of amenable groups. This produces a breaking point in our classification of subgroups of $\mathrm{Diff}_+^{1+bv}(S^1)$. This is the reason that in the next chapter we will follow an "opposite direction" in our study: we will show that because of to some internal (mostly cohomological) properties, certain groups cannot act on the interval or the circle in a nontrivial way.

CHAPTER FIVE

Rigidity via Cohomological Methods

5.1 Thurston's Stability Theorem

The following result corresponds to a weak version of a theorem due to Thurston. For the general version related to the famous Reeb's stability theorem for foliations, we refer to the remarkable article [240]. We also strongly recommend that the reader look at Exercise 5.1.13.

Theorem 5.1.1. *Let Γ be a finitely generated group. If Γ does not admit any nontrivial homomorphism into $(\mathbb{R}, +)$, then every representation $\Phi: \Gamma \to \mathrm{Diff}^1_+([0, 1[)$ is trivial.*[1]

For the proof of this theorem, given a subset B of an arbitrary group Γ and $\varepsilon \geq 0$, we will say that a function $\phi : B \to \mathbb{R}$ is a (B, ε)-***homomorphism*** (into $\mathbb{R}$) if for every g and h in B such that $gh \in B$, one has

$$|\phi(g) + \phi(h) - \phi(gh)| \leq \varepsilon.$$

Fix a finite and symmetric system of generators $\mathcal{G}$ of Γ. We will say that ϕ is ***normalized*** if $\max_{g \in \mathcal{G}} |\phi(g)| = 1$. For simplicity, we let

$$\nabla \phi(g, h) = \phi(g) + \phi(h) - \phi(gh).$$

Notice that a function $\phi : \Gamma \to \mathbb{R}$ is a $(\Gamma, 0)$-homomorphism if and only if $\nabla \phi$ is identically zero, which is equivalent to saying that ϕ is a homomorphism from Γ into $(\mathbb{R}, +)$. Finally, recall that $B_{\mathcal{G}}(k)$ denotes the set of

---

1. Equivalently, for every nontrivial finitely generated subgroup Γ of $\mathrm{Diff}^1_+([0.1])$, the first *cohomology space* $H^1_{\mathbb{R}}(\Gamma)$ is nontrivial.

elements in Γ that may be written as a product of at most k elements in $\mathcal{G}$ (see Appendix B).

Lemma 5.1.2. *If for each $k \in \mathbb{N}$ there exists a normalized $(B_{\mathcal{G}}(k), 1/k)$-homomorphism, then there exists a nontrivial homomorphism from Γ into $(\mathbb{R}, +)$.*

Proof. Given $k \in \mathbb{N}$, let ϕ_k be a normalized $(B_{\mathcal{G}}(k), 1/k)$-homomorphism. It is easy to check that $|\phi_k(g)| \leq k(1 + \varepsilon)$ for every $g \in B_{\mathcal{G}}(k)$. In particular, there exists a subsequence (ϕ_k^1) of ϕ_k such that $\phi_k^1|_{B_{\mathcal{G}}(1)}$ converges (pointwise) to a normalized function from $B_{\mathcal{G}}(1)$ into $\mathbb{R}$. Arguing by induction, for each $i \in \mathbb{N}$, we may find a subsequence $(\phi_k^{i+1})_{k \in \mathbb{N}}$ of $(\phi_k^i)_{k \in \mathbb{N}}$ such that $\phi_k^{i+1}|_{B_{\mathcal{G}}(i+1)}$ converges to a normalized function from $B_{\mathcal{G}}(i + 1)$ into $\mathbb{R}$. The sequence (ϕ_k^k) converges (pointwise) to a function $\phi : \Gamma \to \mathbb{R}$. By construction, ϕ is a homomorphism into $(\mathbb{R}, +)$, which is necessarily nontrivial since it is normalized. $\square$

Now let $\Phi : \Gamma \to \mathrm{Diff}_+^1([0, 1])$ be a nontrivial representation, and let $x \in [0, 1[$ be a point in the boundary of the set of points that are fixed by $\mathcal{G}$ (and hence by Γ). For each $y \in [0, 1[$ that is not fixed by $\mathcal{G}$, we consider the function

$$\phi_y(g) = \frac{1}{C(y)}[\Phi(g)(y) - y],$$

where $C(y) = \max_{g \in \mathcal{G}} |\Phi(g)(y) - y|$. The following lemma shows that if $\Phi(g)'(x) = 1$ for every $g \in \Gamma$, then for y near x (and not fixed by $\mathcal{G}$), the function ϕ_y behaves infinitesimally like a homomorphism from Γ into $(\mathbb{R}, +)$.

Lemma 5.1.3. *Under the preceding conditions, suppose, moreover, that $\Phi(g)'(x) = 1$ for every $g \in \Gamma$. Then for each $n \in \mathbb{N}$ and each $\varepsilon > 0$, there exists $\delta > 0$ such that if $|x - y| < \delta$, then $\phi_y|_{B_{\mathcal{G}}(n)}$ is a normalized $(B_{\mathcal{G}}(n), \varepsilon)$-homomorphism.*

Proof. For $k \in \mathbb{N}$ and $\varepsilon' > 0$, we inductively define

$$\lambda_0(0, \varepsilon') = 0, \qquad \lambda_0(k + 1, \varepsilon') = 1 + \lambda_0(k, \varepsilon')(1 + \varepsilon').$$

Let $\epsilon' > 0$ be small enough that $\varepsilon' \lambda_0(n, \varepsilon') \leq \varepsilon$, and let $\delta' > 0$ be such that $|\Phi(g)'(y) - 1| \leq \varepsilon'$ for all $g \in B_{\mathcal{G}}(n)$ and every point $y \in [0, 1]$ satisfying

$|x - y| \leq \delta'$. Finally, let $\delta \in {]0, \delta'[}$ be such that if $|x - y| \leq \delta$, then $|\Phi(g)(y) - x| \leq \delta'$ for every $g \in B_{\mathcal{G}}(n)$. We claim that this parameter δ verifies the claim of the lemma.

Let us first show that for every $k \leq n$ and every $g \in B_{\mathcal{G}}(k)$,

$$|\phi_y(g)| \leq \lambda_0(k, \varepsilon'). \tag{5.1}$$

Indeed, this inequality is evident for $k = 0$ and $k = 1$. Let us suppose that it holds for $k = i$. If g is an element in $B_{\mathcal{G}}(i + 1)$, then $g = h_1 h_2$ for some $h_1 \in \mathcal{G}$ and $h_2 \in B_{\mathcal{G}}(i)$. Thus

$$|\phi_y(g)| = \frac{1}{C(y)} |\Phi(g)(y) - y|$$

$$\leq \frac{1}{C(y)} |\Phi(h_1)\Phi(h_2)(y) - \Phi(h_2)(y)| + |\phi_y(h_2)|. \tag{5.2}$$

Notice that from the equality

$$\Phi(h_1)\Phi(h_2)(y) - \Phi(h_2)(y) = \int_y^{\Phi(h_2)(y)} [\Phi(h_1)'(s) - 1] ds + [\Phi(h_1)(y) - y]$$

one deduces that

$$\frac{1}{C(y)} |\Phi(h_1)\Phi(h_2)(y) - \Phi(h_2)(y)|$$

$$\leq \max_{|s - x| \leq \delta'} |\Phi(h_1)'(s) - 1| \cdot \frac{1}{C(y)} |\Phi(h_2)(y) - y| + 1.$$

Using (5.2) and the induction hypothesis, we conclude that

$$|\phi_y(g)| \leq \varepsilon' \lambda_0(i, \varepsilon') + 1 + \lambda_0(i, \varepsilon') = \lambda_0(i + 1, \varepsilon'),$$

which completes the proof of (5.1).

Let us now estimate the value of $\nabla \phi_y$. For h_1 and h_2 in $B_{\mathcal{G}}(n)$ such that $h_1 h_2$ belongs to $B_{\mathcal{G}}(n)$, we have

$$\nabla \phi_y(h_1, h_2) = \frac{1}{C(y)} [\Phi(h_1)(y) - y + \Phi(h_2)(y) - y - \Phi(h_1 h_2)(y) + y],$$

that is,

$$\nabla \phi_y(h_1, h_2) = -\frac{1}{C(y)} [\Phi(h_1)\Phi(h_2)(y) - \Phi(h_2)(y) - (\Phi(h_1)(y) - y)]$$

$$= -\frac{1}{C(y)} \int_y^{\Phi(h_2)(y)} [\Phi(h_1)'(s) - 1] ds.$$

Thus

$$
\begin{aligned}
|\nabla \phi_y(h_1, h_2)| &\le \left| \frac{1}{C(y)} [\Phi(h_2)(y) - y] \right| \cdot \sup_{|s-x| \le \delta'} |\Phi(h_1)'(s) - 1| \\
&\le \lambda_0(n, \varepsilon') \varepsilon' \le \varepsilon.
\end{aligned}
$$

Therefore, $\phi_y|_{B_{\mathcal{G}}(n)}$ is a normalized $(B_{\mathcal{G}}(n), \varepsilon)$-homomorphim. □

Proof of Theorem 5.1.1. Suppose that $\Phi : \Gamma \to \mathrm{Diff}_+^1([0, 1[)$ is a nontrivial representation, and let $x \in [0, 1[$ be a point in the boundary of the set of fixed points of Γ. The function $g \mapsto \log(\Phi(g)'(x))$ is a group homomorphism from Γ into $(\mathbb{R}, +)$. By hypothesis, this homomorphism must be trivial (compare Exercise 5.1.14). We are hence under the hypothesis of Lemma 5.1.3, which, together with Lemma 5.1.2, implies that Γ admits a nontrivial homomorphism into $(\mathbb{R}, +)$. □

Notice that the preceding argument involves only the behavior of the maps near a global fixed point. It is then easy to see that Thurston's theorem still holds (with the same proof) for subgroups of the group of germs of diffeomorphism $\mathcal{G}_+^1(\mathbb{R}, 0)$.

Exercise 5.1.4. Prove that Thurston's stability theorem does not hold for non–finitely generated countable groups of diffeomorphisms of the interval.

Hint. Consider the first derived group F' of Thompson's group F, and use the fact that F' is simple [44].

Exercise 5.1.5. Prove that Thurston's stability theorem holds for (perhaps non–finitely generated) groups of germs of *real-analytic* diffeomorphisms.

Hint. Following an argument due to Haefliger (and prior to Thurston's theorem), analyze the coefficients corresponding to the Taylor series expansions of the germs about the origin.

We now give an example (also due to Thurston, but rediscovered some years later by Bergman [17] in the context of orderable groups) showing that Thurston's stability theorem does not extend to actions by homeomorphisms (see also Remarks 2.2.51 and 5.1.8). Let $\tilde{G}$ be the group of presentation $\tilde{G} = \langle f, g, h : f^2 = g^3 = h^7 = fgh \rangle$. We leave to the reader the task of showing that every homomorphism from $\tilde{G}$ into $(\mathbb{R}, +)$ is trivial. (Actually, $\tilde{G} = [\tilde{G}, \tilde{G}]$; that is, $\tilde{G}$ is a *perfect* group.) This is quite natural, since $\tilde{G}$ is the fundamental group of a homology 3-sphere, that is, a closed 3-dimensional manifold with trivial homology but not homeomorphic to the 3-sphere [239, 240].

To construct a nontrivial $\tilde{G}$-action on $[0, 1]$, let us consider the tessellation of the Poincaré disk by hyperbolic triangles of angles $\pi/2$, $\pi/3$, and $\pi/7$. The preimage in $\widetilde{\text{PSL}}(2, \mathbb{R})$ of the subgroup of $\text{PSL}(2, \mathbb{R})$ preserving this tessellation is a group isomorphic to $\tilde{G}$. Since $\widetilde{\text{PSL}}(2, \mathbb{R})$ acts on $\tilde{S}^1 = \mathbb{R}$, viewing the interval $[0, 1]$ as a two-point compactification of $\mathbb{R}$, we obtain a faithful $\tilde{G}$-action on $[0, 1]$. Notice that by Thurston's theorem, the latter action cannot be smooth: the obstruction to the differentiability localizes around each endpoint of the interval.

Exercise 5.1.6. Using the fact that fgh belongs to the center of $\tilde{G}$, as well as Propositions 2.2.45 and 4.2.25, prove that the $\tilde{G}$-action above is not conjugate to an action by C^1 diffeomorphisms of the interval $[0, 1[$ without using Thurston's stability theorem.

Exercise 5.1.7. Following [41], consider the group $\hat{G}$ with presentation

$$\langle f_1, g_1, h_1, f_2, g_2, h_2 :$$
$$f_1^2 = g_1^3 = h_1^7 = f_1 g_1 h_1, f_2^{-1} f_1 f_2 = f_1^2, g_2^{-1} g_1 g_2 = g_1^2, h_2^{-1} h_1 h_2 = h_1^2 \rangle.$$

Show that $\hat{G}$ contains a copy of $\tilde{G}$ and acts faithfully by circle homeomorphisms, but it does not embed into $\text{Diff}_+^1(S^1)$.

Remark 5.1.8. The family of finitely generated groups of homeomorphisms of the interval containing finitely generated subgroups that do not admit nontrivial homomorphisms into $(\mathbb{R}, +)$ is quite large. Indeed, an important result in the theory of orderable groups asserts that a group (of arbitrary cardinality) is *locally indicable* (i.e., all of its finitely generated subgroups admit nontrivial homomorphisms into $(\mathbb{R}, +)$) if and only if it is C-orderable (see § 2.2.6 for the notion of C-order, and see [177] for an elementary proof of this result). Therefore, although all finitely generated groups of interval homeomorphisms are topologically conjugate to groups of Lipschitz homeomorphisms of $[0, 1]$ (cf. Proposition 2.3.15), many of them do not embed into $\text{Diff}_+^1([0, 1[)$ because they fail to be locally indicable. However, this is not the only algebraic obstruction: there exist finitely generated, locally indicable groups without (faithful) actions by C^1 diffeomorphisms of the interval! (see Remark 5.2.33).

A nice consequence of the preceding discussion is that every countable group of C^1 diffeomorphisms of the interval admits a faithful action by homeomorphisms of the interval without crossed elements. Indeed, by Theorem 5.1.1, such a group Γ is locally indicable, and hence by Remark 5.1.8, it is C-orderable. Proposition 2.2.64 (and the comments after its proof) then implies that the dynamical realization of any Conradian ordering on Γ is a subgroup of $\mathrm{Homeo}_+(\mathbb{R})$ without crossed elements (compare Exercise 2.2.21).

Remark 5.1.9. As was cleverly shown by Calegari in [39], the technique of the proof of Thurston's stability theorem (but not its conclusion !) may be used to show that certain actions of locally indicable groups (for example, the free group $\mathbb{F}_2$) on the closed interval are not C^1 smoothable. This is the case, for instance, if the set of fixed points of the generators accumulates at the endpoints and their commutator has no fixed point inside. Let us point out, however, that these actions have (plenty of) crossed elements (cf. Definition 2.2.43).

Remark 5.1.10. Theorem 5.1.1 may also be used to study group actions on higher- dimensional manifolds. For instance, using the idea of its proof, in [42] it is shown that the group of C^1 diffeomorphisms of the closed disk fixing all points in the boundary is orderable.

Exercise 5.1.11. By combining Thurston's stability theorem with the result in Exercise 2.3.23, prove the following result from [203]: if $\Gamma = \Gamma_1 \times \Gamma_2$ is a finitely generated group such that no finite-index normal subgroup of Γ_1 and Γ_2 has a nontrivial homomorphism into $(\mathbb{R}, +)$, then for every action of Γ by C^1 circle diffeomorphisms, the image of either Γ_1 or Γ_2 is finite.

Exercise 5.1.12. A topological group is said to be ***compactly generated*** if there exists a compact subset such that every element may be written as a product of elements therein. Extend Thurston's stability theorem to continuous representations in $\mathrm{Diff}^1_+([0, 1[)$ of compactly generated groups.

Exercise 5.1.13. The goal of this exercise consists in giving an alternative proof of Thurston's stability theorem using a technique from [214] and

[222]. Let us consider a finitely generated subgroup Γ of $\text{Diff}^1_+([0, 1[)$, and to simplify, let us suppose that Γ has no fixed point inside $]0, 1[$ and that all its elements are tangent to the identity at the origin (as we have already seen, the general case easily reduces to this one). Let $\mathcal{G} = \{h_1, \ldots, h_k\}$ be a finite family of generators for Γ. For each $f \in \Gamma$, let us define the *displacement function* Δ_f by $\Delta_f(x) = f(x) - x$. Notice that $(\Delta_f)'(0) = 0$ for every $f \in \Gamma$.

(i) Prove that for every $x \geq 0$ and every f and g in Γ, there exist y and z in $[0, x]$ such that

$$\Delta_{fg}(x) = \Delta_f(x) + \Delta_g(x) + (\Delta_f)'(y)\Delta_g(x),$$
$$\Delta_{f^{-1}(x)} = -\Delta_f(x) - (\Delta_f)'(z)\Delta_{f^{-1}}(x).$$

(ii) Fixing a strictly decreasing sequence of points x_n converging to the origin, for each $n \in \mathbb{N}$, let us choose $i_n \in \{1, \ldots, k\}$ such that $|\Delta_{h_{i_n}}(x_n)| \geq |\Delta_{h_j}(x_n)|$ for every $j \in \{1, \ldots, k\}$. Passing to a subsequence if necessary, we may assume that i_n is constant (say, equal to 1 after reordering the indices), and that each of the k sequences $(\Delta_{h_i}(x_n)/\Delta_{h_1}(x_n))$ converges to a limit ϕ_i (less than or equal to 1) as n goes to infinity. Use the equalities in (i) to show that the map $h_i \longmapsto \phi_i$ extends to a normalized homomorphism from Γ into $(\mathbb{R}, +)$.

Exercise 5.1.14. Following [174] and [245], consider a diffeomorphism φ from $]0, 1[$ into itself such that $\varphi(s) = \exp(-1/s)$ for small-enough $s > 0$. Prove that if $k \geq 0$ and $f : [0, 1[\to [0, 1[$ is a C^k diffeomorphism, then (the extension to $[0, 1[$ of) $\varphi^{-1} \circ f \circ \varphi$ (resp. $\varphi^{-2} \circ f \circ \varphi^2$) is a C^k diffeomorphism with derivative 1 (resp. tangent to the identity up to order k) at the origin.

We close this section by sketching the proof of Theorem 4.3.4. Let Γ be a polycyclic subgroup of $\text{Diff}^{1+\text{Lip}}_+([0, 1])$. By § 4.3, we know that Γ is metabelian. According to Theorem 4.4.5, we need to show that Γ cannot be a semidirect product between $\mathbb{Z}$ and a subgroup of a product of groups of translations as those described in the proof of that theorem. Assume for a contradiction that this is the case. Then it is easy to see that the nilradical $N(\Gamma)$ of Γ is torsion-free Abelian (say, isomorphic to $\mathbb{Z}^k$ for some $k \in \mathbb{N}$) and coincides with the subgroup formed by the elements having fixed points in $]0, 1[$. Moreover, if $\{f_1, \ldots, f_k\}$ is a system of generators of $N(\Gamma)$, then the group Γ is generated by these elements and a certain $f \in \Gamma$ without fixed points in $]0, 1[$.

Let $]a, b[$ be an irreducible component of $N(\Gamma)$; we must necessarily have $[a, b] \subset]0, 1[$. If we identify $N(\Gamma)$ with $\mathbb{Z}^k$ via the map $(m_1, \ldots, m_k) \mapsto g = f_1^{m_1} \cdots f_k^{m_k} \in N$, we readily see that f naturally induces an element f_* in $\mathrm{Hom}(\mathrm{Hom}(\mathbb{Z}^k, \mathbb{R}), \mathrm{Hom}(\mathbb{Z}^k, \mathbb{R}))$, namely, $f_*(\phi)(g) = \phi(fgf^{-1})$. If we view each element in $\mathrm{Hom}(\mathbb{Z}^k, \mathbb{R})$ as the restriction of a homomorphism defined on $\mathbb{R}^k$, the homomorphism f_* yields a linear map $M \in \mathrm{GL}(k, \mathbb{R})$.

Let $\phi_0 : \mathbb{Z}^k \to \mathbb{R}$ be the homomorphism obtained via Thurston's stability theorem applied to the restriction of $N(\Gamma) \sim \mathbb{Z}^k$ to $[a, b[$. From the construction of ϕ_0, it is easy to see that if (g_n) is a sequence of elements in $N(\Gamma)$ whose restrictions to $[a, b]$ converge to the identity in the C^1 topology, then $\phi_0(g_n)$ converges to zero as n goes to infinity. Now recall the following important fact from Exercise 4.1.5: for each $g \in N(\Gamma)$, both sequences $(f^n g f^{-n})$ and $(f^{-n} g f^n)$ restricted to $[a, b]$ converge to the identity in the C^1 topology. Therefore, both $f_*^n(\phi_0)$ and $f_*^{-n}(\phi_0)$ pointwise converge to the zero homomorphism. This means that for the linear map M, there exists a nonzero vector v such that both $M^n(v)$ and $M^{-n}(v)$ converge to zero, which is clearly impossible. This contradiction concludes the proof.

5.2 Rigidity for Groups with Kazhdan's Property (T)

5.2.1 Kazhdan's Property (T)

Let Γ be a countable group, and let $\Psi : \Gamma \to U(\mathcal{H})$ be a unitary representation of Γ in some (real) Hilbert space $\mathcal{H}$. We say that $c : \Gamma \to \mathcal{H}$ is a **cocycle** with respect to Ψ if for every g_1 and g_2 in Γ, one has

$$c(g_1 g_2) = c(g_1) + \Psi(g_1)c(g_2).$$

We say that a cocycle c is a **coboundary** if there exists $K \in \mathcal{H}$ such that for every $g \in \Gamma$,

$$c(g) = K - \Psi(g)K.$$

We denote the space of cocycles by $Z^1(\Gamma, \Psi)$ and the subspace of coboundaries by $B^1(\Gamma, \Psi)$. The quotient $H^1(\Gamma, \Psi) = Z^1(\Gamma, \Psi)/B^1(G, \Psi)$ is the **first cohomology space** of Γ (with values in Ψ).

Definition 5.2.1. A countable group Γ satisfies **Kazhdan's property (T)** (or, to simplify, is a *Kazhdan group*) if for every unitary representation Ψ of Γ, the first cohomology $H^1(\Gamma, \Psi)$ is trivial.

To give a geometric insight into this definition, let us recall that every *isometry* of a Hilbert space is the composition of a unitary transformation and a translation.[2] Indeed, the group of isometries of $\mathcal{H}$ is the semidirect product between $U(\mathcal{H})$ and $\mathcal{H}$.

Let $\Psi : \Gamma \to U(\mathcal{H})$ be a unitary representation, and let $c : \Gamma \to \mathcal{H}$ be a map. If for each $g \in \Gamma$ we define the isometry $A(g) = \Psi(g) + c(g)$, then one easily checks that the equality $A(g_1)A(g_2) = A(g_1 g_2)$ holds for every g_1 and g_2 in Γ if and only if c is a cocycle associated to Ψ. In this case the correspondence $g \mapsto A(g)$ defines an action by isometries.

Suppose now that $K \in \mathcal{H}$ is a fixed point of the isometric action associated to a cocycle $c : \Gamma \to \mathcal{H}$. Then we have, for every $g \in \Gamma$,

$$\Psi(g)K + c(g) = A(g)K = K.$$

Therefore, $c(g) = K - \Psi(g)K$, that is, c is a coboundary. Conversely, it is easy to check that if c is a coboundary, then there exists an invariant vector for the associated isometric action. We then have the following geometric interpretation of Definition 5.2.1: a group has Kazhdan's property (T) if and only if every action by isometries of a Hilbert space has an invariant vector.

Example 5.2.2. Every finite group has Kazhdan's property (T). To see this, it suffices to remark that the mean along any orbit of an action by isometries on a Hilbert space is an invariant vector for the action.

Exercise 5.2.3. A vector $K_{\mathcal{C}} \in \mathcal{H}$ is the *geometric center* of a subset $\mathcal{C}$ of $\mathcal{H}$ if $K_{\mathcal{C}}$ minimizes the function $K \mapsto \sup_{\bar{K} \in \mathcal{C}} \|K - \bar{K}\|$. Show that if $\mathcal{C}$ is bounded, then it has a unique center. Conclude that if $\mathcal{C}$ is invariant under an action by isometries, then its center remains fixed by the action. Use this to prove the following *center lemma* (due to Bass and Tits): an action by isometries has a fixed point if and only if its orbits are bounded. (In particular, a group has property (T) if and only if the orbits associated to its isometric actions on Hilbert spaces are bounded.)

Remark. The same argument applies to spaces on which the distance function satisfies a *convexity property*, for instance, simplicial trees or spaces with nonpositive curvature [28].

---

2. By an isometry of a Hilbert space $\mathcal{H}$ we mean a *surjective* map A from $\mathcal{H}$ into itself satisfying $\|A(K_1) - A(K_2)\| = \|K_1 - K_2\|$ for all K_1 and K_2 in $\mathcal{H}$.

Every group with property (T) is finitely generated, according to a result due to Kazhdan himself (actually, this was one of his motivations for introducing property (T)). We next give an alternative argument due to Serre based on the result from Example 5.2.11.

Let Γ be a countable group, and let $\Gamma_1 \subset \Gamma_2 \subset \ldots \subset \Gamma_n \subset \ldots$ be a sequence of finitely generated subgroups whose union is Γ. Let us consider the oriented simplicial tree $\mathcal{T}$ whose vertices are the left classes of Γ with respect to the Γ_n's, and such that between two vertices $[g]$ in Γ/Γ_n and $[h]$ in Γ/Γ_{n+1} there is an edge (oriented from $[g]$ to $[h]$) if $g \in [h]$. The group Γ naturally acts on $\mathcal{T}$ and preserves the orientation of the edges. By Example 5.2.11, if Γ has property (T), then there exists a vertex $[g]$ that is fixed by this action. If $n \in \mathbb{N}$ is such that $[g]$ represents the class of g with respect to Γ_n, this implies that $\Gamma = \Gamma_n$, and hence Γ is finitely generated.

Example 5.2.4. If Γ has property (T) and is amenable, then it is finite (see Appendix B for the notion of amenability, as well as some notation). In particular, the only Abelian groups having Kazhdan's property are the finite ones. Actually, we will next see that every finitely generated amenable group satisfies *Haagerup's property*, that is, it acts by isometries of a Hilbert space in a *geometrically proper* way (in the sense that $\|c(g)\|_{\mathcal{H}}$ goes to infinity together with $length(g)$). Clearly, such a group cannot have property (T) unless it is finite (see [55] for more on this relevant property).

Let $\mathcal{G}$ be a finite system of generators of Γ, and let (A_n) be a Følner sequence associated to $\mathcal{G}$ such that

$$\frac{card(\partial A_n)}{card(A_n)} \leq \frac{1}{2^n}. \tag{5.3}$$

Let us consider the Hilbert space $\mathcal{H} = \oplus_{n \geq 1} n\ell_{\mathbb{R}}^2(\Gamma)$ defined by

$$\mathcal{H} = \left\{ K = (K_1, \ldots, K_n, \ldots), K_n \in \ell_{\mathbb{R}}^2(\Gamma), \sum_{n \geq 1} n^2 \|K_n\|_{\ell_{\mathbb{R}}^2(\Gamma)}^2 < \infty \right\}.$$

Given g and h in Γ, for $K_n \in \ell_{\mathbb{R}}^2(\Gamma)$, let $\Psi_n(g)K_n(h) = K_n(g^{-1}h)$, and let $\Psi : \Gamma \to U(\mathcal{H})$ be the *regular representation* given by $\Psi(g)K = (\Psi_1(g)K_1, \Psi_2(g)K_2, \ldots)$. Finally, for each $g \in \Gamma$, let $c(g) = (c(g)_1, c(g)_2, \ldots) \in \mathcal{H}$ be defined by

$$c(g)_n = \frac{1}{\sqrt{card(A_n)}} (\mathcal{X}_{A_n} - \mathcal{X}_{g(A_n)}),$$

where $\mathcal{X}$ denotes the characteristic function. By (5.3), for every $g \in \mathcal{G}$, we have

$$
\|c(g)\|_{\mathcal{H}}^2 = \sum_{n \geq 1} n^2 \left\| \frac{\mathcal{X}_{A_n} - \mathcal{X}_{g(A_n)}}{\sqrt{card(A_n)}} \right\|_{\ell_{\mathbb{R}}^2(\Gamma)}^2
$$

$$
\leq \sum_{n \geq 1} n^2 \left(\frac{card(\partial A_n)}{card(A_n)} \right) \leq \sum_{n \geq 1} \frac{n^2}{2^n} < \infty.
$$

The cocycle relation $c(g_1 g_2) = c(g_1) + \Psi(g_1)c(g_2)$ can be easily checked. We claim that the function $c : \Gamma \to \mathcal{H}$ is geometrically proper. To show this, fix an integer k greater than the diameter of $A_n^{-1} = \{h^{-1} : h \in A_n\}$. If an element $g \in \Gamma$ belongs to $B_{\mathcal{G}}(k) \setminus B_{\mathcal{G}}(k-1)$, then $g(A_n) \cap A_n = \emptyset$, and therefore

$$
\left\| \frac{\mathcal{X}_{A_n} - \mathcal{X}_{g(A_n)}}{\sqrt{card(A_n)}} \right\|_{\ell_{\mathbb{R}}^2(\Gamma)}^2 \geq 2.
$$

Thus $\|c(g)\|_{\mathcal{H}}^2 \geq 2n^2$, and this shows our claim.

Example 5.2.5. Group homomorphic images and finite extensions of Kazhdan groups also have property (T). The proof is easy, and we leave it to the reader.

As a consequence of the preceding example, every Kazhdan group satisfies the hypothesis of Thurston's stability theorem. Indeed, the image of a Kazhdan group under a homomorphism into $(\mathbb{R}, +)$ is Abelian and has property (T). Thus it must be finite and hence trivial.

Exercise 5.2.6. Show directly that no nontrivial subgroup Γ of $(\mathbb{R}, +)$ has property (T). To do this, consider the representation of Γ by translations on $\mathcal{L}_{\mathbb{R}}^2(\mathbb{R}, Leb)$ and the associated cocycle $c(g) = \mathcal{X}_{[0,\infty[} - \mathcal{X}_{[g(0),\infty[}$, where $\mathcal{X}$ denotes the characteristic function and $g \in \Gamma$. Show that this cocycle is not a coboundary.

Exercise 5.2.7. Prove directly that every finitely generated subgroup of $\mathrm{Diff}_+^{1+\tau}([0,1])$ with property (T) is trivial, where $\tau > 0$.

Hint. Passing to a quotient if necessary, one may assume that the underlying group Γ has no fixed point in $]0, 1[$. Consider the Radon measure

μ on $]0, 1[$ defined by $d\mu = dx/x$, and let Ψ be the **regular representation** of Γ on $\mathcal{H} = \mathcal{L}^2_{\mathbb{R}}([0, 1], \mu)$ defined by

$$\Psi(g)K(x) = K(g^{-1}(x)) \left[\frac{dg^{-1}}{d\mu}(x) \right]^{1/2}.$$

For each $g \in \Gamma$, let

$$c(g) = 1 - \left[\frac{dg^{-1}}{d\mu}(x) \right]^{1/2}. \tag{5.4}$$

Check the cocycle relation $c(gh) = c(g) + \Psi(g)c(h)$. Using the fact that τ is positive, prove that $c(g)$ belongs to $\mathcal{H}$. Show that if c is cohomologically trivial, then there exists $K \in \mathcal{H}$ such that the measure v on $]0, 1[$ whose density function (with respect to μ) is $x \mapsto [1 - K(x)]^2$ is invariant under Γ. Conclude the using some of the results from § 2.2.5.

Nontrivial examples of Kazhdan groups are lattices in (connected) simple Lie groups of (real) rank greater than 1. (Recall that the (real) **rank** of a Lie group is the dimension of the maximal Abelian subalgebra over which the adjoint representation is diagonalizable (over $\mathbb{R}$), and that a discrete subgroup of a locally compact topological group is a **lattice** if the quotient space has finite Haar measure.) For instance, $SL(3, \mathbb{Z})$ has property (T), since the rank of the simple Lie group $SL(3, \mathbb{R})$ is 2, and $SL(3, \mathbb{Z})$ is a lattice inside. For a detailed discussion of this, see [13].

Different kinds of groups with property (T) have been constructed by many people. In particular, in his seminal work on random groups [106] (see also [260]), Gromov shows that "generic" finitely presented groups have Kazhdan's property. Therefore, a theorem that is true for Kazhdan groups is somehow valid for "almost every group".

To close this section, we give two simple results on obstructions to property (T) that are relevant for us. They concern Thompson's group G and Neretin's groups, to be defined later. The reader who is eager for the connections with groups of circle diffeomorphisms may skip this discussion and pass directly to the next section.

Proposition 5.2.8. *Thompson's group* G *does not have Kazhdan's property* (T).

Proof. We will give a (modified version of a) nice argument due to Farley [77], which actually shows that G has Haagerup's property (cf. Example 5.2.4).

Denote by G_0 the subgroup of G formed by the elements whose restrictions to the subinterval $[0, 1/2]$ of S^1 is the identity. Let us consider the Hilbert space $\mathcal{H} = \ell^2_{\mathbb{R}}(G/G_0)$. The group G naturally acts by isometries of $\mathcal{H}$ by letting $\Psi(g)K([h]) = K([g^{-1}h])$ for all g and h in G and all $K \in \mathcal{H}$.

Given a dyadic interval $I \subset S^1$, choose $g_I \in G$ sending $[0, 1/2]$ into I affinely. Notice that the class $[g_I]$ of g_I modulo G_0 does not depend on this choice. For each $g \in G$, let $c(g) \in \mathcal{H}$ be the function defined by

$$c(g) = \sum \left(\delta_{[g_I]} - \delta_{[gg_I]} \right),$$

where the sum extends over the set of all dyadic intervals I, and $\delta_{[g_I]}$ is the characteristic function of the set $\{[g_I]\}$. Each function $c(g)$ has finite support, since for every $g \in \Gamma$, one has $[g_{g(I)}] = [gg_I]$ for small-enough $|I|$ (the restriction of g to small dyadic intervals is affine). Therefore, $c(g)$ belongs to $\mathcal{H}$. Moreover, it is not difficult to see that c satisfies the cocycle identity with respect to the unitary representation Ψ.

To compute $\|c(g)\|$, notice that from the definition one easily deduces that $\|c(g)\|^2$ equals 2 times the number of dyadic intervals I such that either the restriction of g to I is not affine, or the image of I under g is not a dyadic interval. It easily follows from this fact that $\|c(g)\|$ tends to infinity as $length(g)$ goes to infinity. $\qquad\qquad \square$

In order to introduce Neretin's groups, let us denote the homogeneous simplicial tree of valence $p + 1$ by $\mathcal{T}^p$. Let σ be a marked vertex of $\mathcal{T}^p$, which will be called the **origin**. A (possibly empty) subtree $\mathcal{A}$ of $\mathcal{T}^p$ is **complete** if it is connected and compact and each time two edges in $\mathcal{A}$ have a common vertex, all the edges containing this vertex are included in $\mathcal{A}$. Notice that the complement of a complete subtree is either all of $\mathcal{T}^p$ or a finite union of rooted trees.

Given a pair of complete subtrees $\mathcal{A}$ and $\mathcal{B}$ of $\mathcal{T}^p$, we denote by $\mathcal{N}^p(\mathcal{A}, \mathcal{B})$ the set of bijections from $\overline{\mathcal{T}^p \setminus \mathcal{A}}$ onto $\overline{\mathcal{T}^p \setminus \mathcal{B}}$ sending each connected component of $\overline{\mathcal{T}^p \setminus \mathcal{A}}$ isometrically onto a connected component of $\overline{\mathcal{T}^p \setminus \mathcal{B}}$. If g belongs to $\mathcal{N}^p(\mathcal{A}, \mathcal{B})$, then g induces a homeomorphism of $\partial \mathcal{T}^p$, which we will still denote by g. Notice that $\partial \mathcal{T}^p$ may be endowed with a natural metric: the distance between x and y in $\partial \mathcal{T}^p$ is given by $\partial \text{ist}(x, y) = p^{-n}$, where n is the distance $dist$ (on $\mathcal{T}^p$) between σ and the geodesic joining x and y.

Definition 5.2.9. The group of homeomorphisms of ∂T^p induced by elements in some $\mathcal{N}^p(\mathcal{A}, \mathcal{B})$ is called **Neretin's group** (or the **spheromorphisms group**) and is denoted by $\mathcal{N}^p$.

Roughly, $\mathcal{N}^p$ is the group of "germs at infinity" of isometries of T^p. Notice that if $g_1 \in \mathcal{N}^p(\mathcal{A}, \mathcal{B})$ and $g_2 \in \mathcal{N}^p(\mathcal{A}', \mathcal{B}')$ induce the same element of $\mathcal{N}^p$, then they coincide over $T^p \setminus (\mathcal{A} \cup \mathcal{A}')$. A representative $\tilde{g} \in \mathcal{N}^p(\mathcal{A}, \mathcal{B})$ of $g \in \mathcal{N}^p$ will be said to be **maximal** if its domain of definition contains the domain of any other representative of g. Each element in $\mathcal{N}^p$ has a unique maximal representative. For $g \in \mathcal{N}^p$, let us denote by $\mathcal{A}_g$ (resp. $\mathcal{B}_g$) the closure of the complement of the domain of definition of $\tilde{g}$ (resp. of $\tilde{g}^{-1}$). Notice that $\mathcal{A}_g = \mathcal{B}_{g^{-1}}$. The group $\mathrm{Isom}(T^p)$ of the (extensions to the boundary of the) isometries of T^p is a subgroup of $\mathcal{N}^p$. An element $g \in \mathcal{N}^p$ comes from an element in $\mathrm{Isom}(T^p)$ if and only if $\mathcal{A}_g = \mathcal{B}_g = \emptyset$. Notice, finally, that Thompson's group G may be also be seen as a subgroup of $\mathcal{N}^p$.

Neretin's groups naturally appear in the p-adic context. Indeed, if p is a prime number, then the group of diffeomorphisms of the projective line over $\mathbb{Q}_p$ (that is, of the p-adic circle) embeds into $\mathcal{N}^p$. In a certain sense, $\mathcal{N}^p$ is a combinatorial analog of the group of circle diffeomorphisms. For further information on this, we refer to the excellent survey [191]. Here we will content ourselves with proving the following result from [188].

Proposition 5.2.10. *Let Γ be a subgroup of $\mathcal{N}^p$. If Γ has property* (T), *then there exists a finite-index subgroup Γ_0 of Γ such that the boundary of T^p decomposes into finitely many balls that are fixed by Γ_0, and Γ_0 acts isometrically on each of them.*

Proof. For each vertex $a \neq \sigma$, let A_a be the subtree rooted at a and pointing to the infinity (i.e., in the opposite direction to that of the origin). Let us choose one of the $p + 1$ subtrees of T^p rooted at the origin, and slightly abusing of notation, let us denote it by A_σ. Let $\mathcal{N}_\sigma^p$ be the subgroup of $\mathcal{N}^p$ formed by the elements that fix the boundary at infinity ∂A_σ of A_σ and act isometrically (with respect to the metric $\partial\mathrm{ist}$) on it. The group $\mathcal{N}^p$ has a natural unitary action Ψ on the Hilbert space $\mathcal{H} = \ell_{\mathbb{R}}^2(\mathcal{N}^p/\mathcal{N}_\sigma^p)$, namely $\Psi(g)K([h]) = K([g^{-1}h])$.

To each vertex a of T^p we associate the left-class $\phi_a \in \mathcal{N}^p/\mathcal{N}_\sigma^p$ defined by $\phi_a = [h]$, where $h \in \mathcal{N}^p$ is an element whose maximal representative sends A_σ into A_a isometrically (with respect to the metric $dist$). Given $g \in \mathcal{N}^p$, we let

$$c(g) = \sum g(\delta_{\phi_a}) - \sum \delta_{\phi_b},$$

where δ_{ϕ_a} is the characteristic function of the set $\{\phi_a\}$. Notice that in the preceding expression most of the terms cancel, and only finitely many remain; as a consequence, the function $c(g)$ belongs to $\mathcal{H}$. Moreover, one easily checks that c is a cocycle with respect to Ψ.

For $g \in \Gamma \setminus \text{Isom}(\mathcal{T}^p)$, let $d = d(g)$ be the distance between σ and $\mathcal{A}_g$. Let us consider a geodesic γ joining σ to the vertex in $\mathcal{A}_g$ for which the distance to σ is maximal, and let $a_1, a_2, \ldots, a_{d-1}$ be the vertices in the interior of γ. One easily checks that for every vertex b in $\mathcal{T}^p$ and all $i \in \{1, \ldots, d-1\}$, one has $g(\delta_{\phi_{a_i}}) \neq \delta_b$, from which one deduces that $\|c(g)\|^2 \geq d - 1$.

Now let $g \in \Gamma \cap \text{Isom}(\mathcal{T}^p)$. Let $d' = d'(g)$ be the distance between σ and $\tilde{g}(\sigma)$, and let γ be the geodesic joining these points. Denoting by $a'_1, \ldots, a'_{d'-1}$ the vertices in the interior of $\tilde{g}^{-1}(\gamma)$, one easily checks that for all $i \in \{1, \ldots, d'-1\}$ and every vertex b of $\mathcal{T}^p$, one has $g(\delta_{\phi_{a'_i}}) \neq \delta_b$ (this is because $A_{\tilde{g}(a'_i)}$ does not coincide with the image of $A_{a'_i}$ under $\tilde{g}$, since they point toward different directions !). One thus concludes that $\|c(g)\|^2 \geq d' - 1$.

If Γ has property (T), then the function $g \mapsto \|c(g)\|$ must be bounded. Therefore, there exists an integer $N > 0$ such that $dist(\sigma, \mathcal{A}_g) \leq N$ for all $g \in \Gamma \setminus \text{Isom}(\mathcal{T}^p)$ (and hence $dist(\sigma, \mathcal{B}_g) \leq N$ for all $g \in \Gamma \setminus \text{Isom}(\mathcal{T}^p)$), and such that for all $g \in \Gamma \cap \text{Isom}(\mathcal{T}^p)$, one has $dist(\sigma, \tilde{g}(\sigma)) \leq N$. The proposition easily follows from these facts.   □

5.2.2 The statement of the result

Much evidence suggests that any (reasonable) one-dimensional structure on a space is an obstruction to actions of a Kazhdan group on it. For instance, we will see that for every action of a group with property (T) by isometries of a simplicial tree, there exists a global fixed point. The proof that we present is essentially due to Haglund, Paulin, and Valette. An easy modification allows showing that the same result holds for actions by isometries on real trees. The results are originally due to Alperin and Watatani [2, 250].

Example 5.2.11. Let $\mathcal{T}$ be an oriented simplicial tree (we do not assume that the valence of the edges is finite). Let us denote by $\overrightarrow{edg}(\mathcal{T})$ the set of open (oriented) edges of $\mathcal{T}$. For each $\vec{\Upsilon} \in \overrightarrow{edg}(\mathcal{T})$, let $ver(\vec{\Upsilon})$ be the set of the vertices in $\mathcal{T}$ that are connected to $\vec{\Upsilon}$ by a geodesic whose initial segment is $\vec{\Upsilon}$ (with the corresponding orientation). For each vertex $v \in \mathcal{T}$, let us denote by $\overrightarrow{edg}(v)$ the set of the oriented edges $\vec{\Upsilon}$ for which $v \in ver(\vec{\Upsilon})$.

Let us consider the Hilbert space $\mathcal{H} = \ell^2_{\mathbb{R}}(\overrightarrow{edg}(\mathcal{T}))$. Let Γ be a subgroup of the group of (orientation-preserving) isometries of $\mathcal{T}$. Fix a vertex σ in $\mathcal{T}$, and define a unitary representation Ψ of Γ on $\mathcal{H}$ by letting $\Psi(g)K(\vec{\Upsilon}) = K(g^{-1}(\vec{\Upsilon}))$. For each $g \in \Gamma$, let $c(g) : \overrightarrow{edg}(\mathcal{T}) \to \mathbb{R}$ be defined by

$$c(g) = \mathcal{X}_{\overrightarrow{edg}(\sigma)} - \mathcal{X}_{\overrightarrow{edg}(g(\sigma))},$$

where $\mathcal{X}$ stands for the characteristic function. It is easy to see that $c(g)$ has finite support and hence belongs to $\mathcal{H}$. Moreover, the correspondence $g \mapsto c(g)$ is a cocycle with respect to Ψ.

If Γ has Kazhdan's property (T), then the cocycle c is a coboundary. In other words, there exists a function $K \in \mathcal{H}$ such that for every $g \in \Gamma$, one has $c(g) = K - \Psi(g)K$. In particular, $\|c(g)\| \leq 2\|K\|$. On the other hand, it is easy to see that $\|c(g)\|^2 = 2\,dist(\sigma, g(\sigma))$. One then deduces that the orbit of σ under the Γ-action stays inside a bounded subset of $\mathcal{T}$. We leave to the reader the task of showing that this implies the existence of a vertex that is fixed by the action (see [13, 226] in case of problems; compare also Exercise 5.2.3).

If we have a group action on a general simplicial tree, then by taking the first barycentric subdivision, we may reduce to the oriented case. If the group has property (T), then one concludes that it fixes either a vertex or the middle point of an edge of the original tree.

In § 2.2.3 we have seen that no finite-index subgroup of SL(3, $\mathbb{Z}$) acts by homeomorphisms of the line in a nontrivial way. According to [253] (see also [146]), this is also true for many other lattices in Lie groups having property (T), but it is still unknown whether there exists a nontrivial Kazhdan group action on the line. (Equivalently, it is unknown whether there exist nontrivial, orderable Kazhdan groups.)

More accurate results are known in the case of the circle. In particular, a theorem of Ghys [89] (a closely related version was independently and simultaneously obtained by Burger and Monod [36]) asserts that if $\Phi : \Gamma \to \mathrm{Diff}^1_+(S^1)$ is a representation of a lattice Γ in a simple Lie group of rank greater than 1, then the image $\Phi(\Gamma)$ is finite. For a nice discussion of this result (as well as a complete proof for the case of lattices in $SL(3, \mathbb{R})$), we strongly recommend [88].

Exercise 5.2.12. The main step in Ghys's proof consists in showing that for every action by circle *homeomorphisms* of a lattice Γ in a higher-rank simple Lie group, there is an invariant probability measure. Show that this suffices to prove the theorem (compare Exercise 5.1.11).

Hint. Recall that if there is an invariant probability measure, then the rotation-number function is a group homomorphism into $\mathbb{T}^1$ (see § 2.2.2). Using the fact that Γ has Kazhdan's property, show that the image under this homomorphism is finite. Conclude that the orbit of every point in the support of the invariant measure is finite. Finally, apply Thurston's stability theorem.

Remark 5.2.13. It is not difficult to extend Theorem 2.2.24 and show that finite- index subgroups in $SL(3, \mathbb{Z})$ do not admit nontrivial actions by circle homeomorphisms. Indeed, this result (also due to Witte-Morris) may be proved by combining the claim at the beginning of the preceding exercise and Theorem 2.2.24, although it was originally proved using Margulis's normal subgroup theorem [158]. Let us point out that the proof is quite easy in class C^2. Indeed, the involved lattices contain nilpotent subgroups that are not virtually Abelian. Hence by the Plante-Thurston theorem, their actions by C^2 circle diffeomorphisms have an infinite kernel. By Margulis's normal subgroup theorem, these actions must have finite image.

The following theorem was obtained by the author in [187], inspired by Reznikov's prior work [216].[3] It may be thought of as an analog for Kazhdan groups of the results discussed earlier under a supplementary regularity condition.

Theorem 5.2.14. *Let* $\Phi : \Gamma \to \mathrm{Diff}_+^{1+\tau}(S^1)$ *be a representation of a countable group, with* $\tau > 1/2$. *If* Γ *has Kazhdan's property (T), then* $\Phi(\Gamma)$ *is finite.*

Kazhdan's property may also be considered for nondiscrete groups, mainly for locally compact ones. (In this case the representations and

---

3. In [217] Reznikov proposed an alternative schema of proof that has many gaps (unfortunately, we do not know exactly what he had in mind concerning this).

cocycles involved in the definition should be continuous functions.) The reader can easily check that the technique of the proof of Theorem 5.2.14 still applies in this general context. Another Kazhdan group may then arise as a group of circle diffeomorphisms, namely, SO(2, $\mathbb{R}$) (compare § 1.4).

Notice that Theorem 5.2.14 shows again that Thompson's group G does not have property (T) (see Proposition 5.2.8). Indeed, according to § 1.5.2, G may be realized as a group of C^∞ circle diffeomorphisms (actually, it suffices to use the $C^{1+\text{Lip}}$ realization from § 1.5.1). However, our technique of proof does not lead to Haagerup's property for G shown in § 5.2.1.

Exercise 5.2.15. Show that every Kazhdan subgroup of PSL(2, $\mathbb{R}$) is finite by following these steps (see [13, § 2.6] in case of problems):

(i) Consider the action of the Möbius group on the space M of nonoriented geodesics in the Poincaré disk $\mathbb{D}$. By identifying M with the set of classes PSL(2, $\mathbb{R}$)/G_γ (where G_γ is the stabilizer of a geodesic γ), use the Haar measure on PSL(2, $\mathbb{R}$) to endow M with a Radon measure v.

(ii) For each P and Q in $\mathbb{D}$, let $M_{P,Q}$ be the set of geodesics that "separate" P and Q (i.e., that divide $\mathbb{D}$ into two hyperbolic half-planes so that P and Q do not belong to the same one). Show that there exists a constant C such that the v-measure of $M_{P,Q}$ equals $C \, dist(P, Q)$ for all P, Q, where $dist$ stands for the hyperbolic distance on $\mathbb{D}$.

(iii) Consider the Hilbert space $\mathcal{H} = \mathcal{L}^2_{\mathbb{R}}(M, v)$ and the unitary action Ψ of PSL(2, $\mathbb{R}$) on $\mathcal{H}$ given by $\Psi(f)K(\gamma) = K(f^{-1}\gamma)$. Show that the function c defined on PSL(2, $\mathbb{R}$) by $c(f) = \mathcal{X}_{M_{O,f(O)}}$ belongs to $\mathcal{H}$, where $O = (0, 0) \in \mathbb{D}$ and $\mathcal{X}$ stands for the characteristic function. Show, moreover, that c is a cocycle with respect to Ψ.

(iv) Using the construction in (iii), show that if Γ is a Kazhdan subgroup of PSL(2, $\mathbb{R}$), then all of its elements are elliptic. Conclude that Γ is finite by using the result of Exercise 2.2.34.

5.2.3 Proof of the theorem

To simplify, we will denote by $\bar{g}$ the diffeomorphism $\Phi(g^{-1})$. Recall that the Liouville measure Lv on $S^1 \times S^1$ has density function $S^1 \times S^1$ has density function (see § 1.3.2)

$$(r, s) \mapsto \frac{1}{4\sin^2(\frac{r-s}{2})}.$$

Let $\mathcal{H} = \mathcal{L}_{\mathbb{R}}^{2,\Delta}(S^1 \times S^1, Lv)$ be the subspace of $\mathcal{L}_{\mathbb{R}}^2(S^1 \times S^1, Lv)$ formed by the functions K satisfying $K(x, y) = K(y, x)$ for almost every $(x, y) \in S^1 \times S^1$. Let Ψ be the unitary representation of Γ on $\mathcal{H}$ given by

$$\Psi(g)K(r, s) = K(\bar{g}(r), \bar{g}(s)) \cdot [Jac(\bar{g})(r, s)]^{\frac{1}{2}},$$

where $Jac(\bar{g})(r, s)$ is the Jacobian (with respect to Lv) of the map $(r, s) \mapsto (\bar{g}(r), \bar{g}(s))$:

$$Jac(\bar{g})(r, s) = \frac{\sin^2\left(\frac{r-s}{2}\right)}{\sin^2\left(\frac{\bar{g}(r)-\bar{g}(s)}{2}\right)} \cdot [\bar{g}'(r)\bar{g}'(s)].$$

For each $g \in \Gamma$, let us consider the function (compare (1.5))

$$c(g)(r, s) = 1 - [Jac(\bar{g})(r, s)]^{\frac{1}{2}}. \tag{5.5}$$

One may formally check the relation $c(g_1 g_2) = c(g_1) + \Psi(g_1)c(g_2)$ (compare (1.4)). Indeed, this **Liouville cocycle** c corresponds to the "formal coboundary" of the constant function 1, which does not belong to $\mathcal{L}_{\mathbb{R}}^2(S^1 \times S^1, Lv)$.[4] The main issue here is that if Φ takes values in $\mathrm{Diff}_+^{1+\tau}(S^1)$ for some $\tau > 1/2$, then $c(g)$ is a true cocycle with values in $\mathcal{H}$. This is the content of the following proposition, essentially due to Segal [209] (see also [216]).

Proposition 5.2.16. *If $\tau > 1/2$, then $c(g)$ belongs to $\mathcal{L}_{\mathbb{R}}^{2,\Delta}(S^1 \times S^1, Lv)$ for all $g \in \Gamma$.*

Proof. Notice that for a certain continuous function $K_1 : [0, 2\pi] \times [0, 2\pi] \to \mathbb{R}$, one has

$$\frac{1}{\left|\sin\left(\frac{r-s}{2}\right)\right|} = 2\left[\frac{1}{|r-s|} + K_1(r, s)\right].$$

Therefore, to prove that $c(g)$ is in $\mathcal{L}_{\mathbb{R}}^{2,\Delta}(S^1 \times S^1, Lv)$, we need to verify that the function

$$(r, s) \mapsto \frac{[\bar{g}'(r)\bar{g}'(s)]^{\frac{1}{2}}}{|\bar{g}(r) - \bar{g}(s)|} - \frac{1}{r-s}$$

---

4. The reader will readily notice that the cocycles from Examples 5.2.4, 5.2.6, 5.2.7, and 5.2.11, that of Exercise 5.2.15, and those of the proofs of Propositions 5.2.8 and 5.2.10 also arise as formal coboundaries.

belongs to $\mathcal{L}^2_{\mathbb{R}}(S^1 \times S^1, Leb)$. Now for all r and s in S^1 such that $|r - s| < \pi$, there exists $t \in S^1$ in the shortest segment joining them such that $|\bar{g}(r) - \bar{g}(s)| = \bar{g}'(t)|r - s|$. We then have

$$
\left| \frac{[\bar{g}'(r)\bar{g}'(s)]^{\frac{1}{2}}}{|\bar{g}(r) - \bar{g}(s)|} - \frac{1}{|r - s|} \right|
$$

$$
= \frac{1}{|r - s|\bar{g}'(t)} \cdot \left| [\bar{g}'(r)\bar{g}'(s)]^{\frac{1}{2}} - \bar{g}'(t) \right|
$$

$$
= \frac{1}{|r - s|\bar{g}'(t)} \cdot \frac{|\bar{g}'(r)\bar{g}'(s) - \bar{g}'(t)^2|}{[\bar{g}'(r)\bar{g}'(s)]^{\frac{1}{2}} + \bar{g}'(t)}.
$$

$$
\leq \frac{1}{2\inf(\bar{g}')^2|r - s|} [|\bar{g}'(r) - \bar{g}'(t)|\bar{g}'(s) + \bar{g}'(t)|\bar{g}'(s) - \bar{g}'(t)|].
$$

Since $\bar{g}'$ is τ-Hölder continuous (cf. Example 4.1.18), this gives

$$
\left| \frac{[\bar{g}'(r)\bar{g}'(s)]^{\frac{1}{2}}}{\bar{g}(r) - \bar{g}(s)} - \frac{1}{r - s} \right| \leq \frac{|\bar{g}'|_\tau \sup(\bar{g}')}{2|r - s|\inf(\bar{g}')^2} [|r - t|^\tau + |s - t|^\tau] \leq C|r - s|^{\tau - 1},
$$

where the constant C does not depend on (r, s). The proof is completed by noticing that since $\tau > 1/2$, the function $(r, s) \mapsto |r - s|^{\tau - 1}$ belongs to $\mathcal{L}^2_{\mathbb{R}}(S^1 \times S^1, Leb)$. □

If the group Γ has property (T) and $\tau > 1/2$, then the cocycle (5.5) is a coboundary. In other words, there exists a function $K \in \mathcal{L}^{2,\Delta}_{\mathbb{R}}(S^1 \times S^1, Lv)$ such that for every $g \in \Gamma$ and almost every $(r, s) \in S^1 \times S^1$, one has

$$
1 - [Jac(\bar{g})(r, s)]^{\frac{1}{2}} = K(r, s) - K(\bar{g}(r), \bar{g}(s)) \cdot [Jac(\bar{g})(r, s)]^{\frac{1}{2}},
$$

that is,

$$
[1 - K(\bar{g}(r), \bar{g}(s))]^2 \cdot Jac(\bar{g})(r, s) = [1 - K(r, s)]^2.
$$

We thus conclude the following:

Proposition 5.2.17. *Let* $\Phi : \Gamma \to \mathrm{Diff}^{1+\tau}_+(S^1)$ *be a representation, with* $\tau > 1/2$. *If* Γ *has property* (T), *then there exists* $K \in \mathcal{L}^{2,\Delta}_{\mathbb{R}}(S^1 \times S^1, Lv)$ *such that* Γ *preserves the geodesic current* L_K *given by*

$$
\frac{d L_K}{d Lv} = [1 - K(r, s)]^2.
$$

Since L_K is absolutely continuous with respect to the Lebesgue measure on $S^1 \times S^1 \setminus \Delta$, the following property is evident:

$$L_K([a, a] \times [b, c]) = 0 \quad \text{for all } a < b \le c < a. \tag{5.6}$$

On the other hand, the fact that K is a square integrable function implies that

$$L_K([a, b[\times]b, c]) = \infty \quad \text{for all } a < b < c < a. \tag{5.7}$$

Indeed, if we notice that

$$L_K([a, x] \times [y, c])^{\frac{1}{2}} = \left(\int_a^x \int_y^c [1 - K(r, s)]^2 \, d\, Lv \right)^{\frac{1}{2}}$$

$$\ge \left(\int_a^x \int_y^c \frac{dr \, ds}{4 \sin^2 \left(\frac{r-s}{2} \right)} \right)^{\frac{1}{2}} - \|K\|_2,$$

equality (5.7) easily follows from the relation

$$\int_a^x \int_y^c \frac{dr \, ds}{4 \sin^2 \left(\frac{r-s}{2} \right)} = \log([e^{ia}, e^{ix}, e^{iy}, e^{ic}])$$

and from the fact that the value of the cross-ratio $[e^{ia}, e^{ix}, e^{iy}, e^{ic}]$ goes to infinity as x and y tend to b (with $a < x < b < y < c$).

We will say that a geodesic current satisfying properties (5.6) and (5.7) is **stable**. Proposition 5.2.17 implies the following:

Proposition 5.2.18. *Let $\Phi : \Gamma \to \mathrm{Diff}_+^{1+\tau}(S^1)$ be a representation, with $\tau > 1/2$. If Γ has property (T), then there exists a stable geodesic current that is invariant under Γ.*

The measure L_K is not necessarily fully supported; that is, there may be nontrivial intervals $[a, b]$ and $[c, d]$ for which $L_K([a, b] \times [c, d]) = 0$. This may lead one to think that the group of circle homeomorphisms preserving L_K is not necessarily well behaved. Nevertheless, we will see that the stability properties of L_K lead to rigidity properties for this group.

Lemma 5.2.19. *If a circle homeomorphism preserves a stable geodesic current and fixes three different points, then it is the identity.*

Proof. Suppose that a homeomorphism $f \ne Id$ fixes at least three points and preserves a stable geodesic current L, and let $I =]a, b[$ be a connected component of the complement of the set of fixed points of f. Notice that a and b are fixed points of f. Let $c \in]b, a[$ be another fixed point of f. Since f does not fix any point in $]a, b[$, for each $x \in]a, b[$, the sequence $(f^i(x))$ converges to either a or b. Because both cases are analogous, let us consider only the second one. Then $f^{-i}(x)$ converges to a as i goes to infinity. This yields

$$L([a, x] \times [b, c]) = L([a, f(x)] \times [b, c]),$$

and therefore $L([x, f(x)] \times [b, c]) = 0$. Since $x \in]a, b[$ was arbitrary,

$$L(]a, b[\times [b, c]) = \sum_{i \in \mathbb{Z}} L([f^i(x), f^{i+1}(x)] \times [b, c]) = 0.$$

However, this contradicts (5.7).   □

The next proposition is a direct consequence of the preceding one. The reason for using the notation $\Phi(g)$ instead of $\bar{g}$ here will be clear shortly.

Proposition 5.2.20. *Let* $\Phi : \Gamma \to \text{Diff}_+^{1+\tau}(S^1)$ *be a representation, with* $\tau > 1/2$. *If Γ has property* (T) *and $g \in \Gamma$ is such that $\Phi(g)$ fixes three points, then $\Phi(g)$ is the identity.*

The pretty argument of the proof of the next proposition was kindly communicated to the author by Witte-Morris.

Proposition 5.2.21. *Let* $\Phi : \Gamma \to \text{Diff}_+^{1+\tau}(S^1)$ *be a representation, with* $\tau > 1/2$. *If Γ has property* (T) *and $g \in \Gamma$ is such that $\Phi(g)$ has a fixed point, then $\Phi(g)$ is the identity.*

Proof. Let us consider the 3-fold circle covering $\hat{S}^1$. A degree-3 central extension of Γ acts (by $C^{1+\tau}$ diffeomorphisms) on this covering. More precisely, there exists a group $\hat{\Gamma}$ containing a (central) normal subgroup isomorphic to $\mathbb{Z}/3\mathbb{Z}$, such that the quotient is isomorphic to Γ. Since Γ has property (T) and $\mathbb{Z}/3\mathbb{Z}$ is finite, $\hat{\Gamma}$ also has property (T) (cf. Example 5.2.5). If $g \in \Gamma$ is such that $\Phi(g)$ fixes a point in the original circle, then one of its preimages in $\hat{\Gamma}$ fixes three points in $\hat{S}^1$ by the induced action. Since $\hat{S}^1$ identifies with the circle, by using the preceding proposition, one easily deduces that $\Phi(g)$ is the identity.   □

It is now easy to complete the proof of Theorem 5.2.14. Indeed, by the preceding proposition, the action of the group $\Phi(\Gamma)$ on S^1 is free.

By Hölder's theorem, this group is Abelian. But since $\Phi(\Gamma)$ still satisfies property (T), it is forced to be finite.

Remark 5.2.22. By Exercise 1.3.11, every group of circle homeomorphisms preserving a stable geodesic current is conjugate to a subgroup of the Möbius group. Therefore, if $\tau > 1/2$ and $\Phi : \Gamma \to \mathrm{Diff}_+^{1+\tau}(S^1)$ is a representation whose associated cocycle (5.5) is a coboundary, then the image $\Phi(\Gamma)$ is topologically conjugate to a subgroup of $\mathrm{PSL}(2,\mathbb{R})$. This allows giving an alternative end of the proof of Theorem 5.2.14 by using the result from Exercise 5.2.15.

Exercise 5.2.23. Prove that for every function $K \in \mathcal{L}_\mathbb{R}^{2,\Delta}(S^1 \times S^1, L\upsilon)$, the group Γ_{L_K} of the circle homeomorphisms preserving L_K is uniformly quasi-symmetric (see Exercise 1.3.3).

Remark 5.2.24. A result from [5] allows extending Theorem 5.2.14 to actions by $C^{3/2}$ diffeomorphisms. The theorem is perhaps true even for actions by C^1 diffeomorphisms (compare Exercise 5.2.7 and the comments before it; compare also Remark 5.2.33). However, an extension to general actions by *homeomorphisms* is unclear; see, for instance, Example 5.2.31. A particular case that is interesting in itself is that of piecewise affine circle homeomorphisms.

Remark 5.2.25. By combining Theorem 5.2.14 with the two-dimensional version of Thurston's stability theorem [240], one may show that if $\tau > 1/2$, then every countable group of $C^{1+\tau}$ diffeomorphisms of the closed disk or the closed annulus with property (T) is finite. It is unknown whether this is true for the open disk and/or annulus, as well as for compact surfaces with nonpositive Euler characteristic (notice that the sphere S^2 supports a faithful action of the Kazhdan group $\mathrm{SL}(3,\mathbb{Z})$).

Exercise 5.2.26. Following [115], let $\mathrm{D}^\infty = \mathrm{Diff}_+^\infty(S^1) \setminus int(\rho^{-1}(\mathbb{Q}/\mathbb{Z}))$, where ρ is the rotation-number function. The space D^∞ is closed in $\mathrm{Diff}_+^\infty(S^1)$, and hence it is a Baire space. Prove that the infinite cyclic group generated by a generic $g \in \mathrm{D}^\infty$ is not uniformly quasi-symmetric, and thus the set $\{\|c(g^n)\| : n \in \mathbb{Z}\}$ is unbounded. (Recall that a property is **generic** when it is satisfied on a G_δ-dense set; recall, moreover, that in a Baire space, a countable intersection of G_δ-dense sets is still a G_δ-dense set.)

Hint. The set of $g \in D^\infty$ satisfying $\sup_{[a,b,c,d]=2} \sup_{n \in \mathbb{N}} [g^n(a), g^n(b), g^n(c),$ $g^n(d)] = \infty$ is a G_δ-set. Show that this set contains all infinite-order circle homeomorphisms with rational rotation number having at least three periodic points. Then use the fact that these homeomorphisms are dense in D^∞.

Remark. According to Remark 3.1.10, for every $g \in D^\infty$ with irrational rotation number, the sequence (g^{q_n}) converges to the identity in the C^1 topology, where p_n/q_n is the nth-rational approximation of $\rho(g)$. Moreover, if $k \geq 2$, then for a generic subset of elements g in D^∞, there exists an increasing sequence of integers n_i such that g^{n_i} converges to the identity in the C^k topology. Indeed, if we let $dist_k$ be a metric inducing the C^k topology on $\text{Diff}_+^k(S^1)$, the set of $g \in D^\infty$ satisfying

$$\liminf_{n > 0} \ dist_k(g^n, Id) = 0$$

is a G_δ-set; moreover, this set contains all g for which $\rho(g)$ verifies a Diophantine condition [115, 136, 256]. One thus concludes that for a generic $g \in D^\infty$, the sequence $(\|c(g^n)\|)$ is unbounded, but it has the zero vector as an accumulation point. Furthermore, the orbits of the corresponding isometry $A(g) = \Psi(g) + c(g)$ are unbounded and **recurrent**, in the sense that all their points are accumulation points (see [59, 70] for more on unbounded, recurrent actions by isometries on Hilbert spaces).

Exercise 5.2.27. In class $C^{1+\text{Lip}}$, Kopell's lemma may be shown by using the Liouville cocycle. More precisely, let f and g be commuting $C^{1+\text{Lip}}$ diffeomorphisms of $[0, 1]$. Consider the "Liouville measure" $\overline{Lv}$ over $[0, 1[\times [0, 1[$ whose density function is $(r, s) \mapsto 1/(r - s)^2$.

(i) Show that the functions $c(f)$ and $c(g)$ defined by

$$c(f)(r, s) = 1 - [Jac(f^{-1})(r, s)]^{\frac{1}{2}}, \qquad c(g)(r, s) = 1 - [Jac(g^{-1})(r, s)]^{\frac{1}{2}},$$

are square integrable over all compact subsets of $[0, 1[\times [0, 1[$.

(ii) Show that for every $a \in]0, 1[$, there exists a constant $C = C(a, f)$ such that for every interval $[b, c]$ contained in $[0, a]$, one has

$$\int_b^c \int_b^c \|c(f)\|^2 \leq C|c - b|^2.$$

(iii) Suppose that g fixes a point $a \in]0, 1[$ and that $f(x) < x$ for every $x \in]0, 1[$. Using the preceding inequality and the relation $c(g^k) = c(f^{-n}g^k f^n) = c(f^{-n}) + \Psi(f^{-n})c(g^k) + \Psi(f^{-n}g)c(f^n)$, conclude that the value of

$$\int_{f^2(a)}^a \int_{f^2(a)}^a \|c(g^k)\|^2$$

is uniformly bounded (independently of $k \in \mathbb{N}$).

(iv) By an argument similar to that of the proof of Proposition 5.2.20, show that the restriction of g to $[f^2(a), a]$ (and hence to the whole interval $[0, 1[$) is the identity.

Exercise 5.2.28. Let μ be the (finite) measure on the boundary of T^p giving mass p^{-n} to each ball of radius p^{-n}, where $n \geq 1$ (see § 5.2.1). Given a homeomorphism g of ∂T^p, let us denote by g' its Radon-Nikodym derivative with respect to μ (whenever it is defined). A natural "Liouville measure" Lv on $\partial T^p \times \partial T^p$ is given by

$$dLv(x, y) = \frac{d\mu(x) \times d\mu(y)}{\partial \mathrm{ist}(x, y)^2}.$$

Consider the Hilbert space $\mathcal{H}$ formed by the functions K in $\mathcal{L}_\mathbb{R}^2(\partial T^p \times \partial T^p, Lv)$ satisfying $K(x, y) = K(y, x)$ for almost every $(x, y) \in \partial T^p \times \partial T^p$. Consider the unitary representation Ψ of $\mathcal{N}^p$ on $\mathcal{H}$ given by

$$\Psi(g)K(x, y) = K(g^{-1}(x), g^{-1}(y)) \cdot [Jac(g^{-1})(x, y)]^{1/2},$$

where $Jac(g^{-1})(x, y)$ denotes the Jacobian (with respect to the Liouville measure) of the map $(x, y) \mapsto (g^{-1}(x), g^{-1}(y))$.

(i) Prove that the extension to the boundary of every isometry f of T^p satisfies the equality $\partial \mathrm{ist}(x, y) f'(x) f'(y) = \partial \mathrm{ist}(f(x), f(y))$ for all points x and y in ∂T^p.

(ii) Conclude that for each $g \in \mathcal{N}^p$, the function $c(g) : \partial T^p \times \partial T^p \to \mathbb{R}$ given by

$$c(g)(x, y) = 1 - [Jac(g^{-1})(x, y)]^{1/2}$$

belongs to $\mathcal{H}$.

(iii) Show that c satisfies the cocycle relation with respect to Ψ.

(iv) Prove that if Γ is a subgroup of $\mathcal{N}^p$ satisfying property (T), then there exists $K \in \mathcal{H}$ such that Γ preserves the "geodesic current" L_K given by

$$dL_K = [1 - K(x, y)]^2 dLv.$$

(v) Conclude that there exists a compact subset $\mathcal{C} = \mathcal{C}(K)$ of T^p such that for every element $g \in \Gamma \setminus \mathrm{Isom}(T^p)$, either $\mathcal{A}_g \cap \mathcal{C} \neq \emptyset$ or $\mathcal{B}_g \cap \mathcal{C} \neq \emptyset$.

(vi) Use (v) to give an alternative proof of Proposition 5.2.10.

5.2.4 Relative property (T) and Haagerup's property

Motivated by the preceding section, we may naturally ask whether finitely generated subgroups of $\text{Diff}_+^{1+\tau}(S^1)$ necessarily satisfy Haagerup's property when $\tau > 1/2$ (cf. Example 5.2.4). For instance, this is the case for discrete subgroups of $\text{PSL}(2, \mathbb{R})$ (see Exercise 5.2.15), as well as Thompson's group G (see the proof of Proposition 5.2.8). One of the main difficulties in answering this question is that very few examples of groups satisfying neither Haagerup's nor Kazhdan's property are known. Indeed, most of the examples of groups without Haagerup's property actually satisfy a weak version of property (T), namely, the relative property (T) defined as follows:

Definition 5.2.29. If Γ is a locally compact group and Γ_0 is a subgroup of Γ, then the pair (Γ, Γ_0) has the *relative property* (T) if for every (continuous) representation by isometries of Γ on a Hilbert space, there exists a vector that is invariant under Γ_0.

A relevant example of a pair satisfying the relative property (T) is $(\text{SL}(2, \mathbb{Z}) \ltimes \mathbb{Z}^2, \mathbb{Z}^2)$. The reader will find more examples, as well as a discussion of this notion, in [13, 55]. Let us point out, however, that in most of the examples in the literature, if neither Γ nor Γ_0 has property (T), then Γ_0 (contains a cocompact subgroup that) is normal in Γ (see, however, [58]). Under this hypothesis, the following result may be considered a small generalization of Theorem 5.2.14. Its interest relies on Example 5.2.31.

Theorem 5.2.30. *Let Γ be a subgroup of $\text{Diff}_+^{1+\tau}(S^1)$, with $\tau > 1/2$. Suppose that Γ has a normal subgroup Γ_0 such that the pair (Γ, Γ_0) has the relative property* (T). *Then either Γ is topologically conjugate to a group of rotations or Γ_0 is finite.*

Proof. Let us use again the technique of the proof of Theorem 5.2.14. The Liouville cocycle induces an isometric representation of Γ on $\mathcal{L}_{\mathbb{R}}^{2,\Delta}(S^1 \times S^1, Lv)$. If (Γ, Γ_0) has the relative property (T), then this representation admits a Γ_0-invariant vector, and the arguments of the preceding section (see Remark 5.2.22) show that Γ_0 is topologically conjugate to a subgroup of the Möbius group.

Relative property (T) is stable under finite, central extensions. Using this fact, and by means of the 3-fold covering trick in the proof of

Proposition 5.2.21, one easily shows that Γ_0 is actually topologically conjugate to a subgroup of the group of rotations (see Remark 5.2.22). If Γ_0 is infinite, then it is conjugate to a *dense* subgroup of the group of rotations. Now it is easy to check that the normalizer in $\mathrm{Homeo}_+(S^1)$ of every dense subgroup of $\mathrm{SO}(2, \mathbb{R})$ coincides with the whole group of rotations (compare Exercise 2.2.12). Therefore, if Γ_0 is infinite, then Γ is topologically conjugate to a subgroup of $\mathrm{SO}(2, \mathbb{R})$, which concludes the proof. $\square$

Example 5.2.31. Theorem 5.2.30 does not extend to actions by homeomorphisms: the group $\mathrm{SL}(2, \mathbb{Z}) \ltimes \mathbb{Z}^2$ acts faithfully by circle homeomorphisms. Indeed, starting from the canonical action of $\mathrm{PSL}(2, \mathbb{R})$, one can easily realize $\mathrm{SL}(2, \mathbb{R})$ as a group of real-analytic circle diffeomorphisms. Let $p \in S^1$ be a point whose stabilizer under the corresponding $\mathrm{SL}(2, \mathbb{Z})$-action is trivial. Replace each point $f(p)$ of the orbit of p by an interval I_f (where $f \in \mathrm{SL}(2, \mathbb{Z})$) in such a way that the total sum of these intervals is finite. Doing this, we obtain a topological circle S_p^1 provided with a faithful $\mathrm{SL}(2, \mathbb{Z})$-action (we use affine transformations to extend the maps in $\mathrm{SL}(2, \mathbb{Z})$ to the intervals I_f).

Let $I = I_{id}$ be the interval corresponding to the point p, and let $\{\varphi^t : t \in \mathbb{R}\}$ be a nontrivial topological flow on I. Choose any real numbers u and v that are linearly independent over the rationals, and let $g = \varphi^u$ and $h = \varphi^v$. Extend g and h to S_p^1 by letting

$$g(x) = f^{-1}(g^a h^c(f(x))), \qquad h(x) = f^{-1}(g^b h^d(f(x))),$$

for

$$f = \begin{pmatrix} a & b \\ c & d \end{pmatrix} \in \mathrm{SL}(2, \mathbb{R})$$

and $x \in I_{f^{-1}}$. (For x in the complement of the union of the I_f's, we simply put $g(x) = h(x) = x$.) The reader will easily check that the group generated by g, h, and the copy of $\mathrm{SL}(2, \mathbb{Z})$ acting on S_p^1 is isomorphic to $\mathrm{SL}(2, \mathbb{Z}) \ltimes \mathbb{Z}^2$.

Example 5.2.32. If $\mathbb{F}_2$ is a free subgroup of finite index in $\mathrm{SL}(2, \mathbb{Z})$, then the pair $(\mathbb{F}_2 \ltimes \mathbb{Z}^2, \mathbb{Z}^2)$ still has the relative property (T). Moreover, $\mathbb{F}_2 \ltimes \mathbb{Z}^2$ is an orderable group (actually, it is locally indicable; see Remark 5.1.8), and thus it acts faithfully by homeomorphisms of the interval. (Notice that no explicit action arises as the restriction of the action constructed in the preceding example; however, a faithful action may be constructed by following a similar procedure.) Since $\mathbb{F}_2 \ltimes \mathbb{Z}^2$ is not biorderable, let us formulate the following question: does there exist a finitely generated, biorderable group without Haagerup's property?

Remark 5.2.33. The actions constructed in the preceding examples are not C^1 smoothable. Actually, for every nonsolvable subgroup H of $SL(2, \mathbb{Z})$, the group $H \ltimes \mathbb{Z}^2$ does not embed in either $\text{Diff}^1_+(S^1)$ or $\text{Diff}^1_+(\mathbb{R})$. The proof of this fact does not rely on cohomological properties and is mostly based on dynamical methods; see [178].

5.3 Superrigidity for Higher-Rank Lattice Actions

5.3.1 Statement of the result

Kazhdan's property fails to hold for some higher-rank *semisimple* Lie groups and their lattices. Indeed, some of these groups may act by circle diffeomorphisms. However, these actions are quite particular, as was first shown by Ghys in [92]. To state Ghys's theorem properly, recall that a lattice Γ in a Lie group G is said to be ***irreducible*** if there are no normal subgroups G_1 and G_2 generating G such that $G_1 \cap G_2$ is contained in the center of G (which is supposed to be finite) and such that $(\Gamma \cap G_1) \cdot (\Gamma \cap G_2)$ has finite index in Γ.

Theorem 5.3.1. *Let G be a (connected) semisimple Lie group of rank greater than or equal to 2, and let Γ be an irreducible lattice in G. If Φ is a homomorphism from Γ into the group of C^1 circle diffeomorphisms, then either Φ has finite image, or the associated action is semiconjugate to a finite covering of an action obtained as the composition of the following homomorphisms:*

- *The inclusion of Γ in G.*
- *A surjective homomorphism from G into $PSL(2, \mathbb{R})$.*
- *The natural action of $PSL(2, \mathbb{R})$ on the circle.*

To prove this result, Ghys starts by examining the case of some "standard" higher-rank semisimple Lie groups ($SL(n, \mathbb{R})$, $Sp(2r, \mathbb{R})$, $SO(2, q)$, $SU(2, q)$, and $PSL(2, \mathbb{R}) \times PSL(2, \mathbb{R})$), and then he uses the classification of general semisimple Lie groups [141]. Notice that the first four cases correspond to higher-rank *simple* Lie groups (the involved lattices satisfy Kazhdan's property...). For the last case (which is dynamically more interesting), Ghys proves that up to a semiconjugacy and a finite covering, every homomorphism $\Phi : \Gamma \to \mathrm{Diff}^1_+(S^1)$ factors through the projection of Γ into one of the factors followed by the projective action of this factor on the circle. To understand this case better, the reader should have in mind as a fundamental example the embedding of $\mathrm{PSL}(2, \mathbb{Z}(\sqrt{2}))$ as a lattice in $\mathrm{PSL}(2, \mathbb{R}) \times \mathrm{PSL}(2, \mathbb{R})$ through the map

$$\begin{pmatrix} a_1 + b_1\sqrt{2} & a_2 + b_2\sqrt{2} \\ a_3 + b_3\sqrt{2} & a_4 + b_4\sqrt{2} \end{pmatrix}$$
$$\longmapsto \left(\begin{pmatrix} a_1 + b_1\sqrt{2} & a_2 + b_2\sqrt{2} \\ a_3 + b_3\sqrt{2} & a_4 + b_4\sqrt{2} \end{pmatrix}, \begin{pmatrix} a_1 - b_1\sqrt{2} & a_2 - b_2\sqrt{2} \\ a_3 - b_3\sqrt{2} & a_4 - b_4\sqrt{2} \end{pmatrix} \right).$$

To extend Theorem 5.3.1, we need to generalize the notion of "rank" to an arbitrary lattice. Many tempting definitions have already been proposed (see, for instance, [8]). Here we will deal with perhaps the simplest one, which has been successfully exploited by Shalom [228] (among others). The higher-rank hypothesis corresponds in this general framework to a commutativity hypothesis inside the ambient group. In more concrete terms, the "general setting" that we consider—which is actually that of [228]—is the following one:

- $G = G_1 \times \cdots \times G_k$ is a locally compact, *compactly generated* topological group (see Exercise 5.1.12) with $k \geq 2$, and Γ is a finitely generated cocompact lattice inside.
- The projection $pr_i(\Gamma)$ of Γ into each factor G_i is dense.
- In the case where each G_i is an algebraic linear group over a local field [158], we also allow the possibility that Γ is a noncocompact lattice in G.

We point out that in the last case Γ is necessarily finitely generated. This follows from important theorems due to Kazhdan and Margulis [158]. On the other hand, the second condition is similar to the irreducibility hypothesis in Theorem 5.3.1.

In the introduction of [228] the reader may find many other motivations, as well as some relevant references, concerning the general setting that we consider. We should mention that examples of *nonlinear* lattices satisfying the first two properties have been constructed in [16, 37, 215]. For these lattices, as well as for the *linear* ones (i.e., those that embed into $\mathrm{GL}(n, \mathbb{K})$ for some field $\mathbb{K}$), the next superrigidity theorem for actions on the circle (obtained by the author in [183]; see also [181]) holds.

Theorem 5.3.2. *In the preceding context, let* $\Phi : \Gamma \rightarrow \mathrm{Diff}_+^{1+\tau}(S^1)$ *be a homomorphism such that* $\Phi(\Gamma)$ *does not preserve any probability measure on the circle. If* $\tau > 1/2$, *then either* $\Phi(\Gamma)$ *is conjugate to a subgroup of* $\mathrm{PSL}(2, \mathbb{R})$, *or* Φ *is semiconjugate to a finite covering of an action obtained as a composition of*

- *the injection of* Γ *in* G,
- *the projection of* G *into one of the factors* G_i, *and*
- *an action* Φ *of* G_i *by circle homeomorphisms.*

The hypothesis that $\Phi(\Gamma)$ does not fix any probability measure on S^1 may be suppressed if the first cohomology (with real values) of every finite-index normal subgroup of Γ is trivial. Let us point out that according to [228], this condition is fulfilled if $H^1_{\mathbb{R}}(G)$ is trivial (this is the case, for instance, if the G_i's are semisimple algebraic linear groups over local fields [158]).

Corollary 5.3.3. *Let* Γ *be a finitely generated lattice satisfying the hypothesis of the general setting, and let* $\Phi : \Gamma \rightarrow \mathrm{Diff}_+^{1+\tau}(S^1)$ *be a homomorphism, with* $\tau > 1/2$. *If* $H^1_{\mathbb{R}}(\Gamma_0) = \{0\}$ *for every finite-index normal subgroup of* Γ, *then the conclusion of Theorem 5.3.2 still holds.*

Thanks to the preceding results, to understand the actions of Γ by circle diffeomorphisms, we may use the classification from § 1.4. We can then obtain refined versions under any one of the following hypotheses:

(i) The kernel of Φ is finite, and the orbits by $\Phi(\Gamma)$ are dense.
(ii) The kernel of Φ is finite, and Φ takes values in the group of real-analytic circle diffeomorphisms.
(iii) Every normal subgroup of Γ either is finite or has finite index (that is, Γ satisfies Margulis's normal subgroup theorem [158]).

Let us point out that hypothesis (iii) is satisfied when each G_i is a linear algebraic simple group, as well as by the lattices inside products of groups of isometries of trees constructed in [37] (see [7] for a general version of this fact).

Theorem 5.3.4. *In addition to the hypothesis of Corollary 5.3.3, suppose that* G *is connected and that at least one of the preceding hypotheses* (i), (ii), *or* (iii) *is satisfied. If the image* $\Phi(\Gamma)$ *is infinite, then, up to a topological semiconjugacy and a finite covering,* $\Phi(\Gamma)$ *is a nonmetabelian subgroup of* $\mathrm{PSL}(2, \mathbb{R})$.

One can show that the hypothesis of connectedness for G can be weakened to the hypothesis that no G_i is discrete. (This is not difficult but rather technical; see [35] for a good discussion of this point.) In this general context, hypothesis (i), (ii), or (iii) allows avoiding the degenerate case where G has infinitely many connected components and its action on the circle factors, modulo topological semiconjugacy, through the quotient by the connected component of the identity.

The proofs we give for the results of this section strongly use a superrigidity theorem for isometric actions on Hilbert spaces (due to Shalom) that will be discussed in the next section. This theorem has been generalized in [5] to isometric actions on $\mathcal{L}^p$ spaces. This allows obtaining $C^{1+\tau}$ versions of the preceding results for every $\tau > 0$. Actually, by means of rather different techniques, these results have recently been extended to actions by homeomorphisms by Bader, Furman, and Shaker in [6] (using "boundary theory"), and by Burger in [35] (using bounded cohomology). However, to avoid overloading the presentation, we will not discuss these extensions here.

5.3.2 Cohomological superrigidity

Let us again consider a unitary action Ψ of a group Γ on a Hilbert space $\mathcal{H}$, where Γ now may be nondiscrete but is supposed to be locally compact and compactly generated.

Definition 5.3.5. One says that Ψ **almost has invariant vectors** if there exists a sequence of unitary vectors $K_n \in \mathcal{H}$ such that for every compact subset C of Γ, the value of $\sup_{g \in C} \|K_n - \Psi(g)K_n\|$ converges to zero as n goes to infinity.

Definition 5.3.6. A cocycle $c : \Gamma \to \mathcal{H}$ associated to Ψ is an **almost coboundary** (or is **almost cohomologically trivial**) if there exists a sequence of coboundaries c_n such that for every compact subset C of Γ, the value of $\sup_{g \in C} \|c_n(g) - c(g)\|$ converges to zero as n goes to infinity.

Exercise 5.3.7. Show that the cocycle from Example 5.2.7 is almost cohomologically trivial for every finitely generated subgroup of $\text{Diff}_+^{1+\tau}([0,1])$, where $\tau > 0$.

Hint. Consider the coboundary c_n associated to the function $K_n(x) = \mathcal{X}_{[1/n,1]}(x)$.

The following elementary lemma, due to Delorme [13], appears to be fundamental for studying almost coboundaries.

Lemma 5.3.8. *If Ψ does not almost have fixed vectors, then every cocycle that is an almost coboundary is actually cohomologically trivial.*

Proof. Let $\mathcal{G}$ be a compact generating set of Γ. By hypothesis, there exists $\varepsilon > 0$ such that for every $K \in \mathcal{H}$,

$$\sup_{h \in \mathcal{G}} \|K - \Psi(h)K\| \geq \varepsilon \|K\|. \tag{5.8}$$

Since c is almost cohomologically trivial, there must exist a sequence (K_n) in $\mathcal{H}$ such that $c(g) = \lim_{n \to +\infty}(K_n - \Psi(g)K_n)$ for every $g \in \Gamma$. Inequality (5.8) then gives $M = \sup_{h \in \mathcal{G}} \|c(h)\| \geq \varepsilon \limsup_n \|K_n\|$, and thus $\limsup_n \|K_n\| \leq M/\varepsilon$. Therefore,

$$\|c(g)\| \leq \limsup_{n \to \infty}(\|K_n\| + \|\Psi(g)K_n\|) \leq \frac{2M}{\varepsilon}.$$

Since this holds for every $g \in \Gamma$, the cocycle c is uniformly bounded. By the Bass-Tits center lemma (cf. Exercise 5.2.3), it is cohomologically trivial. $\square$

We now give a version of Shalom's superrigidity theorem [228], which plays a central role in the proof of Theorem 5.3.2. Let us point out, however, that we will not use this result in its full generality. Indeed, in our applications we will always reduce to the case where the corresponding unitary representations do not almost have invariant vectors, and for this case the superrigidity theorem becomes much more elementary.

Theorem 5.3.9. *Let $G = G_1 \times \cdots \times G_k$ be a locally compact, compactly generated topological group, and let Γ be a lattice inside satisfying the hypothesis of our general setting. Let $\Psi : \Gamma \to U(\mathcal{H})$ be a unitary representation that does not almost have invariant vectors. If c is a cocycle associated* *to Ψ that is not an almost coboundary, then c is cohomologous to a cocycle $c_1 + \cdots + c_k$ such that each c_i is a cocycle taking values in a $\Psi(\Gamma)$-invariant subspace $\mathcal{H}_i$ on which the isometric action $\Psi + c_i$ continuously extends to an isometric action of G that factors through G_i.*

This remarkable result was obtained by Shalom with inspiration from the proof of Margulis's normal subgroup theorem. It is based on the general principle that commuting isometries of Hilbert spaces are somehow "degenerate" (Exercises 5.3.2 and 5.3.2 well illustrate this fact). Instead of providing a proof (which may be found in [228]), we have preferred to include two examples where one may appreciate (some of) its consequences (cf. Exercises 5.3.2 and 5.3.2). For the first one, we give a useful elementary lemma for extending group homomorphisms from a lattice to the ambient group. To state it properly, recall that a topological group H is **sequentially complete** if every sequence (h_n) in H such that $\lim_{m,n \to +\infty} h_m^{-1} h_n = id_H$ converges to a limit in H.

Lemma 5.3.10. *Let G and Γ be groups satisfying the hypothesis of our general setting. Let $\Phi : \Gamma \to H$ be a group homomorphism, where H is a sequentially complete, Hausdorff, topological group. Suppose that there exists $i \in \{1, \ldots, k\}$ such that for every sequence (g_n) in Γ satisfying $\lim_{n \to +\infty} pr_i(g_n) = id_{G_i}$, one has $\lim_{n \to +\infty} \Phi(g_n) = id_H$. Then Φ extends to a continuous homomorphism from G into H that factors through G_i.*

Proof. For $g \in G_i$, take an arbitrary sequence (g_n) in Γ such that $pr_i(g_n)$ converges to g. By hypothesis, letting $h_n = \Phi(g_n)$, we have $\lim_{m,n \to +\infty} h_m^{-1} h_n = id_H$. Let us define $\hat{\Phi}(g) = \lim_{n \to +\infty} h_n$. This definition is pertinent because of the hypothesis about the topology of H. Moreover, it does not depend on the chosen sequence, and it is easy to see that the map $\hat{\Phi}$ thus defined is a continuous homomorphism from G that factors through G_i. We leave the details to the reader. $\square$

Exercise 5.3.11. Assuming Theorem 5.3.9, and following the steps below, show the following superrigidity theorem for actions on trees: if Γ is a lattice satisfying the hypothesis of the general setting and Φ is a nonelementary action of Γ by isometries of a simplicial tree $\mathcal{T}$, then there exists a Γ-invariant subtree over which the action extends to G and factors through one of the G_i's (recall that the action is **nonelementary** if there is no vertex, edge, or *point at infinity* that is fixed).

Remark. This result (contained in [228]) still holds for *noncocompact* irreducible lattices. The proof for this general case appears in [170] and uses a superrigidity theorem for bounded cohomology.

(i) By means of a barycentric subdivision, reduce the general case to the one where no element fixes an edge and interchanges its vertices.

(ii) Suppose that there exist a vertex v_0 in T and an index $i \in \{1, \ldots, k\}$ verifying the following condition: for every sequence (g_n) in Γ such that $\lim_{n \to +\infty} pr_i(g_n) = id_{G_i}$, the vertex v_0 is a fixed point of $\Phi(g_n)$ for every large-enough n. Prove that the claim of the theorem is true.

Hint. The set of vertices verifying the preceding condition is contained in a Γ-invariant tree to which Lemma 5.3.10 applies.

(iii) Using Delorme's lemma, prove that the hypothesis that the action is non-elementary implies that the associated regular representation Ψ of Γ on $\mathcal{H} = \ell^2_{\mathbb{R}}(\overrightarrow{edg}(T))$ does not almost have invariant vectors (see Example 5.2.11).

(iv) Let $\mathcal{H}_i$ be the subspace given by Theorem 5.3.9. Define $edg^*(T)$ as the set of edges in T modulo the equivalence relation that identifies $\vec{\Upsilon}_1$ with $\vec{\Upsilon}_2$ if $\psi(\vec{\Upsilon}_1) = \psi(\vec{\Upsilon}_2)$ for every function ψ in $\mathcal{H}_i$. Consider the subset $edg_0^*(T)$ of the classes in $edg^*(T)$ over which at least one of the functions in $\mathcal{H}_i$ is nonzero. Show that every class in $edg_0^*(T)$ is finite.

(v) Let $[\vec{\Upsilon}] \in edg_0^*(T)$ be one of the preceding finite classes, and let Γ^* be its stabilizer in Γ. The subgroup Γ^* fixes the geometric center of the set of edges in $[\vec{\Upsilon}]$, and thus (i) implies that Γ^* fixes a vertex v_0. Show that v_0 satisfies the hypothesis in (ii).

Hint. Given a sequence (g_n) in Γ such that $\lim_{n \to +\infty} pr_i(g_n) = id_{G_i}$, show that g_n belongs to Γ^* for all large-enough n. To do this, argue by contradiction and consider separately the cases where the classes of the g_n's with respect to Γ^* are equal or different, keeping in mind the fact that the functions in $\mathcal{H}_i$ are square integrable and nonzero over the edges in $[\vec{\Upsilon}]$.

Exercise 5.3.12. Recall that there exist nontrivial commuting $C^{1+\tau}$ diffeomorphisms f and g of the interval $[0, 1]$, where $1/2 < \tau < 1$, such that the set of fixed points of f (resp. g) in $]0, 1[$ is empty (resp. nonempty and discrete); see § 4.1.4. Show that the restriction of the Liouville cocycle $c : \Gamma \to \mathcal{L}^{2,\Delta}_{\mathbb{R}}([0, 1] \times [0, 1], \overline{Lv}) = \mathcal{H}$ to the group $\Gamma \sim \mathbb{Z}^2$ generated by f and g is almost cohomologically trivial (see Exercise 5.2.27).

Hint. If we assume the opposite, Shalom's superrigidity theorem provides us with a unitary vector K in $\mathcal{H}$ that is invariant under either $\Psi(f)$ or $\Psi(g)$. Show that the probability measure μ_K on $[0, 1]$ defined by

$$\mu_K(A) = \int_0^1 \int_A |K(x, y)|^2 \, d\overline{Lv}$$

is invariant under either f or g. Then, using the fact that f has no fixed point, show that g preserves μ_K, and that this measure is supported on the set of fixed points of g. Finally, use the fact that the latter set is countable to obtain a contradiction.

Remark. It would be interesting to give an explicit sequence of coboundaries converging to the cocycle c.

Exercise 5.3.13. Given a product $G = G_1 \times G_2$ of compactly generated topological groups, let $A = \Psi + c$ be a representation of G by isometries of a Hilbert space $\mathcal{H}$. Suppose that the isometric representation of G_1 obtained by restriction does not have invariant vectors. Show that the corresponding unitary representation of G_2 almost has invariant vectors.

Hint. Let (g_n) be a sequence in G_1 such that $\|c(g_n)\|$ tends to infinity. Using the commutativity between G_1 and G_2, show that the sequence of unitary vectors $c(g_n)/\|c(g_n)\|$ is almost invariant under $\Psi(G_2)$.

Exercise 5.3.14. Let $\Psi : \mathbb{Z}^2 \to U(\mathcal{H})$ be a unitary representation, and let $c : \mathbb{Z}^2 \to \mathcal{H}$ be an associated cocycle. Show that if both $\Psi((1, 0))$ and $\Psi((0, 1))$ do not almost have invariant vectors, then c is cohomologically trivial.

Hint. Denote by $A_1 = \Psi_1 + c_1$ and $A_2 = \Psi_2 + c_2$ the isometries of $\mathcal{H}$ associated to the generators of $\mathbb{Z}^2$. Check that for all $i \in \{1, 2\}$, all $K \in \mathcal{H}$, and all $n \in \mathbb{Z}$,

$$(Id - \Psi_i)A_i^n(K) = \Psi_i^n(K) - \Psi_i^{1+n}(K) - \Psi_i^n(c_i) + c_i.$$

Using this relation and the commutativity between A_1 and A_2, find a uniform upper bound for $\|(Id - \Psi_1)(Id - \Psi_2)A_1^{n_1} A_2^{n_2}(K)\|$. From the fact that both Ψ_1 and Ψ_2 do not almost have invariant vectors, conclude that the orbits under the isometric action of $\mathbb{Z}^2$ are bounded. Finally, apply the Bass-Tits center lemma (cf. Exercise 5.2.3).

Exercise 5.3.15. Given $C \geq 0$, a subset M_0 of a metric space M is C-*dense* if for every $K \in M$ there exists $K_0 \in M_0$ such that $dist(K_0, K) \leq C$.

(i) Show that there is no isometry of a Hilbert space of dimension greater than 1 having C-dense orbits (compare with the remark in Exercise 5.2.3).

Hint. Following an argument due to Fathi, let $A = \Psi + c$ be an isometry of a Hilbert space $\mathcal{H}$ with a C-dense orbit $\{A^n(K_0) : n \in \mathbb{Z}\}$. Using the equality $A^n(K_0) = \Psi^n(K_0) + \sum_{i=0}^{n-1} \Psi^i(c)$, show that the set $\{(Id - \Psi)A^n(K_0) : n \in \mathbb{Z}\}$ is $2C$-dense in the space $(Id - \Psi)\mathcal{H}$. Using the identity

$$(Id - \Psi)A^n(K_0) = \Psi^n(K_0) - \Psi^{1+n}(K_0) - \Psi^n(c) + c,$$

conclude that the norms of the vectors in this space are bounded from above by $2(\|K_0\| + \|c\|)$, and hence $\Psi = Id$. Thus A is a translation, which implies that $\mathcal{H}$ has dimension 0 or 1.

(ii) As a generalization of (i), show that no action of $\mathbb{Z}^k$ by isometries of a finite-dimensional Hilbert space admits C-dense orbits.

Hint. Let $A_i = \Psi_i + c_i$ be commuting isometries of an infinite-dimensional Hilbert space $\mathcal{H}$, where $i \in \{1, \ldots, k\}$. Assuming that the orbit of $K_0 \in \mathcal{H}$ under the group generated by them is C-dense, show that the set

$$\{(Id - \Psi_1) \cdots (Id - \Psi_k)A_k^{n_k} \cdots A_1^{n_1}(K_0) : n_{i_j} \in \mathbb{Z}\}$$

is $2^k C$-dense in the space $(Id - \Psi_1) \cdots (Id - \Psi_k)\mathcal{H}$. Using a similar argument to that of Exercise 5.3.2, show that this set is contained in the ball in $\mathcal{H}$ centered at the origin with radius $2^k(\|K_0\| + \sum_{i=1}^{k} \|c_i\|)$. Deduce that for at least one of the Ψ_i's, say, Ψ_k, the set $\bar{\mathcal{H}}$ of invariant vectors is an infinite-dimensional subspace of $\mathcal{H}$. Denoting by $\bar{c}_k$ the projection of c_k into $\bar{\mathcal{H}}$, show that the orthogonal projection into the space $\mathcal{H}^* = \langle \bar{c}_k \rangle^\perp \cap \bar{\mathcal{H}}$ induces $k - 1$ commuting isometries generating a group with C-dense orbits. Complete the proof by means of an inductive argument.

Remark. It is not difficult to construct isometries of an infinite-dimensional Hilbert space generating a free group for which all the orbits are dense. The problem of determining what are the finitely generated groups that may act minimally on such a space seems to be interesting. According to (ii) above, such a group cannot be Abelian, and by the next exercise, it cannot be nilpotent either (compare [60]).

Exercise 5.3.16. Suppose that there exist finitely generated nilpotent groups admitting isometric actions on infinite-dimensional Hilbert spaces with C-dense orbits for some $C > 0$. Fix such a group Γ with the smallest possible nilpotence degree, and consider the corresponding affine action $A = \Psi + c$.

(i) Show that the restriction of the unitary action Ψ to the center H of Γ is trivial.

Hint. Consider a point K_0 in the underlying Hilbert space $\mathcal{H}$ with a C-dense orbit, and for each $K \in \mathcal{H}$, choose $g \in \Gamma$ such that $\|A(g)(K_0) - K\| \leq C$. For all $h \in H$, one has

$$
\begin{aligned}
\|A(h)(K) - K\| &\leq \|A(h)(K) - A(hg)(K_0)\| + \|A(hg)(K_0) - A(g)(K_0)\| \\
&\quad + \|A(g)(K_0) - K\| \\
&\leq 2C + \|A(gh)(K_0) - A(g)(K_0)\| \\
&= 2C + \|A(h)(K_0) - K_0\|.
\end{aligned}
$$

In particular, if $K \neq 0$, then, replacing K by λK in the preceding inequality and letting λ tend to infinity, conclude that K is invariant under H.

(ii) By (i), for every $h \in H$, the isometry $A(h)$ is a translation, say, by a vector K_h. Show that K_h is invariant under $\Psi(\Gamma)$. Conclude that the subspace $\mathcal{H}_0$ formed by the $\Psi(\Gamma)$-invariant vectors is not reduced to $\{0\}$ (notice that the action of H on $\mathcal{H}_0$ cannot be trivial). By projecting orthogonally into $\mathcal{H}_0$ and its orthogonal complement $\mathcal{H}_0^{\perp}$, we obtain isometric representations A_0 and $A_0^{\perp}$, both with C-dense orbits. Using the fact that the unitary part of A_0 is trivial (that is, A_0 is a representation by translations) and that Γ is finitely generated, conclude that $A_0^{\perp}$ has infinite dimension. Obtain a contradiction by noticing that over $\mathcal{H}_0^{\perp}$, the affine action of H is trivial, and hence $A_0^{\perp}$ induces an affine action of the quotient group Γ/H whose nilpotence degree is smaller than that of Γ.

Exercise 5.3.17. Give examples of nonminimal, isometric actions on infinite-dimensional Hilbert spaces of finitely generated solvable groups.

5.3.3 Superrigidity for actions on the circle

Let Γ be a subgroup of $\mathrm{Diff}_+^{1+\tau}(S^1)$, where $\tau > 1/2$. According to § 5.2.3 (see Remark 5.2.22), if the restriction of the Liouville cocycle to Γ is cohomologically trivial, then Γ is topologically conjugate to a subgroup of $\mathrm{PSL}(2, \mathbb{R})$. Using Delorme's lemma, we will study the case where this cocycle is almost cohomologically trivial. The next lemma should be compared with Exercise 5.3.2, item (iii).

Lemma 5.3.18. *Suppose that the Liouville cocycle restricted to Γ is an almost coboundary that is not cohomologically trivial. Then Γ preserves a probability measure on the circle.*

Proof. By Delorme's lemma, if c is an almost coboundary that is not cohomologically trivial, then Ψ almost has invariant vectors. Therefore, there exists a sequence (K_n) of unitary vectors in $\mathcal{H} = \mathcal{L}_{\mathbb{R}}^{2,\Delta}(S^1 \times S^1, Lv)$ such that for every $g \in \Gamma$, the value of $\|K_n - \Psi(g)K_n\|$ converges to zero as n goes to infinity. Let μ_n be the probability measure on S^1 defined by

$$\mu_n(A) = \int_{S^1} \int_A K_n^2(x, y) dLv.$$

For every continuous function $\varphi : S^1 \to \mathbb{R}$, we have

$$\begin{aligned}
|\mu_n(\varphi) - g(\mu_n)(\varphi)| &\leq \|\varphi\|_{\mathcal{L}^\infty} \int_{S^1} \int_{S^1} \left| K_n^2 - (\Psi(g)K_n)^2 \right| dLv \\
&\leq \|\varphi\|_{\mathcal{L}^\infty} \|K_n + \Psi(g)K_n\|_{\mathcal{L}^2} \|K_n - \Psi(g)K_n\|_{\mathcal{L}^2} \\
&\leq 2\|\varphi\|_{\mathcal{L}^\infty} \|K_n - \Psi(g)K_n\|_{\mathcal{L}^2}.
\end{aligned}$$

From this it follows that $|\mu_n(\varphi) - g(\mu_n)(\varphi)|$ goes to zero as n goes to infinity. Therefore, if μ is an accumulation point of (μ_n), then μ is a probability measure on S^1 that is invariant under Γ. $\qquad\square$

Proof of Theorem 5.3.2. For each unitary vector $K \in \mathcal{H} = \mathcal{L}_{\mathbb{R}}^{2,\Delta}(S^1 \times S^1, Lv)$, let μ_K be the probability measure of S^1 obtained by projecting on the first coordinate, that is,

$$\mu_K(A) = \int_{S^1} \int_A K^2(x, y) dLv.$$

Let *prob* be the map $prob(K) = \mu_K$ defined on the unit sphere of $\mathcal{H}$ and taking values in the space of probability measures of the circle. This map *prob* is Γ-equivariant in the sense that for all $g \in \Gamma$ and every unitary vector $K \in \mathcal{L}_{\mathbb{R}}^{2,\Delta}(S^1 \times S^1, Lv)$,

$$prob(\Psi(g)K) = \Phi(g)(prob(K)). \tag{5.9}$$

Suppose that $\Phi(\Gamma)$ preserves no probability measure on S^1 and that $\Phi(\Gamma)$ is not conjugate to a subgroup of $PSL(2, \mathbb{R})$. By Lemma 5.3.18 and the discussion before it, Shalom's superrigidity theorem provides us with a family $\{\mathcal{H}_1, \ldots, \mathcal{H}_k\}$ of $\Psi(\Gamma)$-invariant subspaces of $\mathcal{H}$, as well as of co-cycles $c_i : \Gamma \to \mathcal{H}_i$, such that at least one of them is not identically zero, and such that over each $\mathcal{H}_i$, the isometric action associated to c_i extends continuously to G and factors through G_i. Take an index $i \in \{1, \ldots, k\}$ such that $\mathcal{H}_i$ is nontrivial. We claim that the image of the unit sphere in $\mathcal{H}_i$ under the map $prob$ consists of at least two different measures. Indeed, if this image were made of a single measure $prob(K)$, then because of (5.9) and the fact that $\mathcal{H}_i$ is a $\Psi(\Gamma)$-invariant subspace, $prob(K)$ would be invariant under Γ, thus contradicting our hypothesis. Fix an orthonormal basis $\{K_1, K_2, \ldots\}$ of $\mathcal{H}_i$, and define

$$\overline{K} = \sum_{n \geq 0} \frac{|K_n|}{2^n}, \qquad K = \frac{\overline{K}}{\|\overline{K}\|}.$$

The measure μ_K has "maximal support" among those obtained by projecting functions in $\mathcal{H}_i$. Moreover, μ_K has no atom, and it is absolutely continuous with respect to the Lebesgue measure. We denote by Λ the closure of the support of μ_K, which is a compact set without isolated points. Since $\mathcal{H}_i$ is $\Psi(\Gamma)$-invariant, Λ is Γ-invariant, and since $\Phi(\Gamma)$ has no invariant measure, Λ is not reduced to the union of finitely many disjoint intervals.

If Λ is not the whole circle, let us collapse to a point the closure of each connected component of $S^1 \setminus \Lambda$. We thus obtain a topological circle S^1_Λ over which the original action Φ induces an action by homeomorphisms Φ_Λ. However, notice that the orbits of this induced action are not necessarily dense, since Λ might be larger than the (non-empty) minimal invariant closed set of the original action. If Λ is the whole circle, we let $S^1_\Lambda = S^1$. In any case, S^1_Λ is endowed with a natural metric, since it may be parameterized by means of the measure μ_K.

Let K' be a function in the unit sphere of $\mathcal{H}_i$ such that the measure $\mu_{K'}$ is different from μ_K, and let Γ_{μ_K} (resp. $\Gamma_{\mu_{K'}}$) be the group of homeomorphisms of S^1_Λ that preserve the measure (induced on S^1_Λ by) μ_K (resp. $\mu_{K'}$). Notice that Γ_{μ_K} is topologically conjugate to the group of rotations. If (g_n) is a sequence of elements in Γ such that $\lim_{n \to +\infty} pr_i(g_n) = id_{G_i}$, then both $\|\Psi(g_n)K - K\|$ and $\|\Psi(g_n)K' - K'\|$ converge to zero as n goes to infinity. An analogous argument to that of the proof of Lemma 5.3.18 then shows that $(\Phi(g_n))_*(\mu_K)$ (resp. $(\Phi(g_n))_*(\mu_{K'})$) converges to μ_K (resp. $\mu_{K'}$) as n goes to infinity. Using this, one easily concludes that $(\Phi_\Lambda(g_n))$ has accumulation points in $\text{Homeo}_+(S^1_\Lambda)$, all of them contained in $\Gamma_{\mu_K} \cap \Gamma_{\mu_{K'}}$. Since $\mu_{K'}$ is different from μ_K and its support is contained in that of μ_K, the group $\Gamma_{\mu_K} \cap \Gamma_{\mu_{K'}}$ is strictly contained in Γ_K. Since $\Gamma_{\mu_K} \cap \Gamma_{\mu_{K'}}$ is closed in $\mathrm{Homeo}_+(S^1_\Lambda)$, and since every nondense subgroup of the group of rotations is finite, we conclude that $\Gamma_{\mu_K} \cap \Gamma_{\mu_{K'}}$ must be finite.

Let H be the set of elements $h \in \mathrm{Homeo}_+(S^1_\Lambda)$ such that $h = \lim_{n \to +\infty} \Phi_\Lambda(g_n)$ for some sequence (g_n) in Γ satisfying $\lim_{n \to +\infty} pr_i(g_n) = id_{G_i}$. By definition, H is a closed subgroup of $\mathrm{Homeo}_+(S^1_\Lambda)$. Moreover, the preceding argument shows that H is contained in $\Gamma_{\mu_K} \cap \Gamma_{\mu_{K'}}$. It must therefore be finite and cyclic, say, of order d. In the case $d > 1$, we choose a generator h of H, and we notice that the rotation number $\rho(h)$ is nonzero. Let (g_n) be a sequence in Γ such that $\lim_{n \to +\infty} pr_i(g_n) = id_{G_i}$ and $h = \lim_{n \to +\infty} \Phi_\Lambda(g_n)$.

We now show that H is contained in the centralizer of $\Phi_\Lambda(\Gamma)$ in $\mathrm{Homeo}_+(S^1_\Lambda)$. To do this, notice that for each $g \in \Gamma$, the sequence of maps $pr_i(g^{-1}g_n g)$ also tends to id_{G_i} as n goes to infinity. By definition, $(\Phi_\Lambda(g^{-1}g_n g))$ converges to some element $h^j \in H$, where $j \in \{1, \ldots, d\}$. From the equality $\rho(\Phi_\Lambda(g^{-1}g_n g)) = \rho(g^{-1}g_n g) = \rho(g_n) = \rho(\Phi_\Lambda(g_n))$ it easily follows that $j = 1$, which implies that $\Phi_\Lambda(g)$ commutes with h. Since $g \in \Gamma$ was an arbitrary element, the group H centralizes $\Phi_\Lambda(\Gamma)$.

Let $S^1_\Lambda / \sim$ be the topological circle obtained by identifying the points in S^1_Λ that are in the same orbit of H. This circle S^1_Λ is a finite covering of degree d of $S^1_\Lambda / \sim$. Moreover, $\Phi_\Lambda : \Gamma \to \mathrm{Homeo}_+(S^1_\Lambda)$ naturally induces a representation $\tilde{\Phi} : \Gamma \to \mathrm{Homeo}_+(S^1_\Lambda / \sim)$ such that if (g_n) is a sequence in Γ for which $pr_i(g_n)$ tends to id_{G_i}, then $\tilde{\Phi}(g_n)$ tends to the identity (on $S^1_\Lambda / \sim$). Lemma 5.3.10 then allows us to conclude that $\tilde{\Phi}$ extends to a representation $\hat{\Phi} : G \to \mathrm{Homeo}_+(S^1_\Lambda / \sim)$ that factors through G_i. This representation $\hat{\Phi}$ is the one that extends Φ up to a semiconjugacy and a finite covering, as we wanted to show. The proof of Theorem 5.3.2 is thus concluded. $\square$

Now recall that the finite subgroups of $\mathrm{Homeo}_+(S^1)$ are topologically conjugate to groups of rotations, and thus to subgroups of $\mathrm{PSL}(2, \mathbb{R})$. Hence, to prove Corollary 5.3.3, it suffices to show that if $\Phi(\Gamma)$ preserves a probability measure on the circle and $H^1_{\mathbb{R}}(\Gamma_0) = \{0\}$ for every finite-index normal subgroup Γ_0 of Γ, then $\Phi(\Gamma)$ is finite. To do this, notice that if the invariant measure has no atom, then $\Phi(\Gamma)$ is semiconjugate to a group of rotations; in the other case $\Phi(\Gamma)$ has a finite orbit. We claim that this implies that $\Phi(\Gamma)$ must be finite. Indeed, in the case of a finite orbit, this follows from Thurston's stability theorem, whereas in the case of a semiconjugacy to a group of rotations, this follows by taking the rotation-number homomorphism (recall that Γ—and hence Γ_0—is finitely generated). The proof of Corollary 5.3.3 is thus concluded.

Proof of Theorem 5.3.4. Since G is connected, its action $\hat{\Phi}$ on $S^1_\Lambda/\sim$ must factor through homomorphisms into $(\mathbb{R}, +)$, $\text{Aff}_+(\mathbb{R})$, $SO(2, \mathbb{R})$, $\widetilde{PSL}(2, \mathbb{R})$, or $\text{PSL}_k(2, \mathbb{R})$ for some $k \geq 1$ (see § 1.4). In what follows, suppose that $\Phi(\Gamma)$ is infinite, which, according to the proof of Corollary 5.3.3, is equivalent to saying that $\Phi(\Gamma)$ has no invariant probability measure on S^1.

Let us first consider the case of hypothesis (i). The circle S^1_Λ then identifies with the original circle S^1. From the fact that the kernel of Φ is finite, one concludes that there exist sequences (g_n) in Γ such that $pr_i(g_n)$ converges to id_{G_i} and the homeomorphisms $\hat{\Phi}(g_n)$ are two-by-two different. This implies that the Lie group $\hat{\Phi}(G_i)$ cannot be discrete. From the previously recalled classification, the connected component of the identity in this group $\hat{\Phi}(G_i)_0$ corresponds either to $SO(2, \mathbb{R})$, to $\text{PSL}_k(2, \mathbb{R})$ for some $k \geq 1$, or to a subgroup of a product of groups of translations, of affine groups, and/or of groups conjugate to $\widetilde{PSL}(2, \mathbb{R})$, all of them acting on disjoint intervals. The first case cannot arise, since $\phi(\Gamma)$ does not fix any probability measure on S^1. The last case cannot arise either, because the orbits under $\Phi(\Gamma)$ are dense and $\hat{\Phi}(G_i)_0$ is normal in $\hat{\Phi}(G_i)$ (the family of the intervals fixed by $\hat{\Phi}(G_i)_0$ must be preserved by Γ). The group $\hat{\Phi}(G_i)_0$ is therefore conjugate to $\text{PSL}_k(2, \mathbb{R})$ for some $k \geq 1$, and since $\text{PSL}_k(2, \mathbb{R})$ coincides with its normalizer in $\text{Homeo}_+(S^1)$, the same holds for $\hat{\Phi}(G_i)$.

Let us now consider assumption (ii), that is, let us suppose that Φ takes values in the group of real-analytic circle diffeomorphisms. We have already observed that the orbits of the action of Γ on S^1_Λ are not necessarily dense. Let $\dot{\Lambda}$ be the minimal invariant closed set of this action, and let $\dot{\Phi} : \Gamma \to \text{Homeo}(S^1_{\dot{\Lambda}})$ be the action induced on the topological circle $S^1_{\dot{\Lambda}}$ obtained after collapsing the closure of each connected component of $S^1_\Lambda \setminus \dot{\Lambda}$. The orbits under $\dot{\Phi}$ are dense. To apply the arguments used in case (i), it suffices to show that the kernel of $\dot{\Phi}$ is finite. But this is evident, since the fixed points of the nontrivial elements in $\Phi(\Gamma)$ are isolated (the kernel of the restriction of $\dot{\Phi}$ to Γ coincides with the kernel of Φ).

Finally, let us assume hypothesis (iii), that is, that Γ satisfies Margulis's normal subgroup theorem. Once again, we need to show that the kernel of $\dot{\Phi}$ is finite. If this were not the case, then this kernel would have finite index in Γ. This would imply that the orbits of points in $\dot{\Lambda}$ are finite. However, this is impossible, since all the orbits of $\dot{\Phi}$ are dense. The proof of Theorem 5.3.4 is thus concluded. $\square$

Some Basic Concepts in Group Theory

Let Γ_1 and Γ_2 be subgroups of a group Γ. We denote by $[\Gamma_1, \Gamma_2]$ the group generated by the elements of the form $[f, g] = f^{-1}g^{-1}fg$, where $f \in \Gamma_1$ and $g \in \Gamma_2$. The group $\Gamma' = [\Gamma, \Gamma]$ is called the **derived** (or **commutator**) subgroup of Γ. One says that Γ is **metabelian** if $[\Gamma, \Gamma]$ is Abelian, and **perfect** if $[\Gamma, \Gamma] = \Gamma$.

Recall that a subgroup Γ_0 of Γ is **normal** if for every $h \in \Gamma_0$ and all $g \in \Gamma$, one has $ghg^{-1} \in \Gamma_0$. The group Γ is **simple** if the only normal subgroups are $\{id\}$ and Γ itself. Notice that the subgroup $[\Gamma, \Gamma]$ is normal in Γ. Hence if Γ is non-Abelian and simple, then it is perfect. The **center** of a group is the subgroup formed by the elements that commute with all the elements in the group. More generally, given a subset A of Γ, the **centralizer** of A (in Γ) is the subgroup of Γ formed by the elements that commute with all the elements in A.

Given a group Γ, we define inductively the subgroups

$$\Gamma_0^{nil} = \Gamma, \qquad \Gamma_0^{sol} = \Gamma,$$
$$\Gamma_{i+1}^{nil} = [\Gamma, \Gamma_i^{nil}], \qquad \Gamma_{i+1}^{sol} = [\Gamma_i^{sol}, \Gamma_i^{sol}].$$

The series of subgroups Γ_i^{nil} (resp. Γ_i^{sol}) is called the **central series** (resp. **derived series**) of Γ. The group is **nilpotent** (resp. **solvable**) if there exists $n \in \mathbb{N}$ such that $\Gamma_n^{nil} = \{id\}$ (resp. $\Gamma_n^{sol} = \{id\}$). The minimum integer n for which this happens is called the **degree** (also called **order**) of nilpotence (resp. solvability) of the group. From the definitions it easily follows that every nilpotent group is solvable. A group is **virtually nilpotent** (resp. **virtually solvable**) if it contains a nilpotent (resp. solvable) subgroup of finite index.

Exercise A.1.   Prove that the center of every nontrivial nilpotent group is nontrivial. Show that each of the subgroups $\Gamma_i^{\mathrm{nil}}(\Gamma)$ and $\Gamma_i^{\mathrm{sol}}(\Gamma)$ is normal in Γ. Conclude that every nontrivial solvable group contains a nontrivial normal subgroup that is Abelian.

Let P be some property for groups. For instance, P could be the property of being finite, Abelian, nilpotent, free, etc. One says that a group Γ is ***residually*** P if for every $g \in \Gamma$ different from the identity, there exists a group Γ_g satisfying P and a surjective group homomorphism from Γ into Γ_g such that the image of g is not trivial. One says that Γ is ***locally*** P if every finitely generated subgroup of Γ has property P.

Exercise A.2.   Show that a group Γ is residually nilpotent (resp. residually solvable) if and only if $cap_{i \geq 0}\Gamma_i^{\mathrm{nil}} = \{id\}$ (resp. $cap_{i \geq 0}\Gamma_i^{\mathrm{sol}} = \{id\}$).

Exercise A.3.   After reading § 2.2.3 and § 2.2.6, show that every locally orderable (resp. locally biorderable, locally $\mathcal{C}$-orderable) group is orderable (resp. biorderable, $\mathcal{C}$-orderable).

Hint. For a general orderable group Γ, endow $\mathcal{O}(\Gamma)$ with a natural topology and use Tychonov's theorem to prove compactness (see [177] in case of problems with this).

A solvable group Γ for which the subgroups Γ_i^{sol} are finitely generated is said to be ***polycyclic***. If these subgroups are also torsion-free, then Γ is ***strongly polycyclic***. It is easy to check that every subgroup of a polycyclic group Γ is polycyclic and hence finitely generated. It follows that every family of subgroups of Γ has a maximal element (with respect to the inclusion). The maximal element of the family of normal nilpotent subgroups is called the ***nilradical*** of Γ and is commonly denoted by $N(\Gamma)$. Notice that $N(\Gamma)$ is not only normal in Γ but also ***characteristic***, that is, stable under any automorphism of Γ. It is possible to show that there exists a finite-index subgroup Γ_0 of Γ such that $[\Gamma_0, \Gamma_0] \subset N(\Gamma)$ (see [211, Chapter IV]). In particular, the quotient $\Gamma_0/N(\Gamma)$ is Abelian.

Invariant Measures and Amenable Groups

Amenability is one of the most profound concepts in group theory, and it is certainly impossible to give a full treatment in just a few pages. We will therefore content ourselves with exploring a dynamical view of this notion that has been exploited throughout the text. For further information, we refer the reader to [101, 153, 247, 259].

We begin by recalling a classical theorem in ergodic theory due to Bogolioubov and Krylov: every homeomorphism of a compact metric space preserves a probability measure. This result fails to hold for general group actions; consider, for instance, the action of a Schottky group on the circle (see § 2.1.1).

Definition B.1. A group Γ is *amenable* (*moyennable*, in French terminology) if every action of Γ by homeomorphisms of a compact metric space admits an invariant probability measure.

In order to get some insight into this definition (at least for the case of finitely generated groups), let us first recall the strategy of the proof of the Bogolioubov-Krylov theorem. Given a homeomorphism g of a compact metric space M, we fix a probability measure μ on it, and we consider the sequence (ν_n) defined by

$$\nu_n = \frac{1}{n}[\mu + g(\mu) + g^2(\mu) + \cdots + g^{n-1}(\mu)].$$

Since the space of probability measures on M is compact when endowed with the weak* topology, there exists a subsequence (ν_{n_k}) of (ν_n) weakly converging to a probability measure v. We claim that this limit measure v is invariant under g. Indeed, for every k, we have

$$g(v_{n_k}) = v_{n_k} + \frac{1}{n_k}[g^{n_k}(\mu) - \mu],$$

and this implies that

$$g(v) = g\left(\lim_{k \to \infty} v_{n_k}\right) = \lim_{k \to \infty} g(v_{n_k})$$

$$= \lim_{k \to \infty} v_{n_k} + \lim_{k \to \infty} \frac{1}{n_k}[g^{n_k}(\mu) - \mu] = \lim_{k \to \infty} v_{n_k} = v. \qquad (B.1)$$

Let us now try to repeat this argument for a group Γ generated by a finite symmetric family of elements $\mathcal{G} = \{g_1, \ldots, g_m\}$ (recall that *symmetric* means that g^{-1} belongs to $\mathcal{G}$ for every $g \in \mathcal{G}$). For each $g \in \Gamma$, define the *length* of g as the minimal number of (nonnecessarily distinct) elements in $\mathcal{G}$ that are necessary to write g as a product. More precisely,

$$length(g) = \min\{n \in \mathbb{N} : g = g_{i_n} g_{i_{n-1}} \cdots g_{i_1}, \quad g_{i_j} \in \mathcal{G}\}.$$

We call the *ball of radius* n (with respect to $\mathcal{G}$) the set $B_{\mathcal{G}}(n)$ formed by the elements in Γ having length less than or equal to n, and we denote by $L_{\mathcal{G}}(n)$ its cardinal.

Consider now an action of Γ by homeomorphisms of a compact metric space M, and let μ be a probability measure on M. For each $n \in \mathbb{N}$, let us consider the probability measure

$$v_n = \frac{1}{L_{\mathcal{G}}(n-1)} \sum_{g \in B_{\mathcal{G}}(n-1)} g(\mu).$$

Passing to some subsequence (v_{n_k}), we have convergence to some probability measure v. The problem now is that v is not necessarily invariant. Indeed, if we try to repeat the arguments of the proof of equality (B.1), then we need to estimate the value of an expression of the form

$$\frac{1}{L_{\mathcal{G}}(n_k)} \sum g_i g(\mu),$$

where the sum is taken over $g_i \in \mathcal{G}$ and $g \in \Gamma$ such that $length(g) = n_k$. However, this expression does not necessarily converge to zero, since it may happen that the number of elements in $B_{\mathcal{G}}(n_k) \setminus B_{\mathcal{G}}(n_k - 1)$ is not negligible with respect to $L_{\mathcal{G}}(n_k)$. To deal with this problem, the following definitions become natural.

Definition B.2. The *geometric boundary* ∂A of a nonempty subset $A \subset \Gamma$ is defined as

$$\partial A = \bigcup_{g \in \mathcal{G}} (A \triangle g(A)),$$

where $\triangle$ stands for the symmetric difference between sets.

Definition B.3. A *Følner sequence* for a group Γ is a sequence (A_n) of finite subsets of Γ such that

$$\lim_{n \to \infty} \frac{card(\partial A_n)}{card(A_n)} = 0.$$

With this terminology, Bogolioubov-Krylov's argument shows that if Γ is a (finitely generated) group having a Følner sequence, then every action of Γ by homeomorphisms of a compact metric space admits an invariant probability measure. In other words, groups having Følner sequences are amenable. The converse to this is also true, according to a deep result due to Følner.

Theorem B.4. *A finitely generated group is amenable if and only if it admits a Følner sequence.*

It is important to notice that the preceding characterization of amenability is independent of the system of generators. Indeed, it is not difficult to check that the quotient of the length functions with respect to two different systems of generators is bounded (by a constant that is independent of the element in the group). Using this, one easily checks that each Følner sequence with respect to one system naturally induces a Følner sequence with respect to the other one.

Exercise B.5. Show that finite groups are amenable. Show that the same holds for Abelian groups.

Hint. For finitely generated Abelian groups, find an explicit Følner sequence. Alternatively, notice that if f and g are commuting homeomorphisms of a compact metric space, then the argument leading to (B.1) and applied to a probability measure μ that is already invariant under f gives a probability measure ν that is invariant under both f and g.

Exercise B.6. Show that if Γ is a finitely generated group containing a free subgroup on two generators, then Γ is not amenable.

Hint. Consider the natural action of Γ on $\{0, 1\}^{\Gamma}$.

Exercise B.7. Show that amenability is stable under *elementary operations*. More precisely, show the following:

 (i) Subgroups of amenable groups are amenable.
 (ii) Every quotient of an amenable group is amenable.
 (iii) If Γ is a group containing an amenable normal subgroup Γ_0 such that the quotient Γ/Γ_0 is amenable, then Γ itself is amenable.
 (iv) If Γ is the union of amenable subgroups Γ_i such that for all indices i and i' there exists j such that both Γ_i and $\Gamma_{i'}$ are contained in Γ_j, then Γ is amenable.

As an application, conclude that every virtually solvable group is amenable.

Exercise B.8. A finitely generated group Γ has *polynomial growth* if there exists a real polynomial Q such that $L_G(n) \leq Q(n)$ for every $n \in \mathbb{N}$. Show that every group of polynomial growth is amenable.

Remark. A celebrated theorem by Gromov establishes that a finitely generated group has polynomial growth if and only if it is virtually nilpotent [107]. Let us point out that the "easy" implication of this theorem (the fact that the growth of nilpotent groups is polynomial) is prior to Gromov's work and independently due to Bass and Guivarch.

Exercise B.9. A finitely generated group Γ has *subexponential growth* if for every $C > 0$ one has

$$\liminf_{n \to \infty} \frac{L_G(n)}{\exp(Cn)} = 0.$$

Show that finitely generated groups of subexponential growth are amenable.

Hint. Show that for groups of subexponential growth, the sequence of balls in the group contains a Følner sequence.

References

[1] L. Ahlfors. Finitely generated Kleinian groups. *Amer. Journal of Math.* **86** (1964), 413–429, and **87** (1965), 759.

[2] R. Alperin. Locally compact groups acting on trees and property (T). *Mh. Math.* **93** (1982), 261–265.

[3] V. Antonov. Modeling of processes of cyclic evolution type: Synchronization by a random signal. *Vestnik Len. Univ. Mat. Mekh. Astr.* (1984), 67–76.

[4] A. Avila. Distortion elements in $\text{Diff}^{\infty}(\mathbb{R}/\mathbb{Z})$. Preprint (2008), arXiv:0808.2334.

[5] U. Bader, A. Furman, T. Gelander, & N. Monod. Property (T) and rigidity for actions on Banach spaces. *Acta Math.* **198** (2007), 57–105.

[6] U. Bader, A. Furman, & A. Shaker. Lattices, commensurators and continuous actions on the circle. Preprint (2006), arXiv:math/0605276.

[7] U. Bader & Y. Shalom. Factor and normal subgroup' theorems for lattices in products of groups. *Invent. Math.* **163** (2006), 415–464.

[8] W. Ballmann & P. Eberlein. Fundamental groups of manifolds of nonpositive curvature. *J. Diff. Geometry* **25** (1987), 1–22.

[9] W. Ballman, M. Gromov, & V. Schroeder. *Manifolds of nonpositive curvature.* Progress in Mathematics **61**, Birkhäuser (1985).

[10] A. Banyaga. *The structure of classical diffeomorphism groups.* Mathematics and Its Applications, Kluwer Acad. Publ. (1997).

[11] L. Bartholdi & B. Virag. Amenability via random walks. *Duke Math. J.* **130** (2005), 39–56.

[12] A. Beardon. *The geometry of discrete groups.* Graduate Texts in Mathematics **91**, Springer-Verlag, (1995).

[13] B. Bekka, P. de la Harpe, & A. Valette. *Kazhdan's property* (T). New Math. Monographs **11**, Cambridge University Press (2008).

[14] L. Beklaryan. Groups of homeomorphisms of the line and the circle. Topological characteristics and metric invariants. *Russian Math. Surveys* **59** (2004), 599–660.

[15] L. Beklaryan. On analogues of the Tits alternative for groups of homeomorphisms of the circle and the line. *Math. Notes* **71** (2002), 305–315.

[16] N. Benakli & Y. Glasner. Automorphism groups of trees acting locally with affine permutations. *Geom. Dedicata* **89** (2002), 1–24.

[17] G. Bergman. Right-orderable groups that are not locally indicable. *Pacific J. Math.* **147** (1991), 243–248.

[18] C. Bleak. An algebraic classification of some solvable groups of homeomorphisms. *J. Algebra* **319** (2008), 1368–1397.

[19] F. Bonahon. The geometry of Teichmüller space via geodesic currents. *Invent. Math.* **92** (1988), 139–162.

[20] C. Bonatti. Un point fixe commun pour des difféomorphismes commutants de S². *Annals of Math.* **129** (1989), 61–69.

[21] M. Boshernitzan. Dense orbits of rationals. *Proc. of the AMS* **117** (1993), 1201–1203.

[22] R. Botto Mura & A. Rhemtulla. *Orderable groups.* Lecture Notes in Pure and Applied Mathematics **27**, Marcel Dekker (1977).

[23] R. Bowen. A horseshoe with positive measure. *Invent. Math.* **29** (1975), 203–204.

[24] E. Breuillard. On uniform exponential growth for solvable groups. *Pure Appl. Math. Q.* **3** (2007), 949–967.

[25] E. Breuillard & T. Gelander. A topological version of the Tits alternative. *Annals of Math.* **166** (2007), 427–474.

[26] E. Breuillard & T. Gelander. Cheeger constant and algebraic entropy for linear groups. *Int. Math. Res. Not.* **56** (2005), 3511–3523.

[27] E. Breuillard & T. Gelander. On dense free subgroups of Lie groups. *J. Algebra* **261** (2003), 448–467.

[28] M. Bridson & A. Haefliger. *Metric spaces of non-positive curvature.* Grundlehren der mathematischen wissenschaften **319**, Springer-Verlag, (1999).

[29] M. Brin. Elementary amenable subgroups of R. Thompson's group F. *Int. J. Alg. and Comp.* **15** (2005), 619–642.

[30] M. Brin. The ubiquity of Thompson's group F in groups of piecewise linear homeomorphisms of the unit interval. *J. London Math. Soc.* **60** (1999), 449–460.

[31] M. Brin. The chameleon groups of Richard J. Thompson: automorphisms and dynamics. *Publ. Math. de l'IHES* **84** (1996), 5–33.

[32] M. Brin & F. Guzmán. Automorphisms of generalized Thompson groups. *J. Algebra* **203** (1998), 285–348.

[33] M. Brin & C. Squier. Presentations, conjugacy, roots, and centralizers in groups of piecewise linear homeomorphisms of the real line. *Comm. Algebra* **29** (2001), 4557–4596.

[34] M. Brin & C. Squier. Groups of piecewise linear homeomorphisms of the real line. *Invent. Math.* **79** (1985), 485–498.

[35] M. Burger. An extension criterium for lattice actions on the circle. To appear in the Zimmer Festschrift, arXiv:0905.0136.

[36] M. Burger & N. Monod. Bounded cohomology of lattices in higher rank Lie groups. *J. Eur. Math. Soc. (JEMS)* **1** (1999), 199–235.

[37] M. Burger & S. Mozes. Lattices in product of trees. *Publ. Math. de l'IHES* **92** (2000), 151–194.

[38] L. Bursler & A. Wilkinson. Global rigidity of solvable group actions on S^1. *Geometry and Topology* **8** (2004), 877–924.

[39] D. Calegari. Nonsmoothable, locally indicable group actions on the interval. *Algebr. Geom. Topol.* **8** (2008), 609–613.

[40] D. Calegari. Denominator bounds in Thompson-like groups and flows. *Groups, Geometry and Dynamics* **1** (2007), 101–109.

[41] D. Calegari. Dynamical forcing of circular groups. *Trans. of the AMS* **358** (2006), 3473–3491.

[42] D. Calegari. Circular groups, planar groups, and the Euler class. *Geom. Topol. Monogr.* **7** (2004), 431–491.

[43] D. Calegari & N. Dunfield. Laminations and groups of homeomorphisms of the circle. *Invent. Math.* **152** (2003), 149–204.

[44] J. Cannon, W. Floyd, & W. Parry. Introductory notes on Richard Thompson's groups. *L'Enseignement Mathématique* **42** (1996), 215–256.

[45] A. Candel. The harmonic measures of Lucy Garnett. *Contemp. Math.* **176** (2003), 187–247.

[46] A. Candel & L. Conlon. *Foliations I and II.* Graduate Studies in Mathematics **23** and **60**, Amer. Math. Soc. (2000, 2004).

[47] J. Cantwell & L. Conlon. Endsets of exceptional leaves: a theorem of G. Duminy. In *Foliations: geometry and dynamics.* World Sci. Publishing, Warsaw (2000), 225–261.

[48] J. Cantwell & L. Conlon. Leaves of Markov local minimal sets in foliations of codimension one. *Publ. Mat.* **33** (1989), 461–484.

[49] J. Cantwell & L. Conlon. Foliations and subshifts. *Tohoku Math. J.* **40** (1988), 165–187.

[50] J. Cantwell & L. Conlon. Smoothability of proper foliations. *Annales de l'Institut Fourier (Grenoble)* **38** (1988), 219–244.

[51] J. Cantwell & L. Conlon. Poincaré-Bendixson theory for leaves of codimension-one. *Trans. of the AMS* **265** (1981), 181–209.

[52] L. Carleson & T. Gamelin. *Complex dynamics.* Universitext: Tracts in Mathematics, Springer-Verlag (1993).

[53] A. Casson & D. Jungreis. Convergence groups and Seifert fibered 3-manifolds. *Invent. Math.* **118** (1994), 441–456.

[54] M. Chaperon. Invariant manifolds revisited. *Proceedings of the Steklov Institute* **236** (2002), 415–433.

[55] P. Chérix, M. Cowling, P. Jolissaint, P. Julg, & A. Valette. *Groups with the Haagerup property. Gromov's a-T-menability.* Progress in Mathematics **197**, Birkhäuser (2001).

[56] A. Clay. Isolated points in the space of left orderings of a group. *Groups, Geometry, and Dynamics* **4** (2010), 517–532.

[57] P. Conrad. Right-ordered groups. *Michigan Math. J.* **6** (1959), 267–275.

[58] Y. de Cornulier. Relative Kazhdan property. *Annales Sci. de l'École Normale Supérieure* **39** (2006), 301–333.

[59] Y. de Cornulier, R. Tessera, & A. Valette. Isometric group actions on Banach spaces and representations vanishing at infinity. *Transform. Groups* **13** (2008), 125–147.

[60] Y. de Cornulier, R. Tessera, & A. Valette. Isometric group actions on Hilbert spaces: Structure of orbits. *Canad. J. Math.* **60** (2008), 1001–1009.

[61] A. Denjoy. Sur les courbes définies par des équations différentielles à la surface du tore. *J. Math. Pure et Appl.* **11** (1932), 333–375.

[62] B. Deroin. Hypersurfaces Levi-plates immergées dans les surfaces complexes de courbure positive. *Annales Sci. de l'École Normale Supérieure* **38** (2005), 57–75.

[63] B. Deroin & V. Kleptsyn. Random conformal dynamical systems. *Geom. and Functional Analysis* **17** (2007), 1043–1105.

[64] B. Deroin, V. Kleptsyn, & A. Navas. On the question of ergodicity for minimal group actions on the circle. *Moscow Math. Journal* **9** (2009), 263–303.

[65] B. Deroin, V. Kleptsyn, & A. Navas. Sur la dynamique unidimensionnelle en régularité intermédiaire. *Acta Math.* **199** (2007), 199–262.

[66] P. Dippolito. Codimension one foliations of closed manifolds. *Annals of Math.* **107** (1978), 403–453.

[67] S. Druck, F. Fang, & S. Firmo. Fixed points of discrete nilpotent group actions on S^2. *Annales de l'Institut Fourier (Grenoble)* **52** (2002), 1075–1091.

[68] S. Druck & S. Firmo. Periodic leaves for diffeomorphisms preserving codimension one foliations. *J. Math. Soc. Japan* **55** (2003), 13–37.

[69] T. Dubrovina & N. Dubrovin. On braid groups. *Sbornik Mathematics* **192** (2001), 693–703.

[70] M. Edelstein. On non-expansive mappings of Banach spaces. *Proc. Cambridge Philos. Soc.* **60** (1964), 439–447.

[71] S. Egashira. Qualitative theory and expansion growth of transversely piecewise-smooth foliated S^1-bundles. *Erg. Theory and Dynam. Systems* **17** (1997), 331–347.

[72] Y. Eliashberg & W. Thurston. *Confoliations.* University Lecture Series, Amer. Math. Soc. (1998).

[73] A. Erschler. On isoperimetric profiles of finitely generated groups. *Geom. Dedicata* **100** (2003), 157–171.

[74] A. Eskin, S. Mozes, & H. Oh. On uniform exponential growth for linear groups. *Invent. Math.* **160** (2005), 1–30.

[75] H. Eynard. On the centralizer of diffeomorphisms of the half-line. To appear in *Comment. Math. Helvetici.*

[76] B. Farb & J. Franks. Groups of homeomorphisms of one-manifolds III: Nilpotent subgroups. *Erg. Theory and Dynam. Systems* **23** (2003), 1467–1484.

[77] D. Farley. A proper isometric action of Thompson group V on Hilbert space. *Int. Math. Res. Not.* **45** (2003), 2409–2414.

[78] A. Fathi & M. Herman. Existence de difféomorphismes minimaux. *Astérisque* **49** (1977), 37–59.

[79] B. Fayad & K. Khanin. Smooth linearization of commuting circle diffeomorphisms. *Annals of Math.* **170** (2009), 961–980.

[80] W. Feller. *An introduction to probability theory and its applications I and II.* 3rd ed. John Wiley & Sons, (1968, 1971).

[81] D. Fisher. Groups acting on manifolds: around the Zimmer program. To appear in the Zimmer Festschrift, arXiv:0809.4849.

[82] A. Furman. Random walks on groups and random transformations. In *Handbook of dynamical systems*, vol. 1A. North-Holland (2002), 931–1014.

[83] H. Furstenberg. Boundary theory and stochastic processes on homogeneous spaces. *Proc. Sympos. Pure Math.* **26** (1973), 193–229.

[84] D. Gabai. Convergence groups are Fuchsian groups. *Annals of Math.* **136** (1992), 447–510.

[85] L. Garnett. Foliations, the ergodic theorem and Brownian motion. *J. Funct. Anal.* **51** (1983), 285–311.

[86] F. Gëhring & G. Martin. Discrete quasiconformal groups I. *Proc. London Math. Soc.* **51** (1987), 331–358.

[87] K. Gelfert & M. Rams. Geometry of limit sets for expansive Markov systems. *Trans. of the AMS* **361** (2009), 2001–2020.

[88] É. Ghys. Groups acting on the circle. *L'Enseignement Mathématique* **47** (2001), 329–407.

[89] É. Ghys. Actions des réseaux sur le cercle. *Invent. Math.* **137** (1999), 199–231.

[90] É. Ghys. Sur l'uniformisation des laminations paraboliques. In *Integrable Systems and Foliations* (Monpellier, 1995) Progress in Mathematics **145**, Birkhäuser (1997), 73–91.

[91] É. Ghys. Rigidité différentiable des groupes fuchsiens. *Publ. Math. de l'IHES* **78**, 163–185 (1994).

[92] É. Ghys. Sur les groupes engendrés par des difféomorphismes proches de l'identité. *Bol. Soc. Brasileira Mat.* **24** (1993), 137–178.

[93] É. Ghys. L'invariant de Godbillon-Vey. *Astérisque* **177–178** (1989), 155–181.

[94] É. Ghys. Classe d'Euler et minimal exceptionnel. *Topology* **26** (1987), 93–105.

[95] É. Ghys, R. Langevin, & P. Walczak. Entropie géométrique des feuilletages. *Acta Math.* **160** (1988), 105–142.

[96] É. Ghys & V. Sergiescu. Sur un groupe remarquable de difféomorphismes du cercle. *Comment. Math. Helvetici* **62** (1987), 185–239.

[97] É. Ghys & T. Tsuboi. Différentiabilité des conjugaisons entre systèmes dynamiques de dimension 1. *Annales de l'Institut Fourier (Grenoble)* **38** (1988), 215–244.

[98] C. Godbillon. *Feuilletages. Études géométriques.* Progress in Mathematics **98**, Birkhäuser (1991).

[99] C. Goffman, T. Nishiura, & D. Waterman. *Homeomorphisms in Analysis.* Math. Surveys and Monographs **54**, Amer. Math. Soc. (1997).

[100] X. Gómez-Mont & L. Ortiz-Bobadilla. *Sistemas dinámicos holomorfos en superficies.* Aportaciones Matemáticas, Sociedad Matemática Mexicana (1989).

[101] F. Greenleaf. *Invariant means on topological groups and their applications.* Van Nostrand Mathematical Studies **16**, Van Nostrand Reinhold Co. (1969).

[102] R. Grigorchuk. On degrees of growth of *p*-groups and torsion-free groups. *Mat. Sbornik* **126** (1985), 194–214.

[103] R. Grigorchuk. Degrees of growth of finitely generated groups and the theory of invariant means. *Izv. Akad. Nauka* **48** (1984), 939–985.

[104] R. Grigorchuk & A. Maki. On a group of intermediate growth that acts on a line by homeomorphisms. *Math. Notes* **53** (1993), 146–157.

[105] M. Gromov. Entropy and isoperimetry for linear and non-linear group actions. *Groups, Geometry, and Dynamics* **2** (2008), 499–593.

[106] M. Gromov. Random walk in random groups. *Geom. and Functional Analysis* **13** (2003), 73–146.

[107] M. Gromov. Groups of polynomial growth and expanding maps. *Publ. Math. de l'IHES* **53** (1981), 53–73.

[108] A. Haefliger. Foliations and compactly generated pseudo-groups. In *Foliations: geometry and dynamics.* World Sci. Publishing, Warsaw (2000), 275–295.

[109] G. Hall. A C[∞] Denjoy counterexample. *Erg. Theory and Dynam. Systems* **1** (1983), 261–272.

[110] G. Hardy & E. Wright. *An introduction to the theory of numbers.* Clarendon Press, Oxford University Press (1979).

[111] P. de la Harpe. *Topics in geometric group theory.* Univ. of Chicago Press (2000).

[112] J. Harrison. Dynamics of Ahlfors quasi-circles. *Proc. Indian Acad. Sci. Math.* **99** (1989), 113–122.

[113] J. Harrison. Unsmoothable diffeomorphisms on higher dimensional manifolds. *Proc. of the AMS* **73** (1979), 249–255.

[114] G. Hector. Architecture des feuilletages de classe C^2. *Astérisque* **107–108** (1983), 243–258.

[115] M. Herman. Sur la conjugaison différentiable des difféomorphismes du cercle à des rotations. *Publ. Math. de l'IHES* **49** (1979), 5–234.

[116] A. Hinkkanen. Abelian and non discrete convergence groups of the circle. *Trans. of the AMS* **318** (1990), 87–121.

[117] M. Hirsch. A stable analytic foliation with only exceptional minimal set. *Lecture Notes in Math.* **468** (1965), 9–10.

[118] J. Hu & D. Sullivan. Topological conjugacy of circle diffeomorphisms. *Erg. Theory and Dynam. Systems* **17** (1997), 173–186.

[119] S. Hurder. Exceptional minimal sets and the Godbillon-Vey class. Preprint (2005).

[120] S. Hurder. Dynamics of expansive group actions on the circle. Preprint (2004).

[121] S. Hurder. Entropy and dynamics of C^1 foliations. Preprint (2004).

[122] S. Hurder. Dynamics and the Godbillon-Vey class: a history and survey. In *Foliations: geometry and dynamics*. World Sci. Publishing, Warsaw (2000), 29–60.

[123] S. Hurder. Exceptional minimal sets of $C^{1+\alpha}$-group actions on the circle. *Erg. Theory and Dynam. Systems* **11** (1991), 455–467.

[124] S. Hurder. Ergodic theory of foliations and a theorem of Sacksteder. *Lecture Notes in Math.* **1342** (1988), 291–328.

[125] S. Hurder & R. Langevin. Dynamics and the Godbillon-Vey class of C^1 foliations. Preprint (2003).

[126] Y. Ilyashenko. The density of an individual solution and the ergodicity of the family of solutions of the equation $d\eta/d\xi = P(\xi, \eta)/Q(\xi, \eta)$. *Mat. Zametki* **4** (1968), 741–750.

[127] H. Imanishi. On the theorem of Denjoy-Sacksteder for codimension one foliations without holonomy. *J. Math. Kyoto Univ.* **14** (1974), 607–634.

[128] T. Inaba. Examples of exceptional minimal sets. In *A fête of topology*. Academic Press (1988), 95–100.

[129] L. Jiménez. *Grupos ordenables: Estructura algebraica y dinámica*. Master's thesis, Univ. de Chile (2008).

[130] E. Jorquera. *On group actions on 1-dimensional manifolds*. PhD thesis, Univ. de Chile (2009).

[131] E. Jorquera. On the topological entropy of group actions on the circle. *Fundamenta Math.* **204** (2009), 177–187.

[132] V. Kaimanovich. The Poisson boundary of polycyclic groups. In *Probability measures on groups and related structures XI* (Oberwolfach, 1994). World Sci. Publishing, (1995), 182–195.

[133] A. Katok & B. Hasselblatt. *Introduction to the modern theory of dynamical systems.* Encyclopedia of Mathematics and its Applications **54**, Cambridge University Press (1995).

[134] A. Katok & A. Mezhirov. Entropy and growth of expanding periodic orbits for one-dimensional maps. *Fund. Math.* **157** (1998), 245–254.

[135] S. Katok. *Fuchsian groups.* Chicago Lectures in Mathematics, Univ. of Chicago Press (1992).

[136] Y. Katznelson & D. Ornstein. The differentiability of the conjugation of certain diffeomorphisms of the circle. *Erg. Theory and Dynam. Systems* **9** (1989), 643–680.

[137] Y. Kifer. *Ergodic theory of random transformations.* Progress in Prob. and Stat. **10**, Birkhäuser (1986).

[138] V. Kleptsyn. Sur une interprétation algorithmique du groupe de Thompson. Unpublished text (2004).

[139] V. Kleptsyn & M. Nalsky. Convergence of orbits in random dynamical systems on a circle. *Funct. Anal. Appl.* **38** (2004), 267–282.

[140] V. Kleptsyn & A. Navas. A Denjoy type theorem for commuting circle diffeomorphisms with derivatives having different Hölder differentiability classes. *Moscow Math. Journal* **8** (2008), 477–492.

[141] A. Knapp. *Lie groups beyond an introduction.* Progress in Mathematics **140**, Birkhäuser (2002).

[142] V. Kopytov & N. Medvedev. *Right-ordered groups.* Siberian School of Algebra and Logic. Consultants Bureau, (1996).

[143] B. Kolev. Sous-groupes compacts d'homéomorphismes de la sphère. *L'Enseignement Mathématique* **52** (2006), 193–214.

[144] N. Kopell. Commuting diffeomorphisms. In *Global Analysis.* Proc. Sympos. Pure Math. XIV, Berkeley, Calif. (1968), 165–184.

[145] N. Kovačević. Examples of Möbius-like groups which are not Möbius groups. *Trans. of the AMS* **351** (1999), 4823–4835.

[146] L. Lifschitz & D. Witte-Morris. Bounded generation and lattices that cannot act on the line. *Pure Appl. Math. Q.* **4** (2008), 99–126.

[147] P. Linnell. The space of left orders of a group is either finite or uncountable. To appear in *Bull. London Math. Society.*

[148] P. Linnell. The topology on the space of left orderings of a group. Preprint (2006), arXiv:math/0607470.

[149] P. Linnell. Left ordered groups with no nonabelian free subgroups. *J. Group Theory* **4** (2001), 153–168.

[150] P. Linnell. Left ordered amenable and locally indicable groups. *J. London Math. Soc.* **60** (1999), 133–142.

[151] I. Liousse. Unpublished text (2007).

[152] I. Liousse. Nombre de rotation, mesures invariantes et ratio set des homéomorphismes par morceaux du cercle. *Annales de l'Institut Fourier (Grenoble)* **55** (2005), 431–482.

[153] A. Lubotsky. *Discrete groups, expanding graphs and invariant measures.* Progress in Mathematics **125**, Birkhäuser (1994).

[154] M. P. Malliavin & P. Malliavin. An infinitesimally quasi-invariant measure on the group of diffeomorphisms of the circle. Special functions (Okayama, 1990), ICM-90 Satell. Conf. Proc., Springer-Verlag (1991), 234–244.

[155] P. Malliavin. The canonic diffusion above the diffeomorphism group of the circle. *C. R. Acad. Sci. Paris Sér. I Math.* **329** (1999), 325–329.

[156] A. Marden. Universal properties of Fuchsian groups. *Ann. of Math. Studies* **79** (1974), 315–339.

[157] G. Margulis. Free subgroups of the homeomorphism group of the circle. *C. R. Acad. Sci. Paris Sér. I Math.* **331** (2000), 669–674.

[158] G. Margulis. *Discrete subgroups of semisimple Lie groups.* Springer-Verlag (1991).

[159] V. Markovic. Classification of continuously transitive circle groups. *Geometry and Topology* **10** (2006), 1319–1346.

[160] V. Markovic. Uniformly quasisymmetric groups. *Journal of the AMS* **19** (2006), 673–715.

[161] Y. Matsuda. Polycyclic groups of diffeomorphisms of the closed interval. *C. R. Acad. Sci. Paris Sér. I Math.* **347** (2009), 813–816.

[162] S. Matsumoto. Codimension one foliations on solvable manifolds. *Comment. Math. Helvetici* **68** (1993), 633–652.

[163] S. Matsumoto. Measures of exceptional minimal sets of codimension one foliations. In *A fête of topology.* Academic Press (1988), 81–94.

[164] S. McCleary. Free lattice-ordered groups represented as o-2 transitive ℓ-permutation groups. *Trans. of the Ams* **290** (1985), 81–100.

[165] D. McDuff. C^1-minimal sets of the circle. *Annales de l'Institut Fourier (Grenoble)* **31** (1981), 177–193.

[166] W. de Melo & S. van Strien. *One-dimensional dynamics.* Springer-Verlag (1993).

[167] J. Milnor. Growth of finitely generated solvable groups. *J. Diff. Geometry* **2** (1968), 447–449.

[168] H. Minakawa. Classification of exotic circles of $PL_+(S^1)$. *Hokkaido Math. J.* **26** (1997), 685–697.

[169] H. Minakawa. Exotic circles of $PL_+(S^1)$. *Hokkaido Math. J.* **24** (1995), 567–573.

[170] N. Monod & Y. Shalom. Cocycle superrigidity and bounded cohomology for negatively curved spaces. *J. Diff. Geometry* **67** (2004), 1–61.

[171] D. Montgomery & L. Zippin. *Topological transformation groups.* Interscience Publ. (1955).

[172] Y. Moriyama. Polycyclic groups of diffeomorphisms of the half line. *Hokkaido Math. J.* **23** (1994), 399–422.

[173] J. Moulin Ollagnier & D. Pinchon. A note about Hedlund's theorem. *Astérisque* **50** (1977), 311–313.

[174] M. Muller. Sur l'approximation et l'instabilité des feuilletages. Unpublished text (1982).

[175] I. Nakai. Separatrices for non solvable dynamics on $(\mathbb{C}, 0)$. *Annales de l'Institut Fourier (Grenoble)* **44** (1994), 569–599.

[176] A. Navas. A remarkable family of left-ordered groups: central extensions of Hecke groups. *J. Algebra* **328** (2011), 31–42.

[177] A. Navas. On the dynamics of left-orderable groups. *Annales de l'Institut Fourier (Grenoble)* **60** (2010), 1685–1740.

[178] A. Navas. A finitely generated, locally indicable group without faithful actions by C^1 diffeomorphisms of the interval. *Geometry and Topology* **14** (2010), 573–584.

[179] A. Navas. Growth of groups and diffeomorphisms of the interval. *Geom. and Functional Analysis* **18** (2008), 988–1028.

[180] A. Navas. *Grupos de difeomorfismos del círculo.* Ensaios Matemáticos, Brazilian Math. Society (2007).

[181] A. Navas. Reduction of cocycles and groups of diffeomorphisms of the circle. *Bull. Belg. Math. Soc. Simon Stevin* **13** (2006), 193–205.

[182] A. Navas. On uniformly quasisymmetric groups of circle diffeomorphisms. *An. Acad. Sci. Fenn. Math.* **31** (2006), 437–462.

[183] A. Navas. Quelques nouveaux phénomènes de rang 1 pour les groupes de difféomorphismes du cercle. *Comment. Math. Helvetici* **80** (2005), 355–375.

[184] A. Navas. Quelques groupes moyennables de difféomorphismes de l'intervalle. *Bol. Soc. Mat. Mexicana* **10** (2004), 219–244.

[185] A. Navas. Sur les groupes de difféomorphismes du cercle engendrés par des éléments proches des rotations. *L'Enseignement Mathématique* **50** (2004), 29–68.

[186] A. Navas. Groupes résolubles de difféomorphismes de l'intervalle, du cercle et de la droite. *Bull. of the Brazilian Math. Society* **35** (2004), 13–50.

[187] A. Navas. Actions de groupes de Kazhdan sur le cercle. *Annales Sci. de l'École Normale Supérieure* **35** (2002), 749–758.

[188] A. Navas. Groupes de Neretin et propriété (T) de Kazhdan. *C. R. Acad. Sci. Paris Sér. I Math.* **335** (2002), 789–792.

[189] A. Navas & C. Rivas. Describing all bi-orderings on Thompson's group F. *Groups, Geometry, and Dynamics* **4** (2010), 163–177.

[190] A. Navas & C. Rivas, with an appendix by A. Clay. A new characterization of Conrad's property for group orderings, with applications. *Algebr. Geom. Topol.* **9** (2009), 2079–2100.

[191] Y. Neretin. Groups of hierarchomorphisms of trees and related Hilbert spaces. *J. Funct. Anal.* **200** (2003), 505–535.

[192] M. Newman. A theorem on periodic transformations on spaces. *Quart. J. Math.* **2** (1931), 1–8.

[193] P. Nicholls. *The ergodic theory of discrete groups.* London Mathematical Society Lecture Note Series **143**, Cambridge University Press (1989).

[194] A. Norton. Denjoy minimal sets are far from affine. *Erg. Theory and Dynam. Systems* **22** (2002), 1803–1812.

[195] A. Norton. An area approach to wandering domains for smooth surface diffeomorphisms. *Erg. Theory and Dynam. Systems* **11** (1991), 181–187.

[196] A. Norton & B. Tandy. Cantor sets, binary trees and Lipschitz circle homeomorphisms. *Michigan Math. J.* **46** (1999), 29–38.

[197] F. Oliveira & F. da Rocha. Minimal non-ergodic C^1-diffeomorphisms of the circle. *Erg. Theory and Dynam. Systems* **21** (2001), 1843–1854.

[198] Y. Ol'shanskii. On the question of the existence of an invariant mean on a group. *Uspekhi Mat. Nauka* **35** (1980), 199–200.

[199] Y. Ol'shanskii & M. Sapir. Non-amenable finitely presented torsion-by-cyclic groups. *Publ. Math. de l'IHES* **96** (2002), 43–169.

[200] D. Osin. The entropy of solvable groups. *Erg. Theory and Dynam. Systems* **23** (2003), 907–918.

[201] D. Osin. Elementary classes of groups. *Math. Notes* **72** (2002), 75–82.

[202] J. Palis & F. Takens. *Hyperbolicity and sensitive chaotic dynamics at homoclinic bifurcations. Fractal dimensions and infinitely many attractors.* Cambridge Studies in Advanced Mathematics **35**, Cambridge University Press (1993).

[203] K. Parwani. C^1 actions of the mapping class groups on the circle. *Algebr. Geom. Topol.* **8** (2008), 935–944.

[204] D. Pixton. Nonsmoothable, unstable group actions. *Trans. of the AMS* **229** (1977), 259–268.

[205] J. Plante. Subgroups of continuous groups acting differentiably on the half line. *Annales de l'Institut Fourier (Grenoble)* **34** (1984), 47–56.

[206] J. Plante. Solvable groups acting on the line. *Trans. of the AMS* **278** (1983), 401–414.

[207] J. Plante. Foliations with measure preserving holonomy. *Annals of Math.* **102** (1975), 327–361.

[208] J. Plante & W. Thurston. Polynomial growth in holonomy groups of foliations. *Comment. Math. Helvetici* **51** (1976), 567–584.

[209] A. Pressley & G. Segal. *Loop groups*. Oxford Mathematical Monographs (1986).

[210] A. Quas. Non-ergodicity for C^1 expanding maps and g-measures. *Erg. Theory and Dynam. Systems* **16** (1996), 531–543.

[211] M. Raghunathan. *Discrete subgroups of Lie groups*. Ergebnisse der Mathematik und ihrer Grenzgebiete, Band **68**, Springer-Verlag (1972).

[212] J. Rebelo. A theorem of measurable rigidity in $\text{Diff}^w(S^1)$. *Erg. Theory and Dynam. Systems* **21** (2001), 1525–1561.

[213] J. Rebelo. Ergodicity and rigidity for certain subgroups of $\text{Diff}^w(S^1)$. *Annales Sci. de l'École Normale Supérieure* **32** (1999), 433–453.

[214] G. Reeb & P. Schweitzer. Un théorème de Thurston établi au moyen de l'analyse non standard. *Lecture Notes in Math.* **652** (1978), p. 138.

[215] B. Rémy. Construction de réseaux en théorie de Kac-Moody. *C. R. Acad. Sci. Paris Sér. I Math.* **6** (1999), 475–478.

[216] A. Reznikov. Analytic topology of groups, actions, strings and varieties. In *Geometry and Dynamics of Groups and Spaces*. Progress in Mathematics **265**, Birkhäuser (2007), 3–93.

[217] A. Reznikov. Analytic topology. In *European Congress of Mathematics* (Barcelona 2000), vol. 1. Birkhäuser (2001), 519–532.

[218] C. Rivas. On spaces of Conradian orderings. *J. Group Theory* **13** (2010), 337–353.

[219] J. Rosenblatt. Invariant measures and growth conditions. *Trans. of the AMS* **197** (1974), 33–53.

[220] R. Sacksteder. Foliations and pseudogroups. *Amer. Journal of Math.* **87** (1965), 79–102.

[221] R. Sacksteder. On the existence of exceptional leaves in foliations of co-dimension one. *Annales de l'Institut Fourier (Grenoble)* **14** (1964), 221–225.

[222] W. Schachermayer. Addendum: Une modification standard de la démonstration non standard de Reeb et Schweitzer "Un théorème de Thurston établi au moyen de l'analyse non standard." *Lecture Notes in Math.* **652** (1978), 139–140.

[223] A. Schwartz. A generalization of Poincaré-Bendixon theorem to closed two dimensional manifolds. *Amer. J. Math.* **85** (1963), 453–458.

[224] F. Sergeraert. Feuilletages et difféomorphismes infiniment tangents à l'identité. *Invent. Math.* **39** (1977), 253–275.

[225] V. Sergiescu. Versions combinatoires de $\text{Diff}(S^1)$. Groupes de Thompson. Preprint (2003).

[226] J.-P. Serre. *Trees*. Translated from the French by John Stillwell. Springer-Verlag (1980).

[227] Y. Shalom. Bounded generation and Kazhdan's property (T). *Publ. Math. de l'IHES* **90** (2001), 145–168.

[228] Y. Shalom. Rigidity of commensurators and irreducible lattices. *Invent. Math.* **141** (2000), 1–54.

[229] M. Shub & D. Sullivan. Expanding endomorphisms of the circle revisited. *Erg. Theory and Dynam. Systems* **5** (1985), 285–289.

[230] A. Sikora. Topology on the spaces of orderings of groups. *Bull. London Math. Soc.* **36** (2004), 519–526.

[231] V. Solodov. Homeomorphisms of a straight line and foliations. *Izv. Akad. Nauka SSSR Ser. Mat.* **46** (1982), 1047–1061.

[232] M. Stein. Groups of piecewise linear homeomorphisms. *Trans. of the AMS* **332** (1992), 477–514.

[233] S. Sternberg. Local C^n-transformations of the real line. *Duke Math. J.* **24** (1957), 97–102.

[234] D. Sullivan. Discrete conformal groups and measurable dynamics. *Bull. of the AMS* **6** (1982), 57–73.

[235] D. Sullivan. On the ergodic theory at infinity of an arbitrary discrete group of hyperbolic motions. In *Riemann surfaces and related topics.* Proceedings of the 1978 Stony Brook conference. Princeton University Press (1980), 465–496.

[236] D. Sullivan. Cycles for the dynamical study of foliated manifolds and complex manifolds. *Invent. Math.* **36** (1976), 225–255.

[237] G. Szekeres. Regular iteration of real and complex functions. *Acta Math.* **100** (1958), 203–258.

[238] S. Tabachnikov & V. Ovsienko. *Projective differential geometry old and new: From the Schwarzian derivative to the cohomology of diffeomorphism groups.* Cambridge Tracts in Mathematics **165**, Cambridge University Press (2005).

[239] W. Thurston. *Three dimensional geometry and topology.* Princeton University Press (1997).

[240] W. Thurston. A generalization of the Reeb stability theorem. *Topology* **13** (1974), 347–352.

[241] T. Tsuboi. Homological and dynamical study on certain groups of Lipschitz homeomorphisms of the circle. *J. Math. Soc. Japan* **47** (1995), 1–30.

[242] T. Tsuboi. Area functionals and Godbillon-Vey cocycles. *Annales de l'Institut Fourier (Grenoble)* **42** (1992), 421–447.

[243] T. Tsuboi. On the foliated products of class C^1. *Annals of Math.* **130** (1989), 227–271.

[244] T. Tsuboi. Examples of nonsmoothable actions on the interval. *J. Fac. Sci. Univ. Tokyo Sect. IA Math.* **34** (1987), 271–274.

[245] T. Tsuboi. Γ_1-structures avec une seule feuille. *Astérisque* **116** (1984), 222–234.

[246] P. Tukia. Homeomorphic conjugates of Fuchsian groups. *J. für Reine und Angew. Math.* **391** (1989), 1–54.

[247] S. Wagon. *The Banach-Tarski paradox.* Cambridge University Press (1993).

[248] P. Walczak. *Dynamics of foliations, groups and pseudogroups.* Mathematics Institute of the Polish Academy of Sciences, Mathematical Monographs (New Series) **64**, Birkhäuser Verlag (2004).

[249] P. Walters. *An introduction to ergodic theory.* Graduate Texts in Mathematics **79**, Springer-Verlag (1982).

[250] Y. Watatani. Property (T) of Kazhdan implies property FA of Serre. *Math. Japonica* **27** (1981), 97–103.

[251] J. Wilson. On exponential growth and uniform exponential growth for groups. *Invent. Math.* **155** (2004), 287–303.

[252] D. Witte-Morris. Amenable groups that act on the line. *Algebr. Geom. Topol.* **6** (2006), 2509–2518.

[253] D. Witte-Morris. Arithmetic groups of higher $\mathbb{Q}$-rank cannot act on 1-manifolds. *Proc. of the AMS* **122** (1994), 333–340.

[254] J. Wolf. Growth of finitely generated solvable groups and curvature of Riemannian manifolds. *J. Diff. Geometry* **2** (1968), 421–446.

[255] A. Yamada. On Marden's universal constant for Fuchsian groups. *Kodai. Math. J.* **4** (1981), 266–277.

[256] J.-C. Yoccoz. Centralisateurs et conjugaison différentiable des difféomorphismes du cercle. *Astérisque* **231** (1995), 89–242.

[257] J.-C. Yoccoz. Il n'y a pas de contre-exemple de Denjoy analytique. *C. R. Acad. Sci. Paris Sér. I Math.* **298** (1984), 141–144.

[258] K. Yoshida. *Functional analysis.* Classics in Mathematics. Springer-Verlag (1995).

[259] R. Zimmer. *Ergodic theory of semisimple groups.* Monographs in Mathematics, Birkhäuser (1984).

[260] A. Żuk. Property (T) and Kazhdan constants for discrete groups. *Geom. and Functional Analysis* **13** (2003), 643–670.

Index

action
   differentiably expanding, 132
   equicontinuous, 64, 86
   ergodic, xviii, 130–141
   expansive, 64, 75, 86, 102
   faithful, xviii, 21, 39–43, 230, 252
   free, xviii, 10, 43–50, 53, 122, 159,
      182, 192, 200, 218, 248
   minimal, 24
   nonelementary (on a tree), 258
   strongly expansive, 66, 73, 102, 106
affine group, 2–4, 12, 47, 55, 209
Ahlfors Finiteness Theorem, 141
Antonov, 77
Archimedean property, 43, 48, 50, 218

Bass, 272
Birkhoff
   Ergodic Theorem, 89, 104
   sums, 70
Bogolioubov-Krylov's theorem, 37, 70, 269
Bonatti-Crovisier-Wilkinson's lemma, 208
Borel-Cantelli Lemma, 98

Calegari, 231
Cayley graph, 120
center of a group, 48, 56, 192, 208, 230, 262,
   268
central series, 192, 267
centralizer, 77, 155, 156, 159, 165, 187, 208,
   265, 267
characteristic subgroup, 268
coboundary, 233, 257
cocycle, 4, 7, 153, 233
   Liouville, 244, 263

cohomology space, 226, 233
conjugacy, xviii
Conrad, 45
   Conrad property, see group,
      C-orderable
control of distortion, 82, 94, 122, 135, 187
convergence property, 9–11
convolution of measures, 69
crossed elements, 52, 61, 199, 205,
   208, 231

degree
   of a covering, 265
   of a group extension, 247
   of a group of circle
      homeomorphisms, 67, 103
   of a map, 12, 66, 145
   of a power series, 40
   of nilpotence, 194, 267
   of polynomial growth for a group,
      201
   of solvability, 201–225, 267
Delorme's lemma, 257
Denjoy
   counterexamples, 81, 171–182
   generalized theorem, 182–192
   inequality, 89
   theorem, 80–90, 95, 103, 117
density point, 130, 140
derivative
   logarithmic, 3
   Schwarzian, 7, 156
Deroin, 77, 97, 182
diffusion operator, 69, 105
Dippolito, 141

Duminy
    estimates, 92, 124
    first theorem, 110–119, 133
    second theorem, 119–130
    third theorem, xii
dyadic
    interval, 16, 238
    rational number, 13, 19, 90
    rotation, 19
    tree, 15

equicontinuous
    action, *see* action, equicontinuous
    sequence of functions, 171, 207
equivariant family of diffeomorphisms, 173
ergodic
    action, *see* action, ergodic
    decomposition, 105
exceptional minimal set, 24, 38, 64, 75, 80,
    110, 152
expandable point, 132
expanding
    action, *see* action, differentiably
        expanding
    map, 29, 112–116
exponential set of an action, 133

Følner sequence, 224, 235, 271
Farey sequence, 17
Farley, 237
Fathi, 261
first-return map, 114
foliation, 52, 91, 97, 106, 144, 210, 226

geodesic current, 7–10, 245
geometric boundary, 271
germ of a diffeomorphism, 147, 191,
    202, 229
Ghys, 151
    four vertex theorem for the
        Schwarzian derivative, 156
    Ghys-Sergiescu's realization of
        Thompson's groups, 18, 28, 86
    remark on Sacksteder's theorem, 95
    space of orderings, 58
    theorem for lattice actions, 242, 253
    theorem on bounded cohomology,
        xiii
    weak Tits alternative, 64
Gottschalk-Hedlund Lemma, 152

Grigorchuk's examples, 195
Gromov
    random groups, 237
    theorem on polynomial growth
        groups, 196, 272
group
    C-orderable, 58, 62, 230
    amenable, 14, 38, 56, 58, 64, 86, 164,
        223, 269–272
    Baumslag-Solitar, 63
    biorderable, 39, 253
    compact, 2, 11
    compactly generated, 231, 254
    free, xviii, 63–69, 86
    Fuchsian, 26
    Lie, *see* Lie group
    locally indicable, 230, 252
    metabelian, 194, 197, 216, 224, 256,
        267
    modular, 25, 96
    nilpotent, 40, 51, 59, 110, 192, 196,
        202, 262, 267
    orderable, 39, 58, 204, 230, 241, 253,
        268
    perfect, 55, 230, 267
    polycyclic, 58, 209–211, 233, 268
    Schottky, 26, 67, 269
    sequentially complete, 258
    simple, 13, 15, 21, 229, 267
    solvable, 21, 55, 63, 197, 224, 262
    strongly polycyclic, 268
    torsion-free, 13, 39, 59, 202
growth
    of a sequence, 207
    of groups, 195–209, 224, 272
Guivarch, 272

Hölder
    continuity, 172
    derivative, xvii, 92, 188, 208, 245
    inequality, 184
    theorem, 43–49, 53, 96, 102, 159, 163,
        192, 200, 218, 248
Haagerup's property, 235, 237, 243, 251
Haefliger, 92, 229
harmonic function, 72, 107
Hector, 27, 96, 120, 157, 161
Herman, 81, 130, 146, 176
holonomy, 144
homology sphere, 230

Hurder, 97, 144
hyperbolic
    fixed point, 91
    germ, 147
    plane, 4

irreducible component, 158, 164, 193, 217

Jørgensen's inequality, 110, 119

Kakutani
    Fixed Point Theorem, 70
    Random Ergodic Theorem, 104
Katok, 130
Kazhdan's property (T), 233–240
    relative, 251
Klein, 26
    group, 62
    Ping-Pong Lemma, 53, 67, 201
Kleptsyn, 77, 97, 182
Kopell
    generalized lemma, 188
    lemma, 158–164, 179, 193, 197, 221,
        249

Laplacian, 107
lattice, 237, 253, 254
law (in a group), 69
lexicographic ordering, 39, 179, 204
Lie group, 11, 110
    semisimple, 253
    simple, 237
linearization, 87, 147, 165
Linnell, 58, 59
Lipschitz
    conjugacy, 150, 152, 162, 165
    continuity, 172
    derivative, xvii, 81, 92, 123, 168
    homeomorphism, 70, 85, 204, 230
local exceptional set, 92, 120
Lyapunov exponent, 104

Möbius group, 4–12, 25, 243, 248, 251
Magnus, 40
Marden's theorem, 111, 119
Margulis
    inequality, 111, 119
    normal subgroup theorem, 43, 242,
        254, 255
    weak Tits alternative, 63, 86

Markov
    minimal set, 141
    process, 106, 183
martingale, 73, 77
Matsuda, 211
Matsumoto, 210
measure
    Liouville, 9, 243
    quasi-invariant, 3, 49–58, 216, 221
    Radon, 2, 7, 49, 63, 165, 199, 216,
        237, 243
    stationary, 69, 106, 134
Milnor-Wolf's theorem, 197, 201
modulus of continuity, 171
Montgomery-Zippin's theorem, 11
Moriyama, 209

Neretin's groups, 237, 238, 250
Newmann's lemma, 68
nilradical, 210, 232, 268

Oxtoby-Ulam's theorem, 72

piecewise affine homeomorphism, 12, 29, 37,
    41, 89, 103, 220, 248
Plante
    examples of solvable groups, 209
    invariant measure theorem, 51
    Plante-Thurston's theorem, see
        Plante-Thurston's theorem
Plante-Thurston's theorem, 192–196, 202,
    211, 242
Poincaré, 26
    disk, 4, 111, 230
    Recurrence Theorem, 60
    rotation number, see rotation
        number
projective
    group, see Möbius group
    space, 6, 156
    structure, 7
pseudogroup, 90, 115, 119, 142
    compactly generated, 92, 102

quadratic differential, 7
quasisymmetric homeomorphism, 10, 248

Radon-Nikodym derivative, 109, 250
rotation, xviii, 32
    dyadic, see dyadic, rotation

rotation (cont.)
    group, 1–2, 12, 38, 46, 65, 86, 96, 111,
        152, 164, 185, 193, 215, 251
rotation number, 30–37, 66, 80, 95, 146, 184,
    193, 248

Sacksteder
    example, 26
    theorem, 90–110, 123
Schreier graph, 120
Schwartz estimates, 92, 102, 124, 135
Schwarzian derivative, 6
semi-exceptional orbit, 120
semiconjugacy, 24
semigroup, 39, 59, 69, 183
    free, 51, 69, 195–209
Shalom, 256
shift, 72, 142
Sikora, 58
small denominators, 147
Solodov, 47
space of ends, 119
Sternberg
    example, 150
    linearization theorem, 147–152
subexponential set of an action,
    133, 145
Sullivan, 82, 107, 137, 147
superrigidity theorem
    for actions on the circle,
        253–267
    for actions on trees, 259
    for bounded cohomology, 259
    for reduced cohomology, 256
support
    of a function, 154, 241
    of a map, 14, 189
    of a measure, 1, 69
Szekeres's theorem, 165–171

Thompson's groups, 12–22
    m-adic, 22
    F, 63, 223, 229
    G, 18, 27, 55, 87, 90, 96, 238
Thurston, 58
    Example, 229
    Plante-Thurston's theorem, see
        Plante-Thurston's theorem
    realization of Thompson's groups, 15
    stability theorem, 226, 236, 248, 266
Tits
    alternative, 63
    Center Lemma, 234, 257, 260
topological entropy, xii, 72
translation number, 49–58, 63, 201, 221
tree, 15, 130, 195, 234, 238, 241, 259
Tsuboi, 151
    conjecture, 178
    Müller-Tsuboi conjugacy, 209, 232
    Pixton-Tsuboi actions, 179

variation of a function
    α-variation, 185
    total variation, 80, 87, 111, 158

wandering interval, 98, 184, 190
weak* topology, 73, 109, 269
Witte-Morris
    3-fold covering trick, 247, 252
    theorem of nonorderability of
        lattices, 42, 195, 241, 242
    theorem on orderable amenable
        groups, 58, 223
wreath product, 196

Yoccoz, 82, 146, 174

Zazenhäus Lemma, 110
Zorn Lemma, 23, 30, 153